CONSTRUCTION ENGINEERING DESIGN CALCULATIONS AND RULES OF THUMB

CONSTRUCTION ENGINEERING DESIGN CALCULATIONS AND RULES OF THUMB

RUWAN RAJAPAKSE

AMSTERDAM • BOSTON • HEIDELBERG • LONDON
NEW YORK • OXFORD • PARIS • SAN DIEGO
SAN FRANCISCO • SINGAPORE • SYDNEY • TOKYO
Butterworth-Heinemann is an imprint of Elsevier

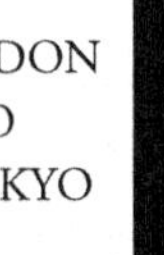

Butterworth-Heinemann is an imprint of Elsevier
The Boulevard, Langford Lane, Kidlington, Oxford OX5 1GB, United Kingdom
50 Hampshire Street, 5th Floor, Cambridge, MA 02139, United States

Notices
Knowledge and best practice in this field are constantly changing. As new research and experience broaden our understanding, changes in research methods, professional practices, or medical treatment may become necessary.

Practitioners and researchers must always rely on their own experience and knowledge in evaluating and using any information, methods, compounds, or experiments described herein. In using such information or methods they should be mindful of their own safety and the safety of others, including parties for whom they have a professional responsibility.

To the fullest extent of the law, neither the Publisher nor the authors, contributors, or editors, assume any liability for any injury and/or damage to persons or property as a matter of products liability, negligence or otherwise, or from any use or operation of any methods, products, instructions, or ideas contained in the material herein.

Library of Congress Cataloging-in-Publication Data
A catalog record for this book is available from the Library of Congress

British Library Cataloguing-in-Publication Data
A catalogue record for this book is available from the British Library

ISBN: 978-0-12-809244-6

Publisher: Joe Hayton
Acquisition Editor: Ken McCombs
Editorial Project Manager: Peter Jardim
Production Project Manager: Vijayaraj Purushothaman
Cover Designer: Matthew Limbert

Typeset by SPi Global, India

CONTENTS

CHAPTER 1

Construction—General Introduction

1.1 HISTORY OF CONSTRUCTION

Early period: It is estimated that cities started to appear in Mesopotamia, India, Egypt, and China approximately 5000 years ago. Egyptians took construction to a very high level by building large-scale pyramids. During later times, construction activities became more diverse. Roman aqua ducts, the Great Wall of China, and large-scale temple structures were built.

Development of the number system: Early engineers used very little mathematics for construction work. When they did use mathematics, it was mainly limited to geometry. Simple computations such as addition, subtraction, multiplication, and division were extremely cumbersome.

A new era of the construction industry began due to the discovery of modern number system by Indian mathematician Aryabhatta around AD 500. One hundred years later Brahmagupta (AD 600) gave arbitrary rules for these four fundamental operations: addition, subtraction, multiplication, and division. Brahmagupta's system was transported to Europe through Arabia and to rest of the world.

Machinery: The next major development to benefit construction work was the use of machinery. Earlier construction workers used wedges, pulleys, and other hand-operated tools. Today construction is mostly an affair of labor and machinery. Backhoes, trucks, loaders, forklifts, cranes, derricks, conveyor belts, jackhammers, compactors, rollers, and compressors are some of the few machines used in construction sites.

Internal combustion engine: The development of the internal combustion engine provided a new dimension to the construction industry. Nicklaus Otto, Gottlieb Daimler, Karl Benz, and James Atkinson invented internal combustion engines around 1870. The internal combustion engine or its variations are needed for today's machinery.

Electricity: The discovery of electricity by Michael Faraday, James Joule, Thomas Edison, and Nicolai Tesla is another major development that affected the construction industry. It is unthinkable to do any construction

Construction Engineering Design Calculations and Rules of Thumb
http://dx.doi.org/10.1016/B978-0-12-809244-6.00001-9

activities without the use of electricity. Power drills, jackhammers, cutting machines, hoists, and conveyor belts are operated using electricity.

Computer: Computers have become an integral part of the construction industry. Scheduling, cost estimating, design, and construction management tasks are highly dependent on computers. Development of the computer by US engineers such as Attanasof, Allen Turing, and Von Neumann brought a new dimension to the construction industry.

1.2 BUILDING CONSTRUCTION

All construction engineers will encounter building construction work eventually. The construction of a building starts with a need. A company may need more office space or new stores. The management of that company would meet an architect and explain their needs. The architect would then come up with a set of architectural drawings, which the owner would look at and make comments. The owner may want the meeting hall to be larger, or more bathrooms be added, or a change in the appearance of the building. Based on owner needs, the architect would redraw the plans. After the architectural drawings are finalized, structural drawings and civil drawings will be prepared by structural engineers and civil engineers. Mechanical drawings and electrical drawings also would be prepared. Today network drawings are prepared for communication and computer networks (Fig. 1.1).

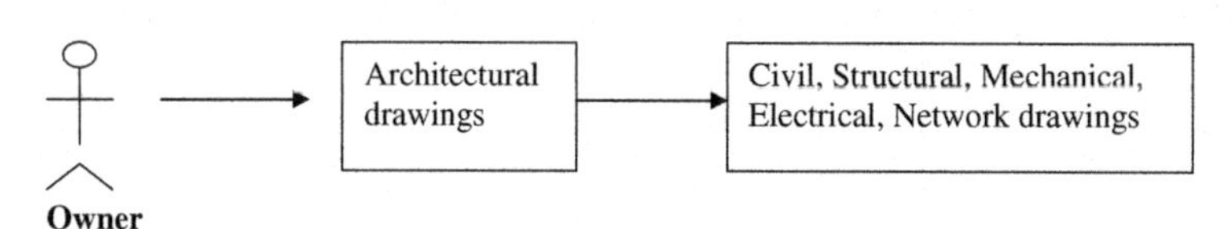

Fig. 1.1 Start of a project.

After all the necessary drawings and specifications are prepared, bids would be called. Based on the bids, a contractor would be selected.

Construction—early phase: Prior to any construction work, the site has to be prepared. Site preparation work includes cutting trees, constructing temporary roads for delivery trucks and concrete trucks, constructing temporary parking lots for workers, dewatering to remove water from excavations, and setting up office trailers, security fences, and erosion control mechanisms such as silt fences, setup temporary phone lines and power (Fig. 1.2).

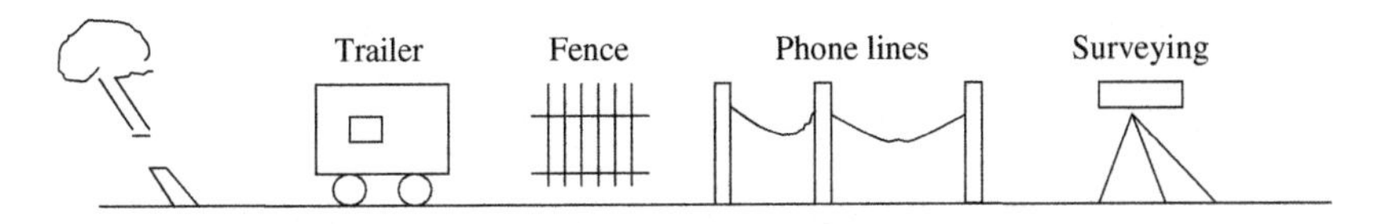

Cutting trees, setting up trailers, security fences, temporary power lines, surveying

Fig. 1.2 Site preparation work.

Soil grading and fill: After site preparation work, the site has to be graded. High ground has to be cut and low ground has to be filled (Fig. 1.3).

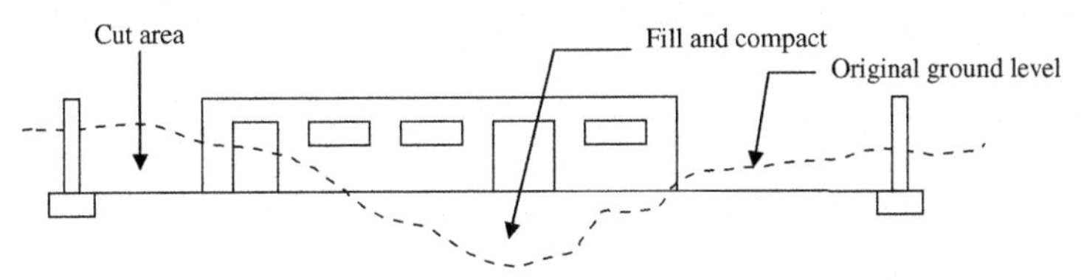

Fig. 1.3 Proposed construction and original ground level.

Soil grading is the process of cutting the original ground to the proposed level (Fig. 1.4).

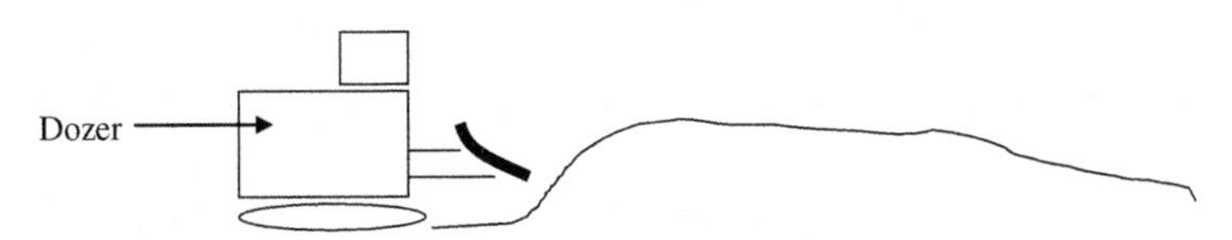

Fig. 1.4 Cutting soil with a dozer.

Fill and compaction: As you see in Fig. 1.4, some areas have to be filled and compacted. Prior to any fill activity, soils that will be used have to be approved by a geotechnical engineer. Usually poorly graded sandy soil or stones are used for fill work. Clay soils and silty soils are considered unsuitable for fill and compaction.

Removal of rock: Rock removal is done using blasting and rock breakers. Generally, igneous rocks are hard and difficult to rip. Sedimentary rocks are relatively easier. Ripability of rock depends on rock type and degree of weathering.

Pile driving: Some buildings require piles when existing ground conditions are not suited for shallow foundations. Piles transfer loads to lower level (Fig. 1.5).

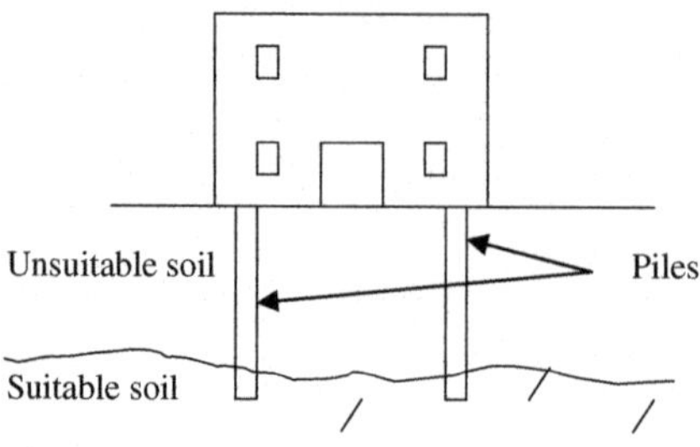

Fig. 1.5 Building on piles.

Excavation: Excavation is needed for basements, shallow foundations, sewer pipes, drainage pipes, manholes, and swimming pools (Fig. 1.6).

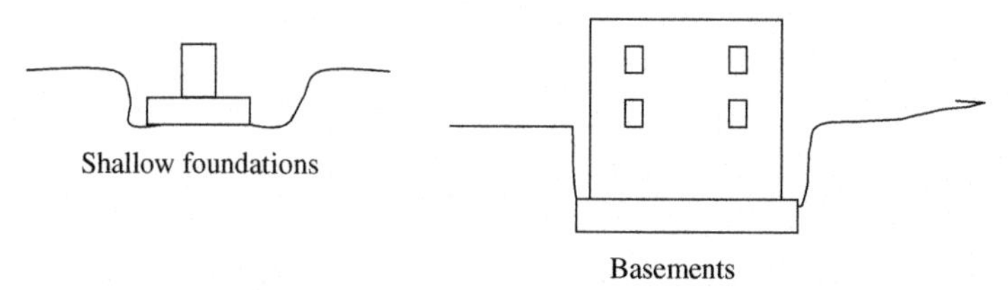

Fig. 1.6 Building on shallow foundations.

Excavation work is usually done using backhoes.

Concreting: After completion of the site preparation work, excavations, and pile driving, concreting work begins.

Concreting work can be divided into:

- formwork preparation,
- installation of reinforcement bars, and
- concreting and curing.

Formwork is needed to hold the concrete in place. Concreting can be done by bringing premixed concrete from a concrete yard or concrete can be prepared on-site using a mixer. Concrete can be pumped or lifted to high elevations.

1.3 STEEL ERECTION

Some buildings are designed using steel beams and columns. Steel members are connected using bolts or welding. Rivets were used in the past and not used anymore (Fig. 1.7).

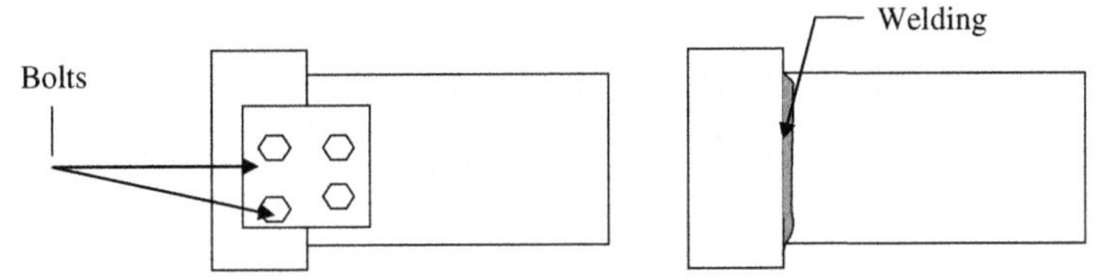

Fig. 1.7 Bolts and welding used to connect steel members.

1.4 MOVING AND LIFTING OF MATERIAL (RIGGING)

Material such as concrete, timber, and steel is brought to the site using trucks. Then cranes are used to move them inside the site. There are many different types of cranes available.

- Tower cranes—Large tower crane can cover a large area. Material on one corner of a site can be moved to the other end in a second.
- Crawler Mounted cranes—These are good for small sites.
- Gantry cranes—These cranes move on rail

CHAPTER 2

Site Work

2.1 PERMANENT AND TEMPORARY SITE WORK

In construction work, many things need to be done outside the main building or structure that is planned. Look at the photograph in Plate 2.1. It shows an undeveloped plot of land.

Plate 2.1 Undeveloped land.

Let us assume a developer is planning to build a shopping mall on this land as shown in Plate 2.2.

Plate 2.2 Shopping mall.

Construction Engineering Design Calculations and Rules of Thumb
http://dx.doi.org/10.1016/B978-0-12-809244-6.00002-0

Can you prepare a list of work items that needs to be done other than construction of the building?

First, the developer needs to cut the tress and clean the top soil. This is not a complicated activity. However, there are many state and federal permits that have to be obtained prior to cutting trees. Plate 2.2 also shows a small water logged area. This may be considered as a wetland. The developer needs to find out from relevant authorities whether he can fill that area.

Site work can be divided into permanent site work and temporary site work. A sheet pile wall may be erected to hold back the soil during construction. This could be considered as temporary retaining wall. In addition, some retaining walls can be a permanent structure that is part of the design. A contractor may decide to have a quick and dirty drainage lines during construction. Later he may build the permanent drainage lines to the site. Some areas may be temporarily paved for delivery trucks to come and go.

Some temporary site work:

- site clearing or grubbing
- demolition of abandoned structures
- fill depressed areas
- excavations
- cut and fill
- breaking rock
- mass grading
- fine grading
- compaction of soil
- removal of existing utilities
- install temporary lighting, water or gas supply
- provide temporary drainage to the site
- provide temporary paving
- temporary retaining walls, coffer dams, sheet pile walls
- temporary sediment and erosion control structures (rip rap, silt fences)
- soil stabilization (vibroflotation, dynamic compaction, soil surcharging)

Some permanent site work:

- retaining walls
- roads
- parking lots
- construct permanent utilities (water supply, electricity, gas, cable, communication)
- planting trees
- ponds and canals
- landscaping

2.2 SITE CLEARING

Site clearing is also known as "grubbing." Site clearing involves cutting trees, removal of bushes, and the removal of top soil. Typically, backhoes, dozers, and tree cutting machines are used for site clearing (Plate 2.3).

Some specialized equipment used for site clearing:

- stump splitters
- stump pullers
- clearing rakes
- grapples

Plate 2.3 Cleared site.

2.3 DEMOLITION OF EXISTING STRUCTURES AND UTILITIES

In many situations, old abandoned structures need to be demolished. Typically, demolition involves demolition of concrete, steel structures, fences, masonry structures, and roofs.

2.4 MASS GRADING

Grading is the process of attaining the required ground elevation. Depressed areas have to be filled and high areas have to be cut. After site clearing and demolition is completed, mass grading is done. Mass grading is done using dozers, excavators, and loaders. Dozers are good to cut through soil. But not a good machine to transport soil. Dozers are efficient when transportation of soil is kept to a minimum. Loader can transport soil in the bucket.

Other widely used equipment is the scraper. Scrapers have an underbelly to transport soil and are much more efficient in transporting soil than loaders. More details regarding these machines are provided under cut and fill chapter.

2.5 FINE GRADING

The blade of a grader is not as robust as the blade of a dozer. Typically, blades of graders are at the center between wheels. On the other hand, a dozer blade is much more robust and located in front.

2.6 TEMPORARY DRAINAGE

Once the trees and grass is removed, water tends to make the site muddy. Working becomes highly inefficient in a muddy site. Hence, temporary drainage should be provided.

Temporary drainage is provided thru backfilling, gravel beds, perforated pipes, and trenches. Typically, one has to locate low areas where ponding could occur. These areas can be backfilled to bring it up. In addition, gravel could be placed and a perforated pipe can be installed.

2.7 EROSION AND SEDIMENT CONTROL

Many states require an erosion and sediment control plan be submitted prior to start of any construction work. Silt fences are used to stop soil eroding away. Near riverbeds, riprap is provided. Hay bales and geo-fabrics also can be used to stop erosion (Fig. 2.1).

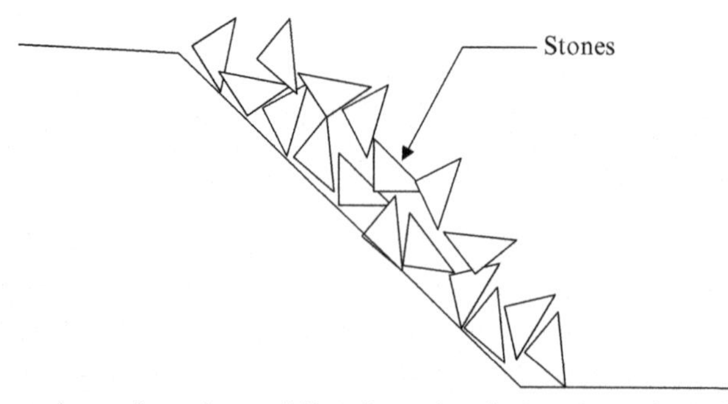

Fig. 2.1 Riprap (stones) used to stabilize slopes by placing them along the slope.

Stone size in riprap changes depending upon the velocity of water. Larger stones are used on slopes near high velocity rivers and streams. Small stones can be used for low velocity streams.

2.8 SURVEYING

Horizontal control: Design documents would specify that the coordinates should be obtained using a monument nearby. Surveyors need to use the monuments provided and establish control points near the site. In many instances these control points get runover by machines and new control points need to be installed. It is important to make sure that the control points are protected. If not the building would be constructed at wrong coordinates.

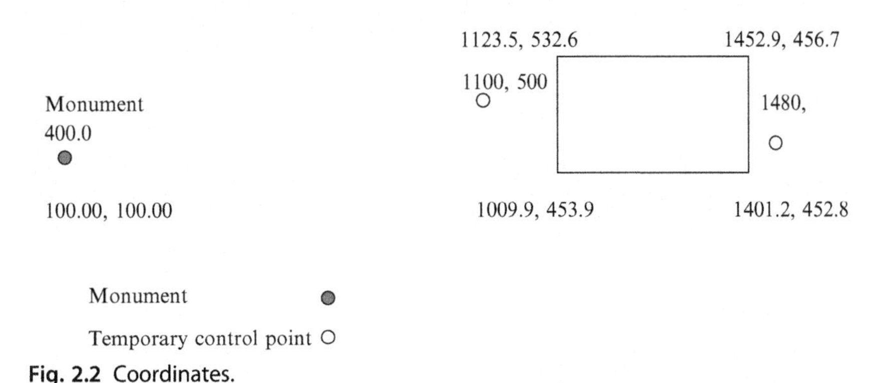

Fig. 2.2 Coordinates.

Fig. 2.2 shows a proposed building, monument, and two control points. The control points need to be closer to the site. However, when they are too close they get damaged due to construction activities.

Vertical control: Similar to horizontal control, design documents would indicate the elevation and the datum used. Surveyors need to establish temporary benchmarks near and around the site to be used for grading, establish footing and slab elevations, and elevations of utility pipes.

Manhole construction: Manholes are required to clean out pipes. Manholes can be temporary or permanent.

2.9 SHEET PILES

It is very rare to see a construction site without sheet piles. Sheet piles can be used as retaining walls, excavation support, trenching, cofferdams, and temporary bearing platforms (Plate 2.4).

Plate 2.4 Sheet pile walls.

2.10 SOIL STABILIZATION

Some sites are not suitable to have shallow foundations. In such cases, piles need to be driven. Piles generally come with a heavy price tag. Instead, one can try to improve the soil bearing strength. This is known as soil stabilization. Many methods are used for soil stabilization. Vibroflotation, dynamic compaction, surcharging, Wick Drains, and pressure grouting are some processes used for soil stabilization.

2.11 SITE WORK—PERMANENT CONSTRUCTION

Any building needs roads and parking lots. Roads and parking lots are constructed by bringing the soil to the required grade and then providing a gravel base. This is known as a sub-base. After the gravel layer (sometimes crushed stone is also used), an asphalt base course is provided. On top of the asphalt base course, asphalt surface layer is provided. Surface course is designed to have better friction between tires and asphalt. Base course is designed to provide rigidity to the road.

2.12 PERMANENT DRAINAGE

Drainage is provided through the installation of storm water pipes and manholes. Trenches need to be dug to install storm water pipes. Excavation support is provided with trench boxes or shoring.

2.13 CONSTRUCTION OF UTILITIES (WATER PIPES, SEWER PIPES, ELECTRICAL CONDUITS)

Any facility requires water, electricity, sanitary, cable, and communication lines. These utilities are part of site work.

Other site work items include retaining walls, storm water detention systems, playgrounds, and landscaping.

Storm water detention pond: During storms, all the storm water ends up in storm pipes. This could create overflow of manholes and flooding. Hence, large sites are required to maintain storm water detention ponds. During a storm, water is drained to the storm water detention pond. Later when the storm is over, detention pond would discharge to the storm water pipes.

2.14 LANDSCAPING

Landscaping is the process of creating an aesthetic and natural environment around the facility. Generally, trees, flowers, water ponds, and plants are used to create a pleasant environment.

CHAPTER 3

Concrete Construction

Concrete is a product made of cement, sand, stone, and water. Sand is known as fine aggregates and stone is known as coarse aggregates. Chemical compounds, known as admixtures, are also added to concrete to obtain special properties (Fig. 3.1).

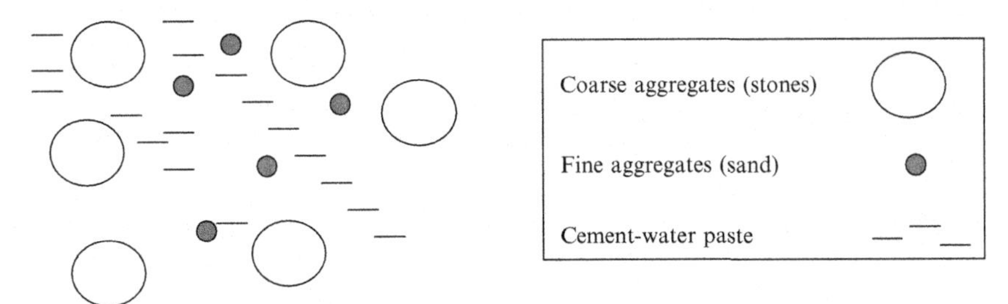

Fig. 3.1 Cement, sand, stones, and water (ingredients of concrete).

3.1 CEMENT, FINE AGGREGATES, AND COARSE AGGREGATES

Cement water mixture acts as the binder between coarse and fine aggregates. In general, the higher the cement content, the higher the strength of the concrete. It has been found that high water content leads to less strength. Hence, to achieve high strength, one should minimize the water content. However, if the water content is reduced, the concrete may not be workable. Because, in many cases, concrete needs to be pumped, a certain amount of flowability is needed in the concrete. Chemical admixtures can be added to increase the workability without reducing the strength of the concrete for situations where high workability and strength need to be maintained.

$$\text{High water content} \rightarrow \text{Low strength} + \text{High workability}$$
$$\text{Low water content} \rightarrow \text{High strength} + \text{Low workability}$$
$$\text{Low water content} + \text{Admixtures} \rightarrow \text{High strength} + \text{High workability}$$

Cement types: Five major types of cements are available in the US market.

Construction Engineering Design Calculations and Rules of Thumb
http://dx.doi.org/10.1016/B978-0-12-809244-6.00003-2

Type I cement: Type I cement is known as general-purpose cement and is widely used. This is the cheapest type of cement.
Type II cement: Type II cement generates less heat than type I cement. This property can be useful for mass concretes. When a large mass of concrete is poured (eg, dams, large footings, retaining walls) heat that is generated inside the core may not be able to escape. High temperatures give rise to low strength. In such situations, type II cement can be used. Another property of type II cement is its resistance to sulfate attack. Sulfates are present in some soils and groundwater.
Type III cementrun: Type III cement is known as high early strength cement. High early strength is required to remove forms and move forward. Type III cement is more expensive than type I cement. Hence, one may have to consider cost versus schedule benefits when recommending type III cement. Typically, type III cement will achieve the 28-day strength of type I cement in 7 days. Eventually they both will have the same strength assuming other ingredients are the same (Fig. 3.2).

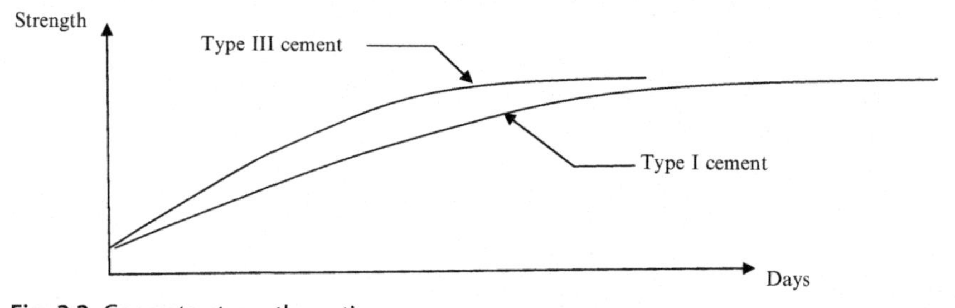

Fig. 3.2 Concrete strength vs. time.

Type IV cement: Type IV cements generates much less heat during hydration. Hence, this cement is used for very large concrete structures. These cements have the lowest heat generation. Type IV cements are not readily available in the market.
Type V cement: Type V cements are known as sulfate resistance cements. These cements are used when the groundwater or the soil contain large concentrations of sulfates.

3.2 POZZOLANS

Pozzolans are known as supplementary cementitious material. Pozzolans can be mixed with cement to reduce cost without reducing its strength. Pozzolans react with by-products of cement hydration. One of the main by-product of cement hydration is calcium hydroxide. Pozzolans react with this by-product to generate strength. Hence, pozzolans do not contribute to initial strength. However, later pozzolans react with by-products of cement hydration and generate additional strength.

Other than the cost benefit, there is another benefit of pozzolans. Since most pozzolans are materials that are wastes from other processes, owners can gain LEEDs environmental points by using them in the mixture.

Three widely used pozzolans are

1. Fly ash
2. Blast furnace slag
3. Micro silica or silica fume

3.2.1 Fly Ash

Coal is burned to generate electricity. It has been reported that 50% of electricity generated in US comes from coal plants. Fly ash is a by-product of coal burning plants. In the past, fly ash was sent to landfills. Recently it has been found that fly ash could be used as a supplement to cement without affecting the strength. Fly ash is the most commonly used supplementary cementitious material.

Since fly ash particles are more spherical in shape than cement particles, workability and pumpability can be improved by adding fly ash. Some fly ashes cause low early strength. This can be a problem when strength needs to be attained sooner. Adding fly ash usually improves the resistance against sulfate attack. Another property of fly ash is to reduce the air content in concrete. In freezing and thawing conditions, air entrained concrete is preferred. In such situations, fly ash should be avoided.

3.2.2 Blast Furnace Slag

Slag is produced in blast furnaces that produce iron and steel. Slag can also be used as a supplementary cementitious material. Slag tends to improve resistance for sulfate attack. In addition, it has been reported that slag develops higher long-term strength.

3.2.3 Silica Fume

Silica fume is a by-product of the silica alloy industry. Silica fume is also known as micro silica. Micro silica particles are 100 times smaller than cement particles. The main advantage of silica fume is high durability. Since micro silica particles are extremely small, permeability of concrete is reduced. This is an important quality since rebars are better protected from water permeation.

3.3 CONCRETE ADMIXTURES

Chemical admixtures are widely used to improve the required properties of concrete.

3.3.1 Air Entraining Admixtures

Concrete that is subjected to repeated freezing and thawing develops cracks. This can be avoided by increasing entrapped air. A wide array of chemicals is used for air entrainment. Vinsol resin is the most popular air entrainment admixture.

3.3.2 Water-Reducing Admixtures

Less water in the concrete generates higher strengths. On the other hand, there needs to be enough water for workability. Water-reducing agents can maintain a low water content while maintaining workability.

3.3.3 Accelerating Admixtures

Accelerating agents are used to accelerate the concrete's setting process. Some accelerators are capable of increasing the early strength of concrete as well.

3.3.4 Super Plasticizers (Commonly Known as Super Ps)

Super plasticizers are used to maintain high workability while at the same time maintaining strength. When concreting highly reinforced structures, concrete has to be able to flow freely. Super plasticizers can be used in such situations to increase the flowability without compromising strength.

Concrete retarders: Concrete retarders are added to delay the setting of concrete. The following situations require that the setting of concrete is delayed.

(1) Concrete has to be transported longer distances
(2) Provide more time for the workers to carve grooves, curves, and architectural features
(3) To avoid cold joints

3.4 CONCRETE SLUMP TEST

A concrete slump test should be done as described in ASTM C 143. A slump test is widely used to check the workability and consistency of concrete (Fig. 3.3).

Slump test procedure:

STEP 1: Obtain fresh concrete from the truck.

STEP 2: Fill one-third of the slump cone and tamp 25 times with a rod. (This is commonly called rodding.)

STEP 3: Fill another one-third of the slump cone and tamp 25 times. Fill the last one-third and tamp 25 times.

STEP 4: Lift the cone.

STEP 5: Measure the slump.

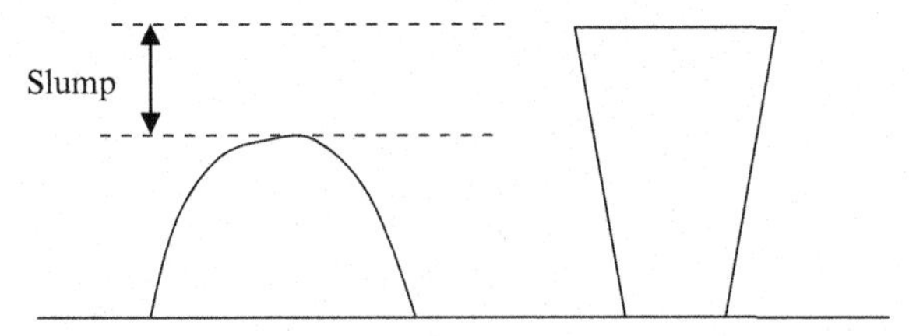

Fig. 3.3 Concrete slump.

3.5 CONCRETE CYLINDERS

Concrete cylinders are taken to conduct compressive strength tests. Concrete compressive strength tests need to be done as per ASTM C 31 and ASTM C 39. Concrete cylinders are 6 in. diameter and 12 in. high. Small size cylinders (4 in. diameter and 8 in. tall) are also used.

Procedure to obtain concrete cylinders:

STEP 1: Fill one-third of the cylinder and tamp 25 times with a rod.

STEP 2: Fill another one-third of the cylinder and tamp 25 times. Finally, fill the last one-third of the cylinder and tamp 25 times.

Concrete cylinders should be placed in the job site in a controlled environment at temperature 60°F to 80°F. Cylinders should be transported to the lab within 48 h.

Acceptance criteria: Typically, three cylinders are taken. One cylinder is broken after 7 days. The strength of the 7-day test should be ~65–70% of the 28-day strength. The 7-day break is used for informational purposes only.

Two other cylinders are broken after 28 days. Average of these two cylinders should be equal to or more than the required strength. In addition, none of the 28 days tests should fall below the required strength by more than 500 psi.

It is a good practice to obtain four cylinders. If the 28-day breaks do not reach the required strength, the fourth cylinder can be tested later.

Concrete cylinders should be placed in concrete curing boxes when on-site. During wintertime, these boxes need to be powered and heat should be provided to attain proper temperatures (Plate 3.1).

Plate 3.1 Concrete cylinders.

3.6 SPLITTING TENSILE STRENGTH TEST

Tensile strength of concrete is not utilized for design, and is rarely used. The tensile strength of concrete is found using a splitting tensile strength test.

The cylinder is tested by placing it horizontally. There are correlations between tensile strength and the compressive strength.

3.7 MIXING, TRANSPORTATION, AND PLACEMENT OF CONCRETE (ACI 304R)

It is important to make sure that the design mix has not undergone major changes when the concrete truck arrives at a job site. One major problem that occurs during transportation is segregation. Common sense dictates that heavy particles, such as aggregates, tend to settle at the bottom if given the chance. Hence, it is important that the drums of concrete trucks rotate during transportation.

Another major parameter that affects the strength of concrete is the water content. Higher water content gives rise to low strength and low durability.

It is not possible to reduce the water content indefinitely, since workability will be reduced. It is very difficult to get a smooth finish if the water content is too low.

3.7.1 Concrete Plants

Concrete is a mixture of cement, sand, coarse aggregates, admixtures, and water. As previously discussed, fly ash and various other pozzolans are also added in some cases. The general functioning of a concrete plant is shown in Fig. 3.4.

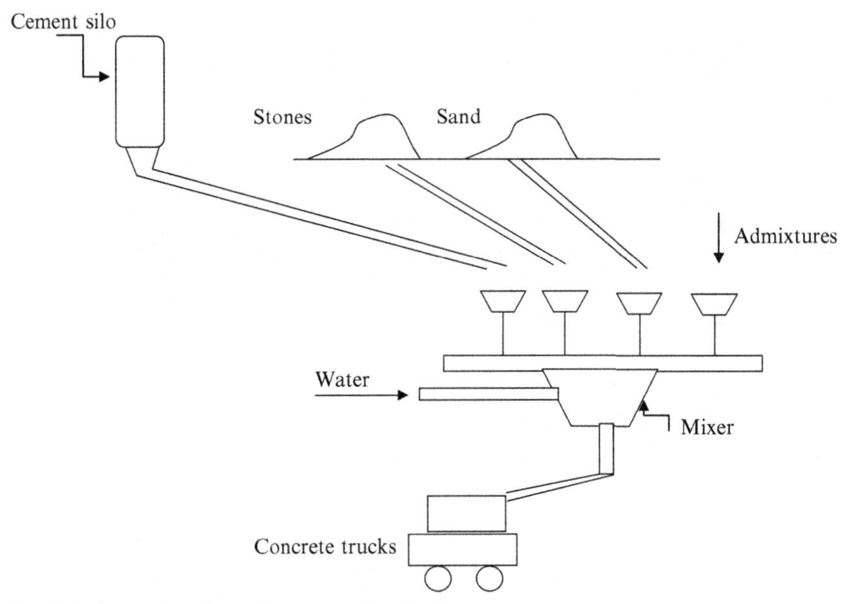

Fig. 3.4 General outline of concrete batch plant.

A concrete plant needs to have storage facilities for cement, fine aggregates, coarse aggregates, chemical admixtures, and water. These ingredients need to be transported to the mixer. Typically, conveyor belts are used to transport materials in a concrete plant. Modern mixers are computer controlled. The operator input the proportions of material that need to be mixed. This data is used by a control system and the correct amount of material is transported to the mixer. Concrete trucks are placed just under the mixer. In some plants, the mixing is done inside the truck while it is traveling.

3.7.2 Storage of Material

Some concrete plants tend to store sand and coarse aggregates outside where they are exposed to the elements. Indoor silos for sand and coarse aggregates are expensive. However storing sand and coarse aggregates in the open allows dust and other contaminants to get into stones and sand. In addition, stones and sand will freeze in the winter time.

3.7.3 Cement Silos

The storage of cement should be carefully planned. Cement tends to absorb moisture. Hence, it is important to make sure that cement silos are free of moisture. As per ACI 304R, a cement silo should have a minimum bottom slope of 50 degrees from the horizontal for a circular silo and 55–60 degrees for a rectangular silo.

3.8 CONCRETE MIXING

Concrete mixing can be done in the plant or in the truck. If the concrete has to travel far, then it is advantageous to mix the concrete in the truck. A well-proportioned mix is dropped into the truck and the mixer inside the truck combines the concrete while going to the job site. This can save time. On the other hand, mixing can be done in the plant. Either way, it is important for the concrete truck to have a rotating drum to make sure segregation does not occur.

ACI 304R prefers all water to be added at the plant so that the water content can be properly controlled. Some water can be added at the job site to obtain the correct slump.

Concrete mixing at the job site: For small projects, mixing of concrete can be done at the job site. Portable mixers of various sizes are available.

3.9 CONCRETE PLACEMENT

Once the concrete arrives at a jobsite, concrete placement can be done with much different types of equipment.

3.9.1 Concrete Buckets

Concrete trucks dump the concrete into buckets. Concrete buckets are then lifted by a crane to the proper elevation for concreting.

It is important to clean the concrete bucket at the end of each day so that the openings are not obstructed by hardened concrete. ACI 304R recommends the side slopes of the bucket to be at least 60 degrees from the horizontal (Plate 3.2).

Plate 3.2 Concreting with a bucket.

3.9.2 Concrete Buggies

Concrete buggies are used for transporting concrete horizontally. Concrete tends to segregate during the transportation of concrete using buggies. Hence, it is not a very good way of transporting concrete. ACI limits the concrete transportation distance to 200 ft for manually operated buggies and 1000 ft for power buggies. To minimize segregation, rails should be provided for the buggies. Hence, transportation can be made smooth. If rails cannot be provided, the surface should be made smooth as possible.

3.9.3 Concrete Chutes

Concrete chutes are typically used to transport concrete from a higher elevation to a lower elevation. ACI does not give a maximum allowed length that can be used to transport concrete using chutes.

Concrete chutes should have rounded corners. The slope should be steep enough for the concrete to travel freely.

3.10 CONCRETE PUMPING

Concrete pumping has become very popular in large projects. Concrete pumping requires less labor and is easy to control. The height of pumping is dependent on the size of the pump. Concrete pump design mixes typically have a higher slump compared to regular concrete mix. To obtain a high slump, one needs to increase the water content. Increasing the water content affects the strength. Hence, water-reducing admixtures are used to obtain a higher slump without increasing the water content.

Concrete pumping pipes can be rigid or flexible. Rigid pipes will have fewer problems during pumping. The major disadvantage of rigid piping is that it is difficult to handle, whereas workers can take the flexible pipes to the placement location more easily. Concrete pumps have a maximum rate of flow and a maximum pressure. Both cannot be achieved at the same time. One foot of additional vertical height is equal to 3–4 ft of additional horizontal distance (Plates 3.3 and 3.4).

Plate 3.3 Concrete pumping.

Plate 3.4 Concrete pump truck. The hoses are folded when not being used.

3.10.1 Pumping During Cold Weather

Major problems can occur when pumping concrete during the winter months. Some common occurrences when pumping in cold weather conditions are freezing pipes, valves, and concrete, among other problems.

As per ACI, the maximum size of angular aggregates should be less than one-third of the internal diameter of the pipe. The maximum size of round aggregates can be two-fifths of the internal diameter of the pipe. If coarse aggregate sizes are larger than these recommended values, concrete blockages can occur.

Fine aggregates also play a major role in concrete pumping. Fine aggregates, water, and sand combines together to make a paste. If this paste consists of large particles, pumping is difficult. Therefore it is important to control the size of fine aggregates (sand).

ACI recommendations on size of fine aggregates:

- At least 15–30% should pass through a No. 50 sieve and
- 5–10% should pass through a No. 100 sieve

Pumping of lightweight concrete: Lightweight concrete is needed to reduce the weight on metal decks, bridge decks, and high-rise building slabs. The density of lightweight concrete is between 115 and 120 pcf while the density of normal weight concrete is around 145 pcf. Lightweight concrete is achieved by using light material. Density of cement and water cannot be changed. Lightweight concrete is achieved by using lightweight aggregates and lightweight sand.

Lightweight aggregates: Lightweight aggregates are typically manufactured by burning shale or clay and then fragmenting them into pieces. Clay or shale tend to expand when they are burned. Hence, the density of the material decreases. Some naturally occurring sands may be less dense than normal sand. In many cases lightweight sand needs to be transported from far away quarries. Since the transportation cost is a factor, lightweight sand may be costlier than normal sand.

It is important to make sure that the aggregates are fully saturated, prior to pumping lightweight concrete. If the aggregates are not saturated, they tend to rise to the top of the pipe. This creates segregation of the materials (Fig. 3.5).

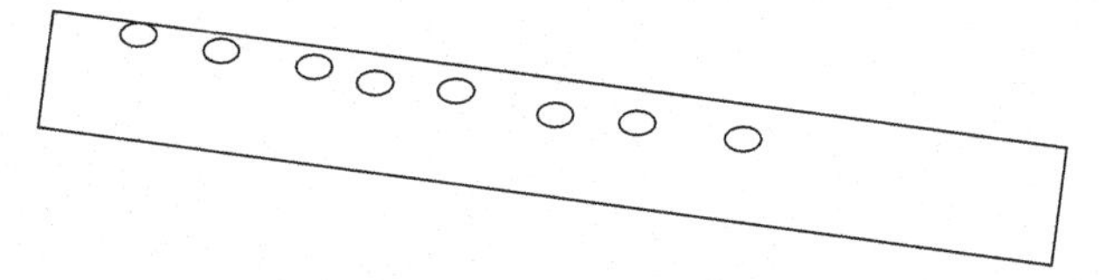

Fig. 3.5 Lightweight aggregates tend to rise to top of the pipe. Hence, light weight aggregates need to be fully saturated prior to mixing.

3.11 TREMIE PIPES

Tremie pipes are used to concrete piers, caissons, and bridge substructures. ACI recommends that tremie pipes are between 8 and 12 in. diameter. Tremie pipes should be embedded 3–5 in. in fresh concrete (Fig. 3.6).

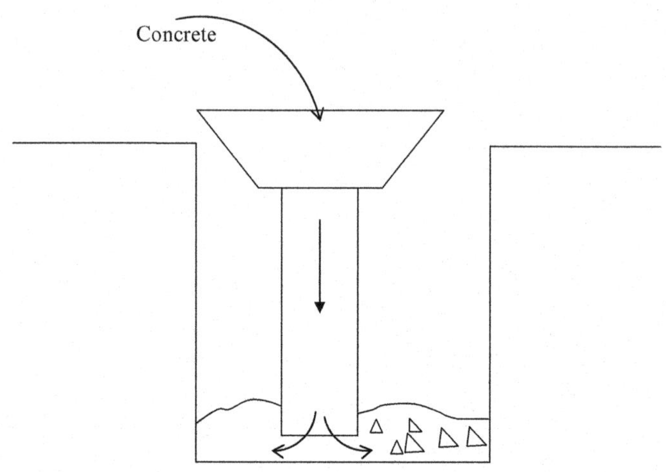

Fig. 3.6 Free fall of concrete through a Tremie pipe.

3.11.1 Free Fall of Concrete

ACI does not limit the height from which concrete can free fall. AS per ACI 304R, as long as the free fall path of concrete does not have rebars or any other obstructions, segregation is not an issue.

> *The stream of concrete should not be separated by falling freely over rods, spacers, reinforcement, or other embedded materials. If forms are sufficiently open and clear so that the concrete is not disturbed in a vertical fall into place, direct discharge without the use of hoppers, trunks or chutes is favorable. Concrete should be deposited at or near its final position because it tends to segregate when it has to be flowed laterally into place.*
>
> ***ACI 304R***

3.12 CONCRETE VIBRATION (CONCRETE CONSOLIDATION) ACI 309

ACI (American Concrete Institute) document 309 deals with concrete consolidation or vibration.

The following is a list of benefits that can be obtained due to the proper vibration of concrete:

- Higher compressive strength
- Higher bond between rebars and concrete
- Increase bond at cold joints
- Reduction of honeycombing and air pockets inside concrete
- Avoidance of segregation of cement paste and aggregates

3.12.1 External Vibrators

In some cases, it is not possible to insert a regular vibrator into the concrete. In such situations, external vibrators are used. External vibrators are attached to the formwork and a motor inside the mechanism creates vibration.

Adverse conditions due to inadequate vibrating;

Inadequate vibration can cause:

- Honeycombing
- Voids due to high air entrapment—Proper vibration allowing air to escape.
- Sandstreaking—Sandstreaking is the loss of cement water paste due to excessive bleeding. When cement water paste bleeds out between the form and mass of concrete, sand lines are exposed. Proper vibration can be useful in avoiding sandstreaking. Not enough fines in the concrete mix could also lead to sandstreaking.
- Placement lines—When concrete is not placed with one truck, there can be a time lag of up to a half hour in some cases between trucks. Placement lines will be visible if concrete is not properly vibrated.

3.13 CONCRETE FINISHING

Concrete finishing, also known as concrete screeding, is the process of achieving a smooth concrete surface. The most common screeding apparatus is the bull float. A bull float can be used for average sized concrete slabs.

Using bull floats and hand trowels may be time consuming for larger slabs. In such situations, concrete trowel machines are used. Trowel machines come in various sizes.

In some instances, concrete is finished with a broom to get a rough surface. This type of finish is required when epoxy or other layers are to be placed on the concrete. The rough surface that is obtained will be useful in providing friction so that whatever is going on the top will properly adhere to the concrete.

3.13.1 Ride-on Trowel Machines

Large projects typically use ride-on trowel machines. A person can ride these machines and trowelling can be done faster.

3.14 CONCRETE GRINDING

Not all projects proceed smoothly. It may be necessary to grind the concrete to the correct elevation if a concrete slab is placed at a higher elevation than that which is specified. In such situations, concrete grinding machines are used. Concrete grinding machines have a cutter that rotates in order to grind the concrete to the correct elevation.

3.15 CONCRETE SCARIFIERS

To obtain a rough surface in the concrete, scarification is used. Some situations require rough surfaces, these include certain topping slabs, to avoid tripping, and certain epoxies, etc. In addition, clarifiers are used to remove concrete coatings.

3.16 TOLERANCES

Concrete elements can never be constructed to the exact dimensions. For example, column may be slightly off from the vertical or a slab may not be 100% flat. ACI 117 provides guidelines for tolerances for concrete construction.

Tolerance for vertical walls: As per ACI 117, a wall height of <83 ft 4 in. should have a tolerance of <1 in. or 0.3% of the height. For walls taller than 84 ft 4 in., the tolerance is lesser of 0.1% times the height or 6 in.

Practice Problem 3.1

Find the tolerance at the top for the vertical wall shown in Fig. 3.7.

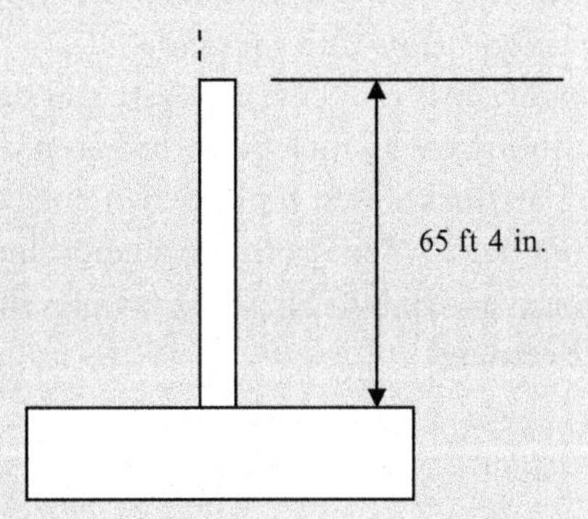

Fig. 3.7 Tolerance of a vertical wall.

Solution

This wall is less 83 ft 4 in. Hence, the tolerance is <1 in. or 0.3% of the height.

$$0.3\%\ \text{of height} = 0.3/100 \times (65.33)$$
$$= 0.196\,\text{ft} = 2.35\,\text{in.}$$

One inch is <2.35 in.

Tolerance at the top of wall = 1 in.

Tolerance for horizontal distance between concrete elements: In most cases, 1-in. tolerance is recommended for distance between concrete elements (Fig. 3.8).

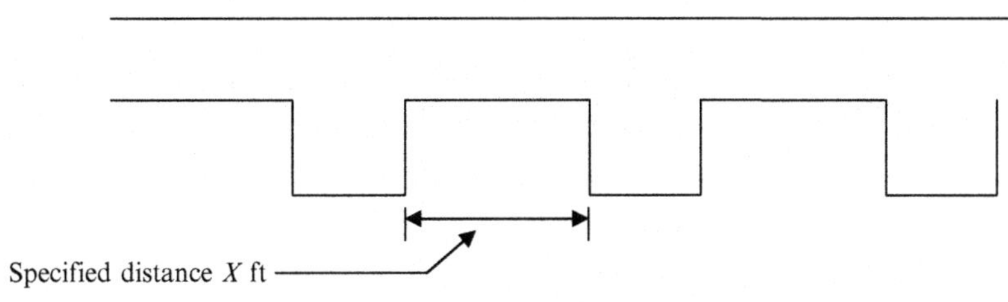

Fig. 3.8 Horizontal tolerance.

$$\text{Acceptable distance} = X\,\text{ft} \pm 1\,\text{in.}$$

Tolerance for soil grading below slabs on grade: Soil grading below slabs on grade is important. If the soil were at a higher elevation, then the slab thickness would be reduced. ACI 117 recommends three-fourths of an inch of tolerance for soil elevation below slab on grade.

Tolerance for slab thickness: If one core is taken, the slab thickness should not be less than three-fourths of an inch from the given value. ACI 117 does not mention of larger slab thicknesses. If more than one core sample is taken, the thickness should not be <3/8 in. from the given value. At least one sample should be taken for every 10,000 SF. Core samples should be taken after 7 days of concrete placement.

Practice Problem 3.2

Three core samples were taken for a 14-in. slab. The thicknesses of the three core samples are 13.25, 13.12, and 13.32 in. Is this slab within the tolerance limit?

Solution

$$\text{Average thickness} = (13.25 + 13.12 + 13.32)/3 = 13.23$$

$$\text{Deviation} = 14 - 13.23 = 0.77\,\text{in.} \quad (\text{Tolerance limit} = 3/8\,\text{in. or } 0.375\,\text{in.})$$

The slab exceeds the tolerance limit and corrective measures should be taken.

3.17 COLD WEATHER CONCRETING

The process of cement reacting with water to create a binder is called hydration. The hydration process can slow down when the temperature is low. Hence, cement water binder could take longer to form or might not form at all. Another cold weather problem is freezing water. When water freezes, the intended chemical reactions do not take place.

ACI 306 defines cold weather as follows:

Cold weather is defined as a period when, for more than 3 consecutive days, the following conditions exist:

(1) The average daily air temperature is <40°F (5°C) and
(2) The air temperature is not >50°F (10°C) for more than one-half of any 24-h period.

The average daily air temperature is the average of the highest and the lowest temperatures occurring during the period from midnight to midnight.

For these reasons, concrete inspectors must pay attention to the temperature to determine if cold weather conditions, as defined by ACI 306, exist.

Solutions for cold weather:

- Heat the aggregates
- Use hot water for mixing
- Lay blankets and sprinkle hot water
- Provide heat to formwork
- Provide heat to metal deck or other steel elements attached to wet concrete
- Provide heaters to heat the ambient air

3.18 HOT WEATHER CONCRETING (ACI 305)

Concreting during hot weather can lead to many problems. ACI 305 deals with hot weather concreting.

As per ACI 305, the following problems can be expected when concreting in hot weather.

- *Increased rate of slump loss and corresponding tendency to add water at the job site:* As per ACI, adding water at the job site is not a very good practice since it is difficult to control the water cement ratio properly.
- *Increased rate of setting:* Hydration reaction (or concrete setting) will take place at a faster rate during hot weather. This creates difficulty in handling, compacting, and finishing. Another problem of fast setting is the formation of cold joints.
- *Increased tendency for shrinkage cracking:* When the temperature in the surrounding goes down, concrete shrinks. Shrinkage of concrete can generate cracks. It has been reported that concrete poured during hot weather conditions are highly vulnerable to shrinkage cracking.
- *Increased difficulty in controlling entrained air content:* High air content gives rise to low strength and low air content gives rise to cracking during freezing and thawing. Controlling air content is extremely difficult during hot weather conditions.
- *Decreased 28-day and later strengths resulting from high temperature:* Concreting during hot weather can result in low strength concrete.
- *Cold joint formation:* Since the concrete sets fast when poured in high temperatures, cold joints could form in between arrival of concrete trucks (Fig. 3.9).

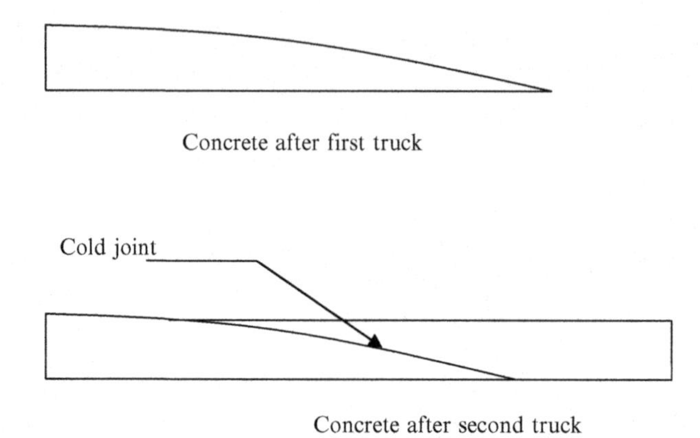

Fig. 3.9 Concrete cold joint.

The formation of cold joints reduces the strength and durability of concrete.

Definition for hot weather concreting: Hot weather is defined as any combination of the following conditions:

- High ambient temperature,
- High concrete temperature,
- Low relative humidity,
- Wind speed, and
- Solar radiation.

It is obvious that high ambient (air) temperature gives rise to high concrete temperature. In addition, it is possible that the temperature of the concrete was high during mixing. Low relative humidity causes water loss and have the same effect as high temperature in many cases.

Solutions for hot weather concreting: If concreting is planned during the summer months, the contractor may need to submit a hot weather-concreting plan. In that plan, the contractor needs to address the issue of hot weather concreting.

Widely used solutions during hot weather:

- *Use ice with mixing water*: Ice absorbs large amount of heat during the melting phase. This keeps the concrete from reaching high temperatures. As per ACI 305, a maximum of 20°F can be reduced by mixing ice with water.
- *Use low hydrating cement:* If the hydration rate of cement is lower, then heat generation is lower at the initial stages.
- *Use of fly ash and other pozzolans:* Pozzolans react with by-products of cement hydration. Hence, the cement amount can be reduced and

pozzolans can be introduced. This decreases the heat generation during hydration.

- *Use of retarding admixtures:* Retarding admixtures increase the setting time. Hence, peak temperatures of the concrete is reduced.

Protecting the concrete from hot weather during curing: Precautions should be taken to protect the concrete from high temperatures during the curing period. Concrete can be kept moist by continuously spraying water. Wet blankets can also be used to keep the concrete at moderate temperatures.

3.19 CONCRETE ELEMENTS

Typically, there are many different types of concrete elements in a structure. Some of the concrete elements are beams, columns, slabs on grade, structural slabs, slabs on metal decks, concrete walls, retaining walls, concrete piles, pile caps, topping slabs, piers, footings, curbs, stairs, equipment pads, conduit encasements, and concrete filled metal pan stairs.

3.20 CONCRETE ACCESSORIES

Reinforcement supports: Rebars in slabs and beams must be supported to obtain proper concrete cover requirements. A concrete cover is needed to protect rebars from water and outside chemicals (Fig. 3.10).

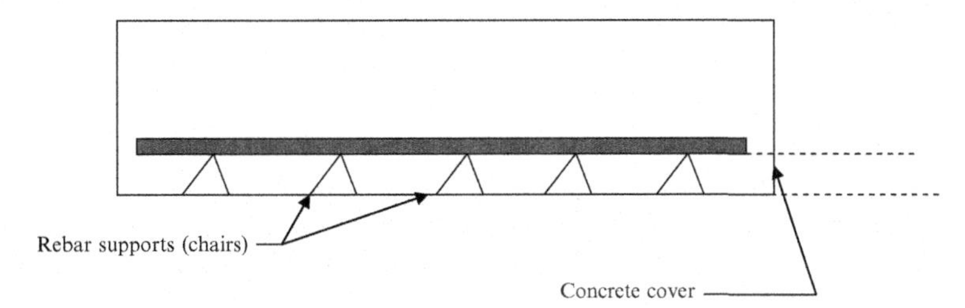

Fig. 3.10 Rebar chairs.

Different types of rebar supports available in the market. Bricks are the most commonly used rebar supports.

3.21 CURING COMPOUNDS

Curing is the process of concrete hardening or hydration. The hardening process needs water. The concrete must be kept moist while it is curing.

This can be achieved by sprinkling or flooding the concrete with water during the curing process. Other methods include wet blankets and curing compounds (Fig. 3.11).

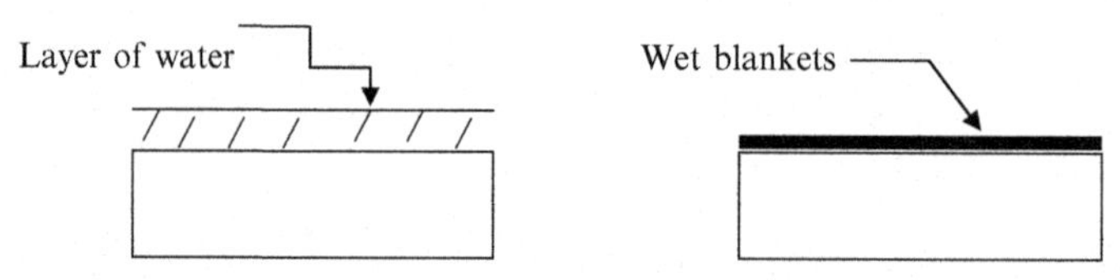

Fig. 3.11 Concrete curing. Left: flooding the concrete with water and right: place a wet blanket and sprinkle water.

Most contractors prefer to apply a curing compound to the concrete. Curing compounds are liquids that seal the concrete surface and keep the water inside. After the concrete is hardened, the curing compound breaks down and the sealing effect will disappear.

Curing compounds essentially act as a sealant during curing. It will not allow water to evaporate from the concrete.

3.22 BONDING ADMIXTURE (BONDING AGENT)

Fresh concrete does not bond well to old concrete. Bonding agents are used to bond new concrete to old concrete, metal to concrete, and concrete to topping slabs.

3.23 WATERSTOPS

Waterstop is an expanding material that is installed in footings and in walls to stop water from entering the concrete. Waterstops are made of PVC, stainless steel, and swellable clays (Fig. 3.12).

Fig. 3.12 shows a concrete footing and a concrete wall installed on top of the footing. The footing is constructed first with rebars sticking out. See Figs. 3.13–3.15.

Most waterstop materials expand when they encounter wet concrete. Due to the expansion, they create a barrier for water. Water coming through the gap is stopped by the water stop material. If only one side is exposed to water, then only one waterstop material is needed.

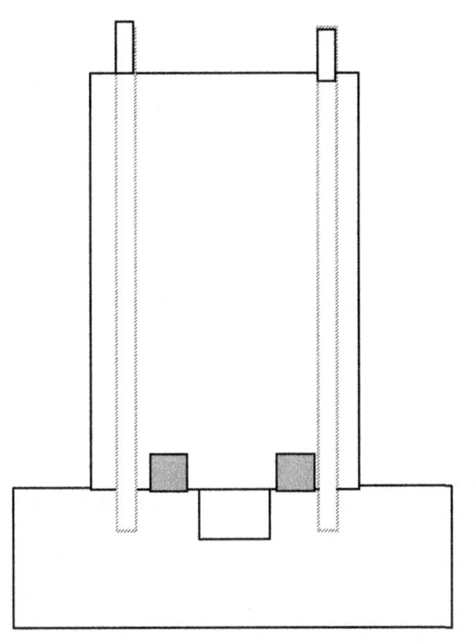

Fig. 3.12 Waterstop material placed between a wall and a footing.

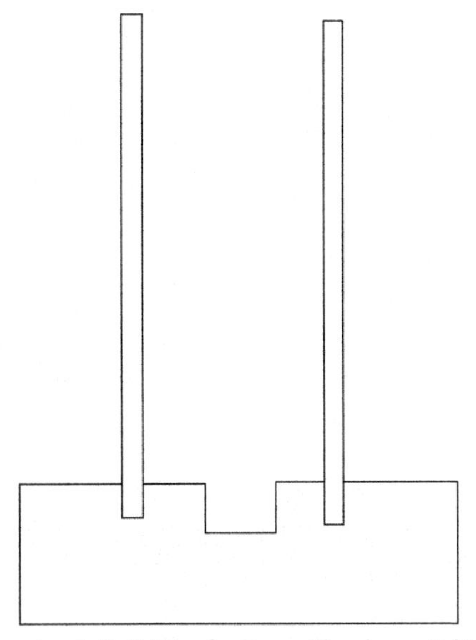

Fig. 3.13 Construction step 1: Build the footing with rebars sticking out. The keyway is installed to provide a good shear bond between wall and the footing.

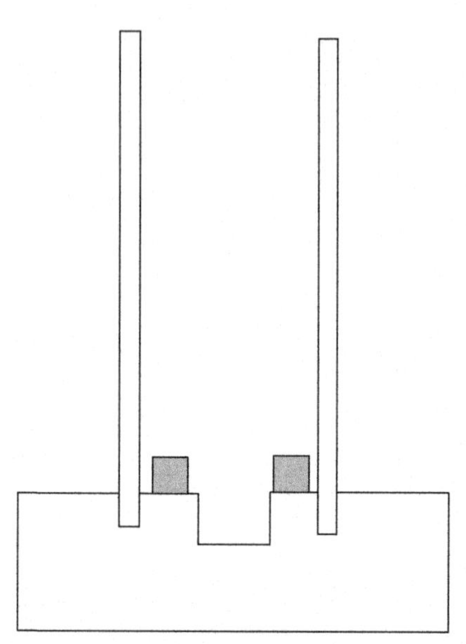

Fig. 3.14 Construction step 2: Install the waterstop material. Waterstop material is a special plastic that expands when soaked with water. Do not place the waterstop material inside the keyway. The keyway need to have a good bond between the wall and the footing against lateral forces.

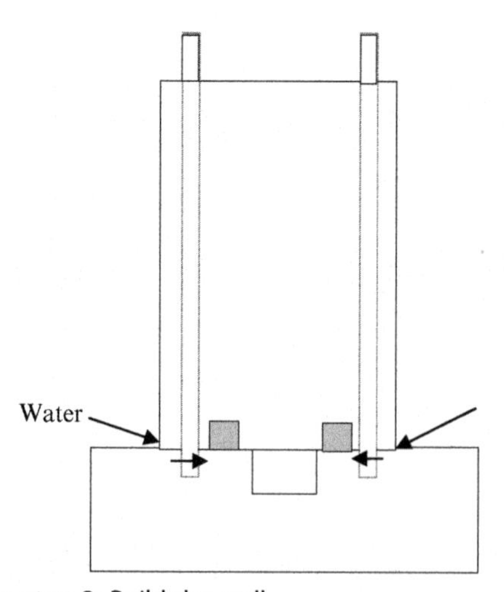

Fig. 3.15 Construction step 3: Build the wall.

3.24 CONCRETE FORMWORK

In many projects, the concrete formwork cost exceeds the cost of concreting. Still, it is vital to pay proper attention to the formwork. In the past, formwork was mostly constructed using timber. Timber is still widely used today. However, timber formwork, though cheaper than metal or plastic formwork, cannot be reused several times due to damage. Hence, other products can be more economical than timber formwork.

3.24.1 Wall Formwork

Wall formwork using timber is shown in Figs. 3.13–3.15.

The main items in a timber wall formwork system are (Fig. 3.16):

(1) Sheathing
(2) Wales
(3) Posts
(4) Buttresses

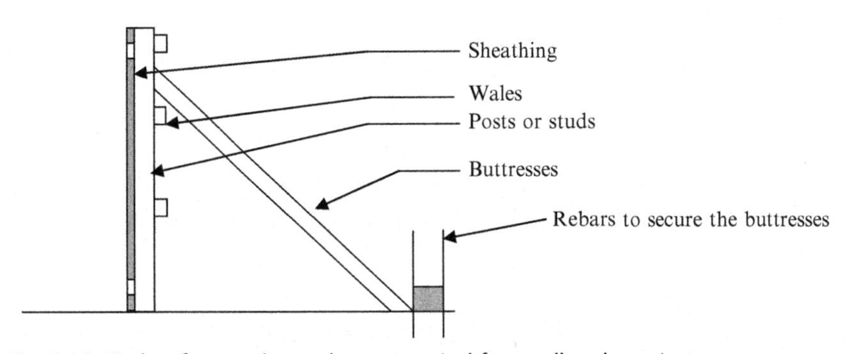

Fig. 3.16 Timber formwork may be economical for small-scale projects.

3.24.2 Prefabricated Formwork

For larger projects, it is more economical to rent or buy prefabricated formwork. In addition, it may be faster to install them than constructing timber formwork (Plate 3.5, Figs. 3.17 and 3.18).

Plate 3.5 The photo shows rebars for a wall. On the left hand side, metal formwork is shown to hold the concrete. Formwork on the right side is not installed yet. Once all the rebars are installed, formwork on right hand side will be installed. Then the wall will be concreted.

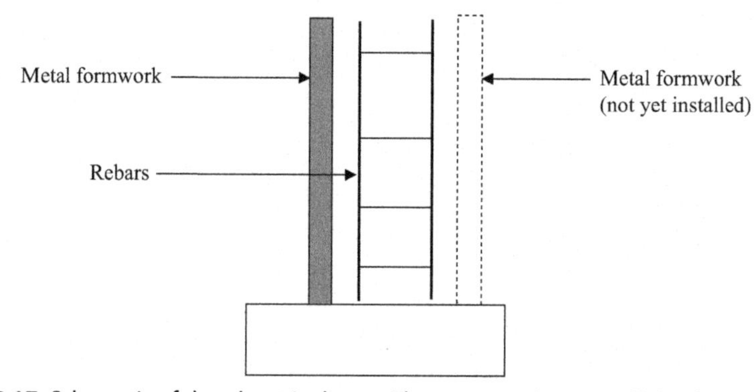

Fig. 3.17 Schematic of the photo is shown. The next step is to install the formwork on the right hand side.

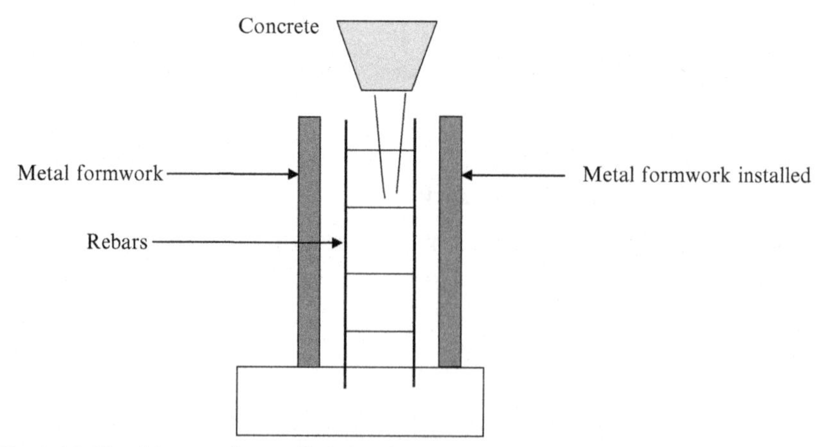

Fig. 3.18 Metal formwork installed. Ready to concrete.

3.24.3 Formwork Material

Plywood forms: Plywood is the cheapest material used to make forms. Plywood forms can be reused many times.

Steel forms: Steel forms can be reused up to 300 times. The cost of steel forms is much higher than plywood forms. Steel forms come in different sizes and shapes, which helps to reduce the work in the field.

Aluminum forms: Since aluminum is lighter than steel, forms made of this material can be erected with fewer men.

Lateral concrete pressure in formwork: Exam questions are highly likely in computing the pressure in formwork during concreting. When concrete is poured, the pressure in concrete starts to build up. When the concrete starts to harden, lateral pressure starts to reduce and becomes zero after concrete is fully hardened.

Lateral concrete pressure assuming concrete to be a liquid (hydrostatic force): Concrete flows freely and at the time of pouring, concrete can be considered to act as a liquid. However, within hours, concrete starts to harden. When concrete is semi hardened, lateral force diminishes. When concrete is fully hardened, lateral force disappears completely and the formwork is not needed.

The following formula can be used to find the hydrostatic pressure:

$$P = \gamma \cdot h$$

where P is the lateral pressure, γ is the concrete density, and h is the concrete height.

Practice Problem 3.3

Find the hydrostatic concrete pressure at the bottom of the formwork. Assume the concrete density to be 145 pcf.

(a) Draw the concrete pressure distribution diagram.

(b) Find the total force acting on the formwork per linear ft (Fig. 3.19).

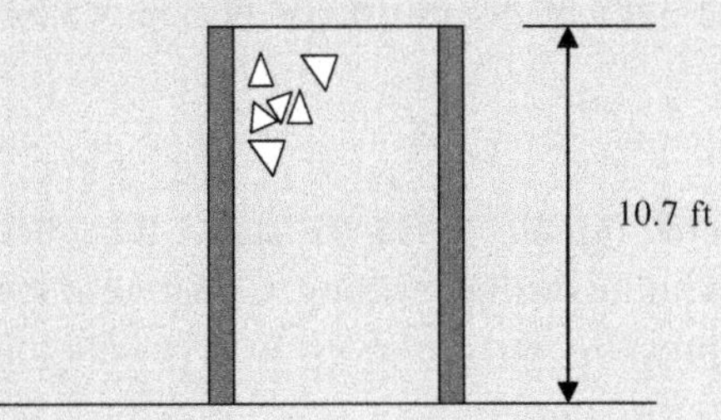

Fig. 3.19 Hydrostatic pressure.

Continued

Solution

$$\gamma = 145\text{pcf},\ h = 10.7$$

$$P = \gamma \cdot h = 145 \times 10.7 = 1551.5\text{psf}$$

Concrete pressure distribution diagram (Fig. 3.20)

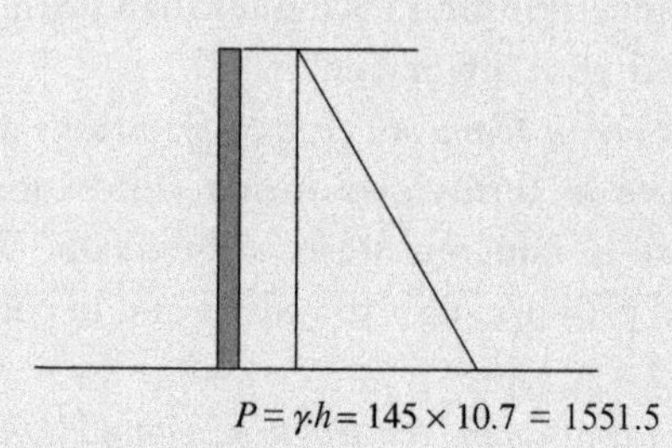

$P = \gamma . h = 145 \times 10.7 = 1551.5$

Fig. 3.20 Hydrostatic pressure diagram.

(b) Total force per linear ft = area of the pressure diagram
Total force per linear ft = 1551.5 × 10.7/2 (area of the pressure triangle) = 8300.5 lbs per linear ft of formwork.

Practice Problem 3.4

If a single form is 10.7 ft high and 3 ft wide, what is the total force on the form in the previous example.

Solution

$$\text{Total force per linear ft} = 8300.5\text{ lbs}$$

$$\text{Total force per 3 ft wide formwork} = 8300.5\text{ lbs} \times 3.0 = 24{,}901.6\text{ lbs}$$

ACI (American Concrete Institute) recommendations: It has been found that concrete does not stay a liquid during pouring. Concrete at lower depths starts to harden and lateral pressure starts to drop. ACI has the following two equations to address this situation.

Equations for wall forms: For wall forms filled at a rate <7 ft/h and total height <14 ft:

$$\boxed{p_{max} = C_W \cdot C_c(150 + 9000 \times R/T)}$$

where p_{max} is the maximum lateral pressure, R is the rate of pouring (ft/h), T is the temperature in degrees (°F), C_w is the unit weight coefficient (see Table 3.1 for ACI 347), C_c is the chemistry coefficient (lateral pressure on formwork depends on concrete type and concrete chemistry), C_c is obtained from ACI 347 Table 3.2.

For wall forms filled at a rate >7 ft/h but <10 ft/h, wall height less or higher than 14 ft:

$$\boxed{p_{max} = C_W \cdot C_c(150 + 43,400/T + 2800)R/T}$$

Two rules for the previous equation:

- The maximum pressure value obtained from the previous equation should not be lower than $600C_W$. If it is $<600C_W$, use $600C_W$ as the maximum pressure. Use Table 3.1 to find C_W.
- The maximum pressure value obtained from the previous equation should not exceed "*w·h*." If it exceeds "*w·h*," use "*w·h*" as the maximum pressure (w is the unit weight of concrete in pcf and h is the height of concrete in ft).

Table 3.1 Unit weight coefficient C_W (ACI 347-04)

Unit weight of concrete (w)	C_W
<140 pcf	$C_W = 0.5[1 + (w/145)]$ w is the unit weight of concrete given in pcf or lbs/ft^3 C_W should not be <0.80
140–150 pcf	1.0
>150 pcf	$C_W = w/145$

Table 3.2 Chemistry coefficient C_c (ACI 347-04)

Cement type	C_c (chemistry coefficient)
Cement Types I, II, and III without retarding admixtures	1.0
Cement Types I, II, and III with retarding admixtures	1.2
Other types of cements or blends containing <70% slag or 40% fly ash without retarding admixtures	1.2
Other types of cements or blends containing <70% slag or 40% fly ash with retarding admixtures	1.4
Blends containing more than 70% slag or 40% fly ash	1.4

Practice Problem 3.5

(a) Find the maximum concrete pressure acting on the formwork shown below using the ACI recommended equation. The rate of pouring is 6 ft/h and temperature of concrete is 75°F. Density of concrete is 145 pcf. C_w and C_c coefficients are 1.0.
(b) What is the highest point where maximum pressure occurs?
(c) Draw the pressure diagram
(d) What is the force acting on one linear ft of formwork?
(e) Find the total force acting on the form, if the form is 3 ft wide (Fig. 3.21).

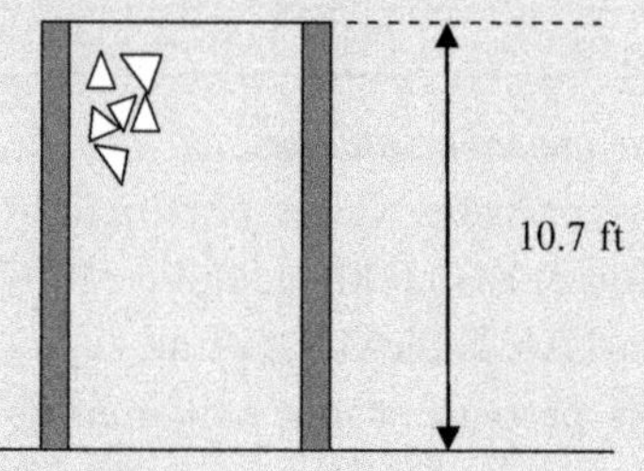

Fig. 3.21 Total force on a formwork.

Solution

(a) Since the rate is <7 ft/h, the following equation should be used:

$$p_{\max} = C_w \cdot C_c(150 + 9000 \times R/T)$$

C_w and C_c are given to be 1.0

$$\begin{aligned} p_{\max} &= 150 + 9000 \times 6/75 \text{ psf} \\ &= 870\text{psf. (This is the maximum pressure as per ACI guidelines).} \end{aligned}$$

(b) Find the highest depth where the maximum pressure occurs.

$$870 = \gamma \cdot h = 145 \text{ h}$$

$$h = 870/145 = 6.0\text{ft}$$

(c) *Draw the pressure diagram* (Fig. 3.22)

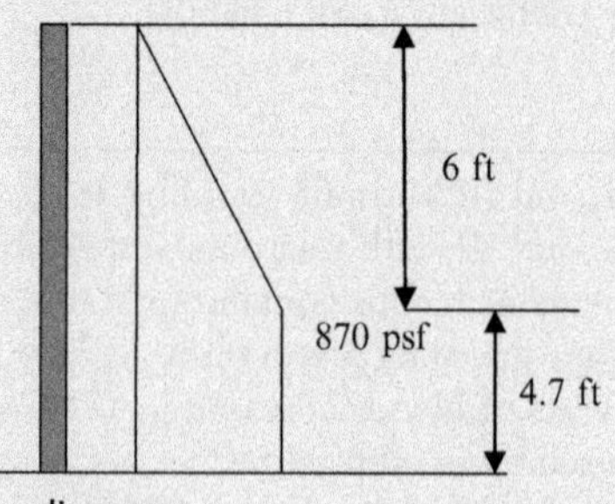

Fig. 3.22 Concrete pressure diagram.

(d) *Find the force per one linear ft of formwork*: Total force is given by the area of the pressure diagram.

$$\text{Area of the pressure diagram} = (870 \times 6/2) + (870 \times 4.7)$$
$$= 6699 \text{lbs per 1 ft wide form}$$

(e) Total force on a 3ft wide form $= 3 \times 6699 \text{lbs} = 20,097 \text{lbs}$

Note that hydrostatic force for a 3 ft wide form was 24,901.6 lbs. (See the previous problem.) (Fig. 3.23)

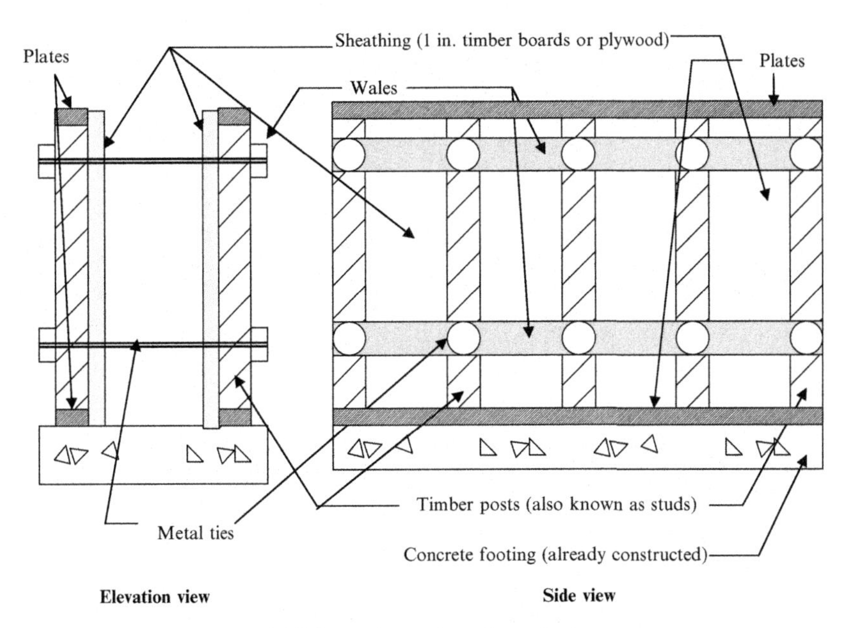

Fig. 3.23 Wall forms.

Sheathing: Sheathing keeps the concrete from falling. Sheathing could be 1 in. thick timber planks or three-quarter in plywood.

Studs (vertical timber posts): Vertical timber posts have to be erected to hold the sheeting. The function of the studs is to hold the sheeting in place. Studs (timber posts) could be 2 × 4 in. timber. Studs are attached to the sheeting using nails.

Wales (horizontal timber): Horizontal timber is erected to hold the studs. These could be 2 × 4 in. or 2 × 2 in. timber depending upon the height. Nails are used to attach wales to the studs.

Plates: Timber placed on the top and bottom of the studs are known as plates. Typically, 2 × 4 in. timber is used for plates. Studs are constructed on top of plates (Fig. 3.24).

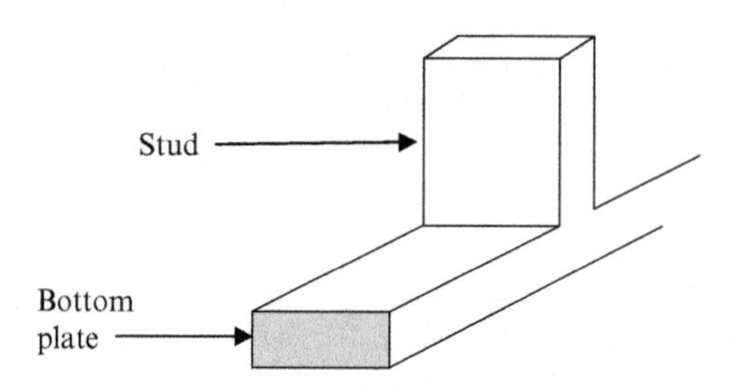

Fig. 3.24 Plates.

Metal ties: Metal ties are used to hold the wales together. They usually break off when the concrete hardens (Fig. 3.25).

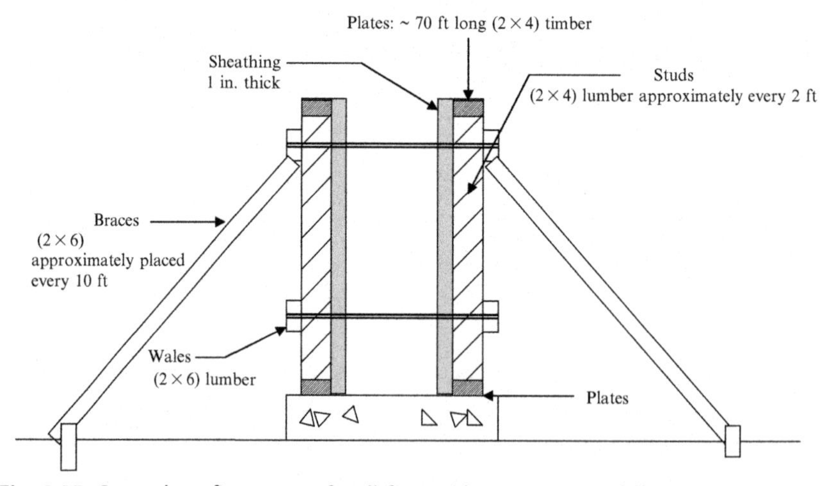

Fig. 3.25 General configuration of wall forms (dimensions are different based on the situation).

Load on sheathing: The load on sheathing due to concrete is considered to be a uniformly distributed load for computation purposes (Fig. 3.26).

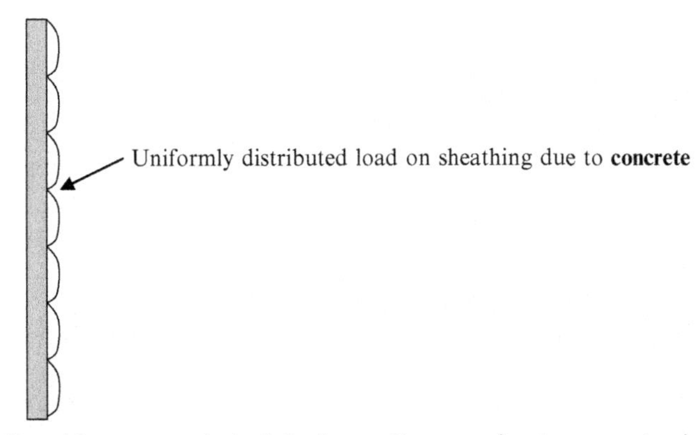

Fig. 3.26 Sheathing or wood plank is shown. Pressure due to concrete also shown.

Load on studs: The load on studs due to sheathing can also be approximated to a uniformly distributed load (Fig. 3.27).

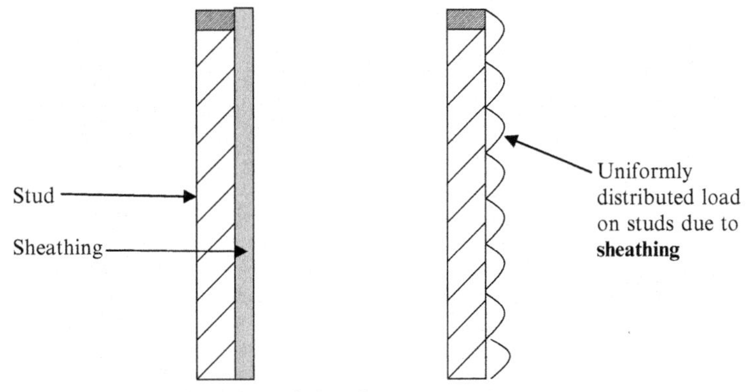

Fig. 3.27 Wall formwork studs and sheathing.

Load on wales due to studs: The loads on wales due to studs are point loads. For practical purposes, load on Wales also considered to be uniformly distributed (Figs. 3.28 and 3.29).

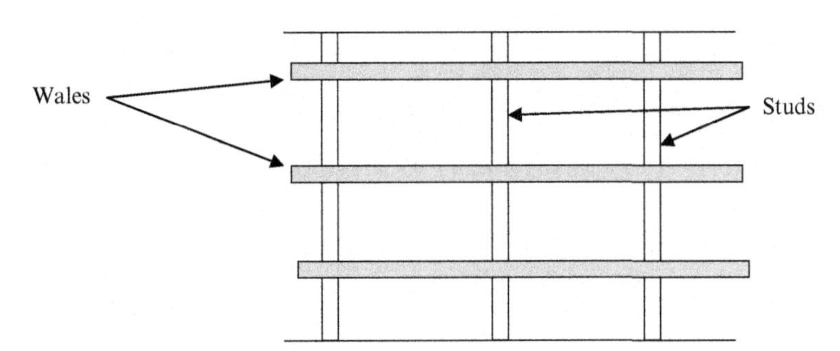

Fig. 3.28 Wales and studs.

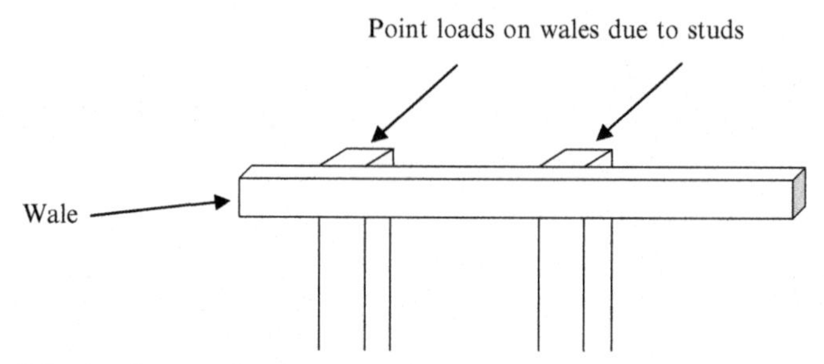

Fig. 3.29 Point loads on wales due to studs.

Formwork for columns: The formwork for columns are built using 1 in. sheathing or three-quarter in plywood. The typical arrangement is shown in Fig. 3.30.

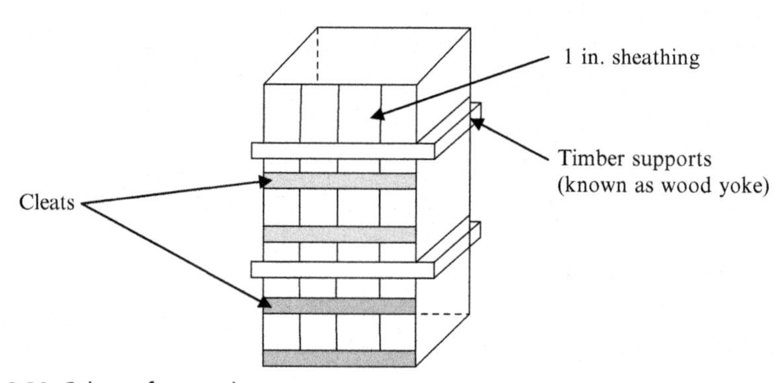

Fig. 3.30 Column formwork.

Elements in a column formwork:

1 in. sheathing: 1 in. sheathing comes in vertical strips as shown.

Cleats: Cleats are nailed to the sheathing. Typically 1 in. thick and 4 in. wide timber is used for cleats.

Timber supports (yoke): The yoke holds the structure together. The yoke is built using 4 × 4 in. timber. Typically, timber supports are held together with bolts as shown in Fig. 3.31, Plates 3.6 and 3.7.

Column configuration (plan view):

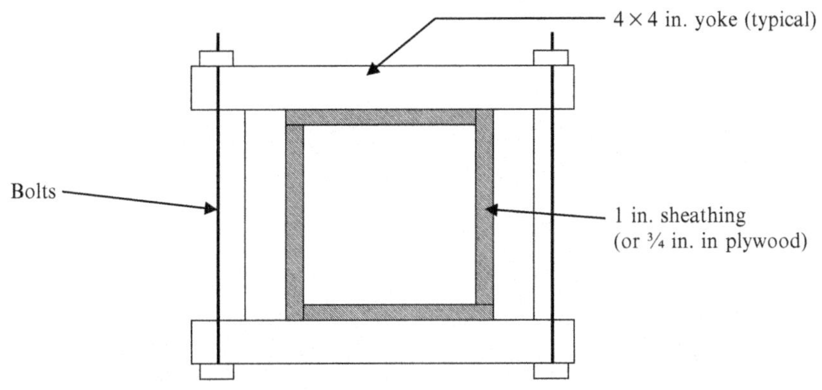

Fig. 3.31 Column formwork—plan view.

Plate 3.6 Column formwork.

Plate 3.7 Columns ready to concrete.

3.24.4 Formwork for Curbs

Curbs are a common element in any building. Formwork for a curb is shown in Fig. 3.32 and Plate 3.8.

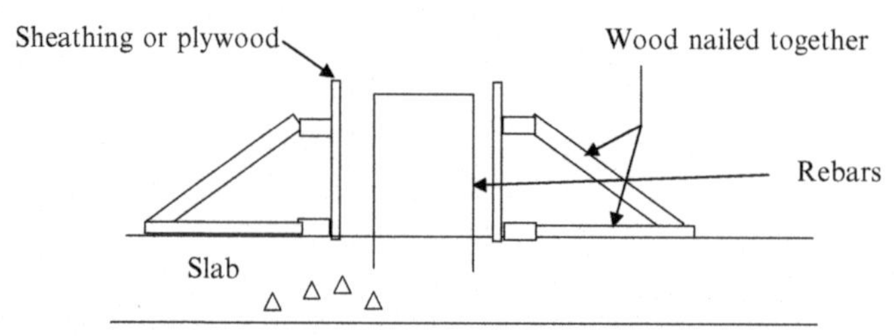

Fig. 3.32 Formwork for a curb.

Plate 3.8 Formwork for a curb is shown. Rebars are extended for a wall on top of the curb.

3.25 SHORING AND RESHORING

Shoring is the process of supporting wet concrete slabs or walls until they harden. Usually shores are placed below the formwork (Fig. 3.33).

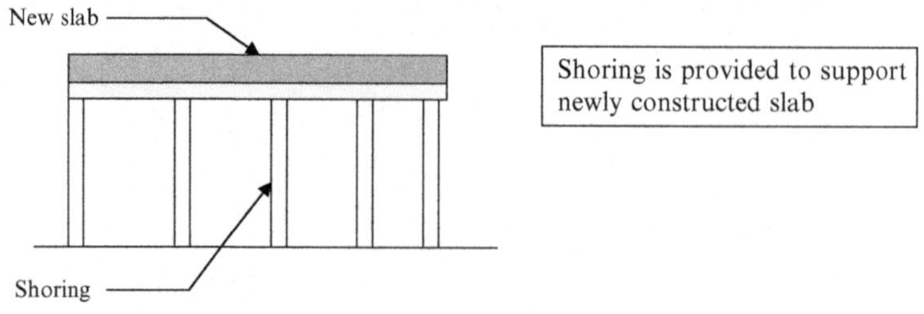

Fig. 3.33 Shoring.

Newly formed slabs are not capable of supporting themselves. With time, concrete will gain strength. In many cases, maximum strength is achieved in ~28 days. During this time, concrete slabs need to be supported.

Lateral support: There are many situations where shoring has failed due to a lack of lateral support. Lateral loads can be significant due to wind, motorized buggies on top of the slab, and vibrations. It is important to consider adequacy of a lateral support during the design of shoring. Another major reason for failure is the removal of shoring prior to a slab gaining adequate strength.

Load transfer during shoring: Shoring should be designed to carry the total slab load + load due to the form + any workers and equipment on top of the slab. It is no secret that concrete is not very strong during the time of pouring.

Reshoring: Reshoring is the process of supporting multiple floor slabs. During the construction of high-rise buildings, multiple floor slabs will be supported by shores. The original shoring in lower floors needs to be removed and reshored to support the higher-level slabs. If the original shoring were not removed, the entire load coming from the top floors would go to the original shoring at the lowest level. Hence, shoring at the lower levels needs to be removed to allow the hardened slab to take its own weight. In addition, the slab should be allowed to deflect during the curing process. If the slab is not allowed to deflect, it could lead to cracks (Plate 3.9).

Plate 3.9 Top most slab is about to be concreted. Wet concrete slab will be supported by shoring. The vertical posts are the shoring. Load from above will be transferred to four floors down using vertical posts.

The following demonstration will show how a load is transferred to shoring and reshores:

First floor slab construction (wet slab):

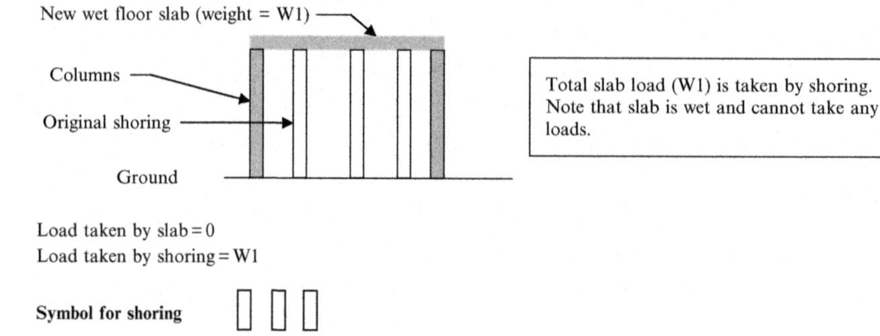

Load taken by slab = 0
Load taken by shoring = W1

Symbol for shoring

Symbol for reshores

Hardened slab:
Hardened floor slab (weight = W1)

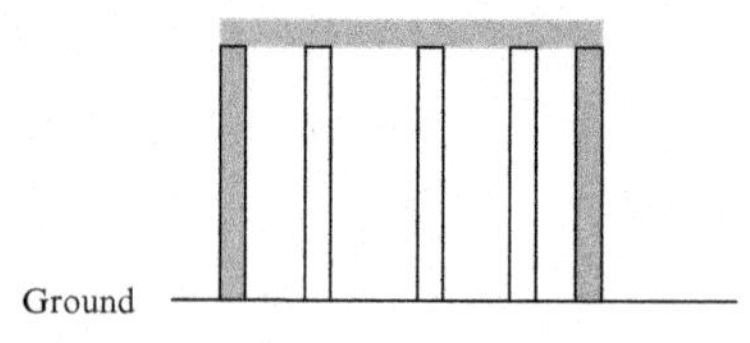

Now the slab is hardened. Total slab load (W1) is still taken by shoring. Still no load is taken by the slab.

Load taken by slab = 0
Until shoring is removed, slab cannot take its own weight. Once the shoring is removed, slab will slightly deflect and take its own weight.
Load taken by shoring = W1

Shoring removed:

Hardened floor slab (weight = W1)

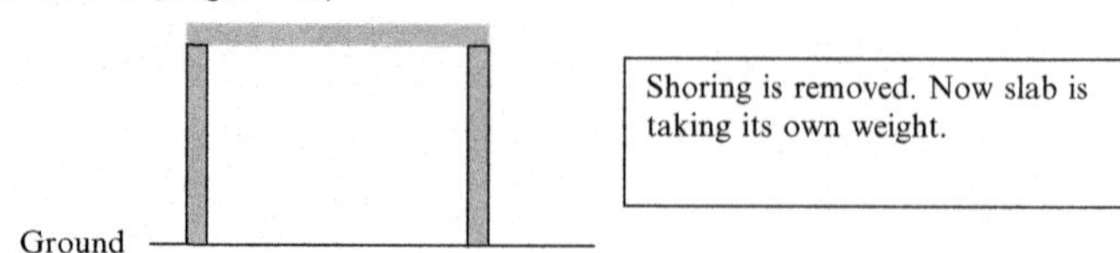

Load taken by slab = W1

Reshores installed:

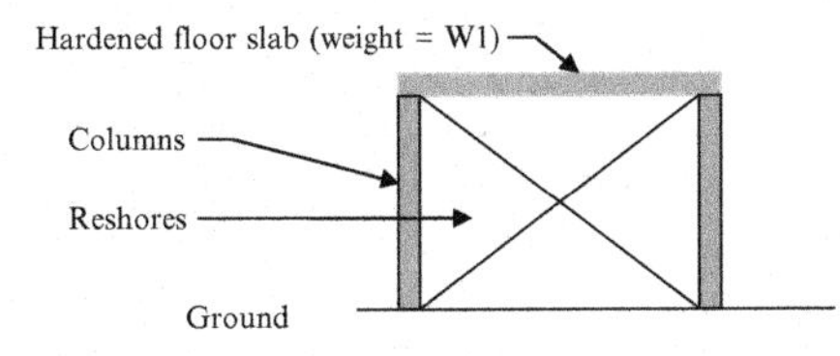

Next, reshores are installed. Reshores are snug fitted. Snug fitted means reshores are not loaded. They are just placed under the slab touching the slab.

Load taken by slab = slab dead weight = W1

Load taken by reshores = 0 (reshores are not loaded)

Construct second floor slab:

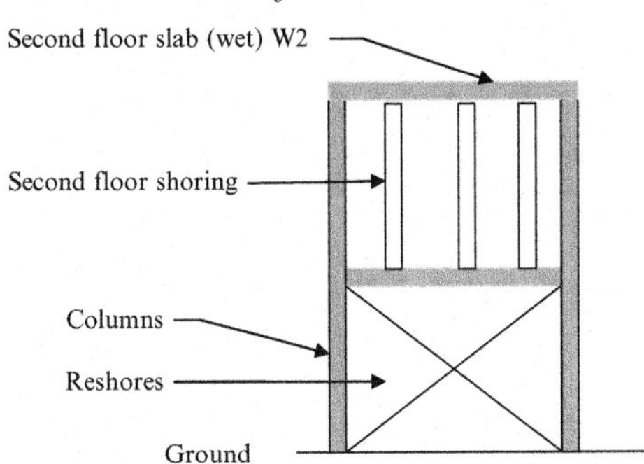

Next, second floor slab is constructed. Second floor slab is wet and will not take any load. It would pass the load to shoring below. Shoring will transfer the load to first floor slab. The first floor slab will try to deflect but it cannot. Reshores would immediately get loaded and absorb the load.

Load taken by second floor slab = 0
Load taken by second floor shoring = W2
Load taken by first floor slab = first floor slab dead weight = W1
Note that first floor slab is carrying its own weight (W1).
W2 weight coming from top will be transferred to the reshores below.
Load taken by reshores = W2 (reshores are now loaded)

Second floor slab hardened. Remove second floor shoring:

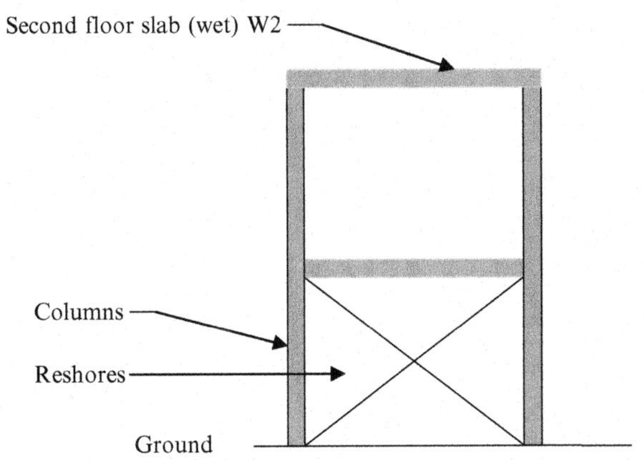

Load taken by second floor slab = W2

Load taken by second floor shoring = 0 (shoring removed)

Load taken by first floor slab = Slab dead weight = W1

Load taken by reshores = 0 (W2 load is taken by the second floor slab)

Second floor slab hardened. Install second floor reshores:

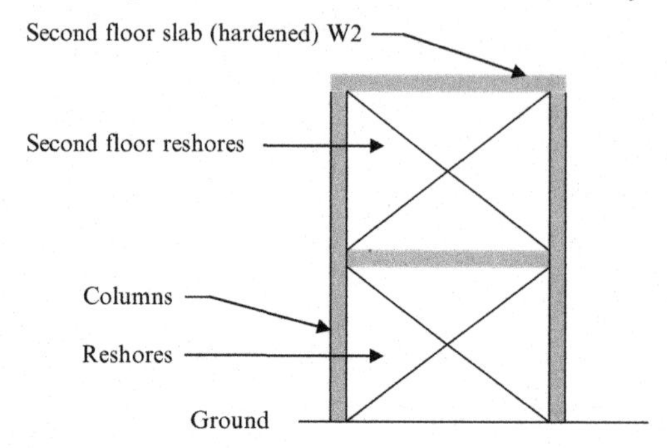

Load taken by second floor slab = W2
Load taken by second floor reshores = 0 (reshores are snug fitted)

Load taken by first floor slab = Slab dead weight = W1
Load taken by first floor reshores = 0 (W2 load is taken by the second floor slab)

Construct the third floor slab (W3):

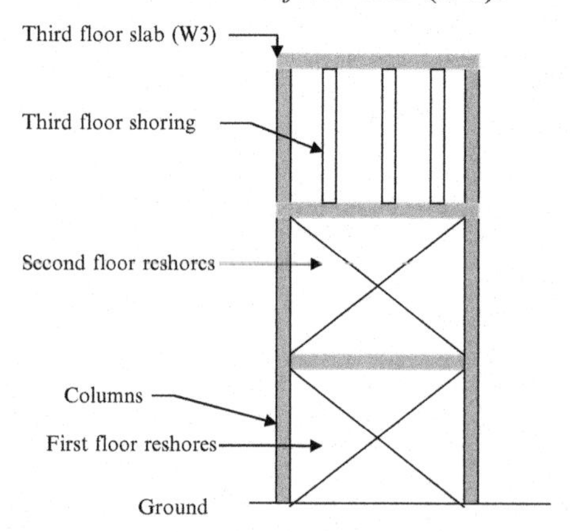

Load taken by third floor wet slab = 0 (slab is wet)
Load taken by third floor shoring = W3
Load taken by second floor slab = W2 (second floor slab takes only its own weight)
Load taken by second floor reshores = W3 (W3 load coming from top will be transferred to the reshores)
Load taken by first floor slab = Slab dead weight = W1
Load taken by first floor reshores = W3
Basically new slab weight (W3) is transferred to third floor shoring to second floor reshores to first floor reshores and finally to the ground. Slabs will not take any of W3 load. Slabs will be taking only their own dead weight.

Remove third floor shoring and install third floor reshores. Also remove the first floor reshore:

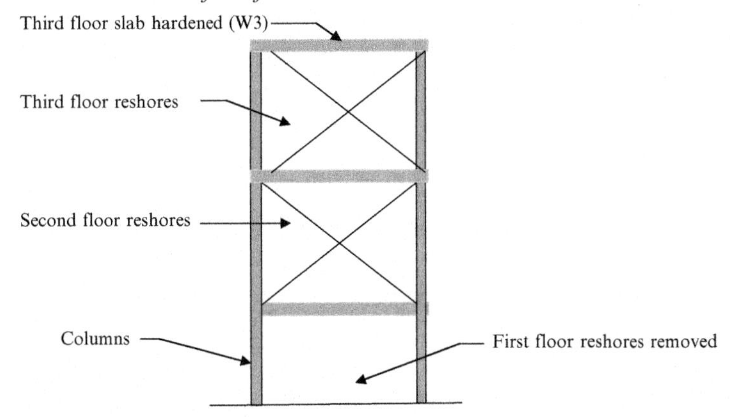

Load taken by third floor hardened slab = W3
Load taken by third floor reshores = 0 (Reshores are snug fitted).

Load taken by second floor slab = W2 (2nd floor slab takes only its own weight)
Load taken by second floor reshores = 0 (W3 load is taken by the hardened third floor slab)

Load taken by first floor slab = slab dead weight = W1
first floor reshores are removed.

Construct the fourth floor slab: (*VERY IMPORTANT STEP*):

This step is very important and different from previous steps. Note that new load is NOT transferred to the ground.

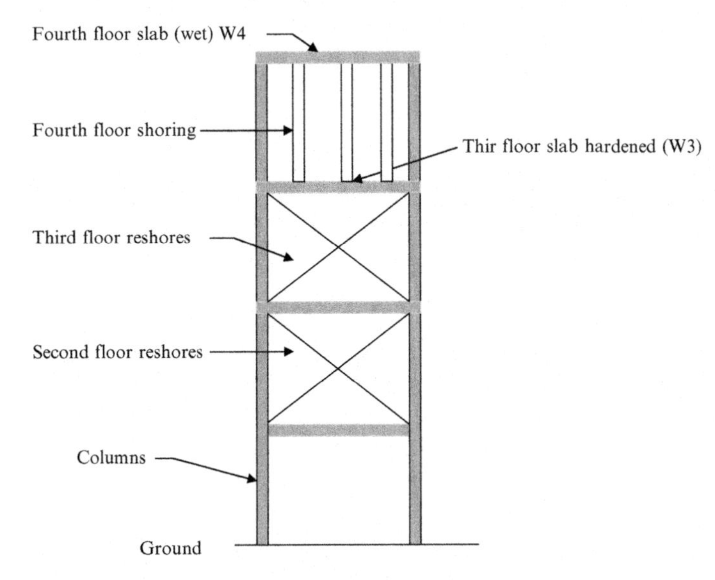

Load taken by Fourth floor slab (wet) = 0
Load taken by Fourth floor shoring = W4

The fourth floor slab is wet and will not take any load. It will transfer its weight to the shoring below.

In the previous instances, the load was transferred to the ground. Now the first floor reshores are removed, there is no way to transfer the W4 weight to ground.

W4 weight has to be shared by three slabs. (First, second, and third floor slabs.)

Each of these slabs will take a load of W4/3.

$$\text{Load taken by third floor hardened slab} = W3 + W4/3$$

Now what is the load taken by the third floor reshores? W4 load is coming from top. W4/3 was taken by the third floor slab. Hence, the remaining slab weight will be transferred to the third floor reshores.

$$\text{Load taken by third floor reshores} = 2W4/3$$

$$\text{Load taken by second floor slab} = W2 + W4/3$$

We will transfer the W4 load equally among three slabs. Hence, the second floor slab will take its own weight and W4/3.

Now what is the load transferred to second floor reshores?

The load coming from third floor reshores is 2 W4/3. Out of this load, W4/3 was taken by the second floor slab. The remaining load need to be taken by the second floor reshores.

$$\text{Load taken by second floor reshores} = W4/3$$

$$\text{Load taken by first floor slab} = W1 + W4/3$$

The W4 load is fully transferred between the three slabs. Subsequently, this process is repeated for upper floors.

This step is very important since each slab is taking its own weight plus a share of the fourth floor slab. If a slab were to fail, it would happen in this step. Previously, slabs were taking only their own weights.

Loads taken by slabs and reshores are shown below;

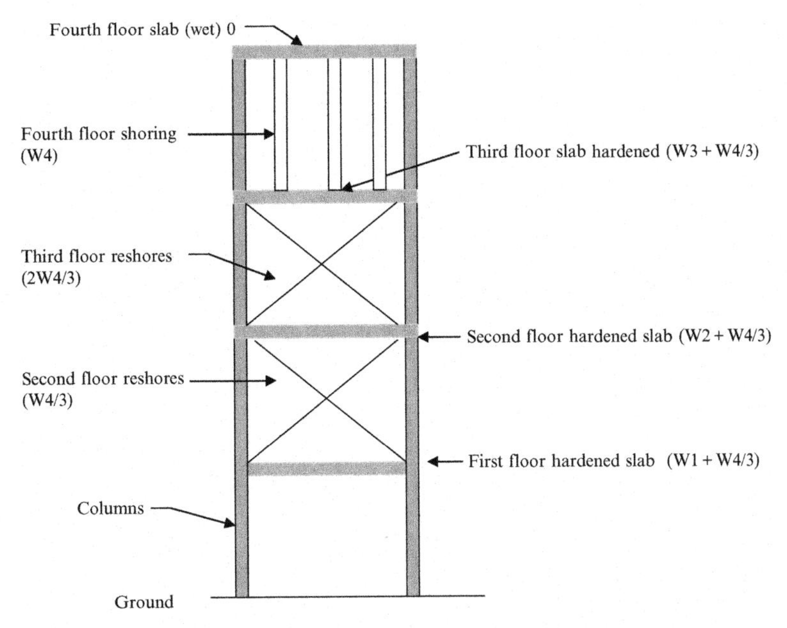

ACI 347 example:

Now we are ready to discuss the method given in ACI 347. This is similar to the method given previously.

Building slabs are shored/reshored with one level of shoring and two levels of reshoring.

Following loads are given:

Weight of each slab = D

Construction live load = $0.4D$

Shore and form weight = $0.1D$

Reshore weight is ignored

Construction live load is the load due to workers and their tools and vehicles.

Symbol for shoring

Symbol for reshoring

Stage 1: Slab is wet and will not carry any load. Total load will be taken by shoring.
Total load taken by shoring = $1.5D$ ($1D + 0.4D + 0.1D$)
Load taken by wet slab = 0

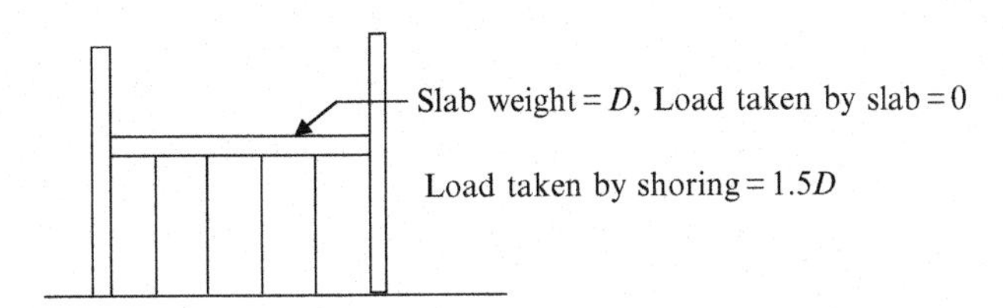

Stage 2: Once the slab is hardened, shoring is removed and reshores are installed. Reshores are lightly fitted or snug fitted. Hence reshores are not loaded. Once the slab is hardened there is no need of having any construction load on top of the slab. Hence construction live load = 0. Slab will take its own weight.

Load taken by slab = D

Load taken by reshores = 0

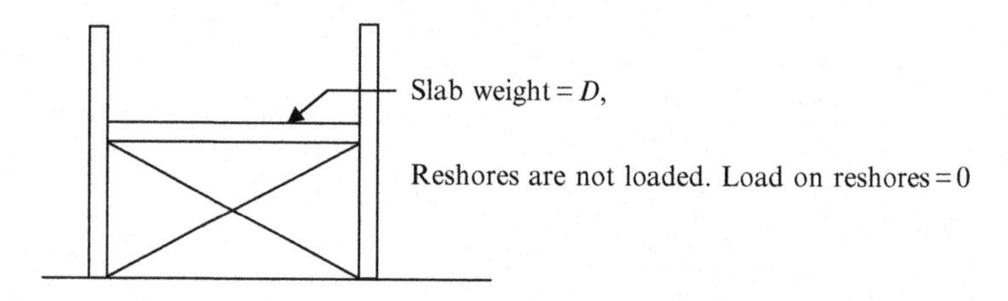

Stage 3: Shores are installed on top of hardened first floor slab and second floor slab is constructed.
Load taken by second floor slab = 0 (This slab is still wet)
Load taken by shores between first and second floor = $1.5D$ ($1D + 0.4D + 0.1D$)
Load taken by first floor slab = D (First floor slab will carry its own weight. Any new load will be passed to the reshores below).
Load taken by first floor reshores = $1.5D$

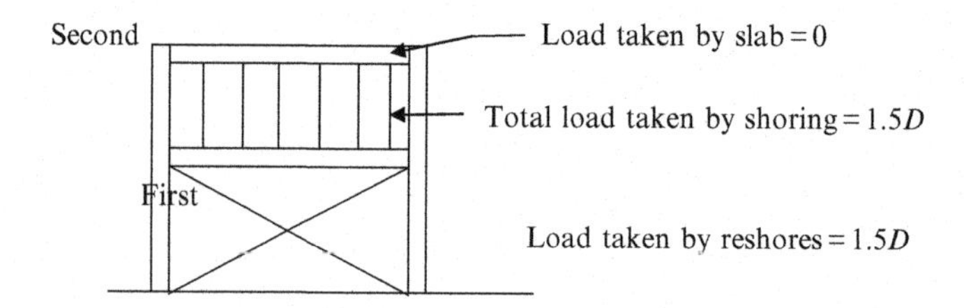

Stage 4: Once the second floor slab is hardened, shoring between first and second floor is removed and reshores are installed. Construction live load and shoring is gone.

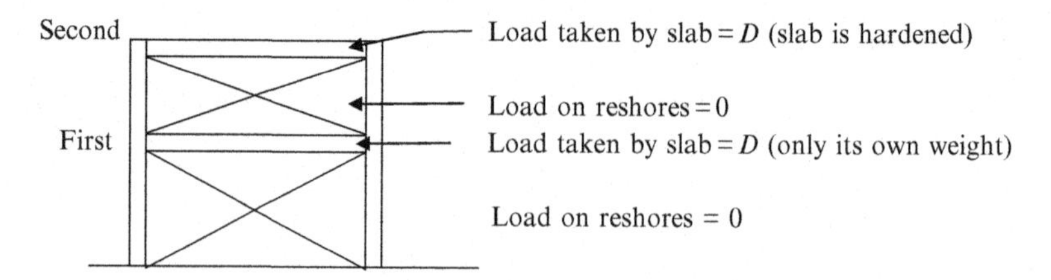

Stage 5: Install shoring on second floor slab and build the third floor slab

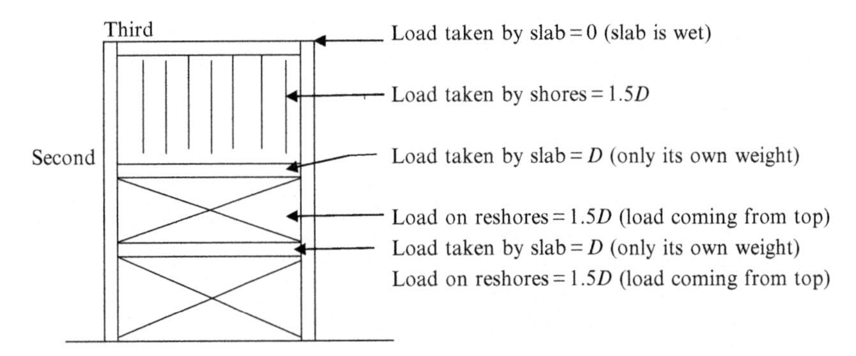

The load coming from the third floor will be taken by shores and then pass to reshores below. Finally the load is transferred to the ground. Slabs will not take any load other than its own weight since slabs are not allowed to deflect.

Stage 6: Remove shoring below third floor and install reshores. Third floor slab is now hardened. Construction live load is gone. In addition, remove reshores on ground floor.

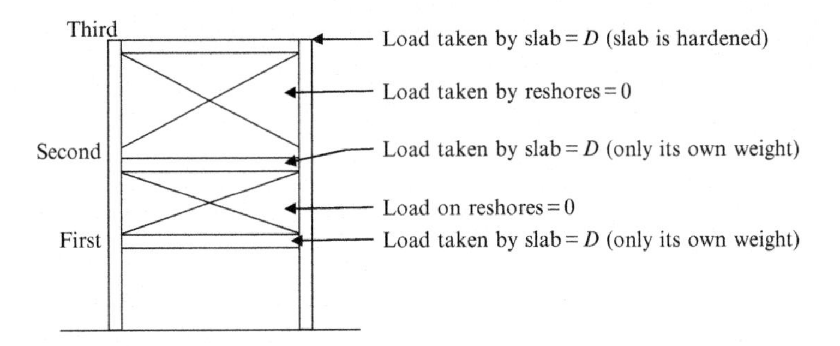

Stage 7: (*Very important step*): Install shoring on third floor and build the fourth floor slab. Now this situation is slightly different from previous cases. If you notice reshoring on ground floor is removed. Hence load coming from top cannot be transferred to ground. Total load on fourth floor is $1.5D$. Where does this load end up? In previous cases, $1.5D$ was directly transferred to ground. Now this $1.5D$ cannot be transferred to ground. It has to be taken by three slabs. (Third, second, and first floor slabs). $1.5D$ is equally distributed among three slabs.

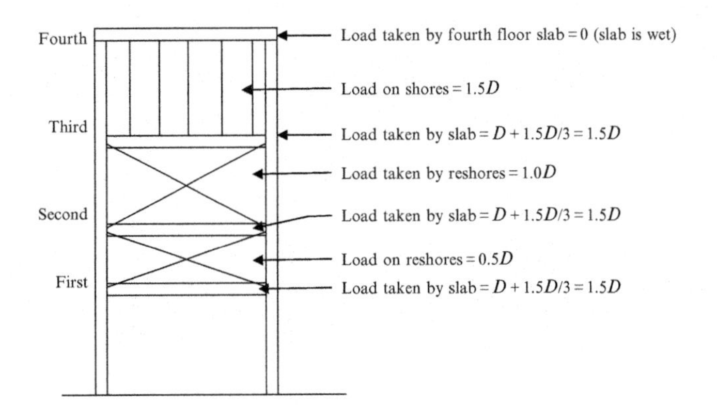

The fourth floor has a load of 1.5D including shoring weight and construction live load. This 1.5D will be taken by shoring below the fourth floor. The third floor slab will take one-third of the load that is coming from top. One-third of 1.5D is 0.5D.

Hence, the third floor slab will take 0.5D plus its own weight of 1.0D. The third floor slab will take a total of 1.5D.

Now let us see how much loading will be taken by reshores below third floor.

The fourth floor loading was 1.5D. 0.5D was already taken by the third floor slab. What is remaining is 1.0D.

Hence, reshores below the third floor will take 1.0D.

We agreed to distribute the 1.5D, equally among three slabs. Hence, loading taken by the second floor slab will be $1.0D+0.5D=1.5D$.

Now how much loading will be taken by reshores below the second floor?

The load coming from reshores above is 1.0D. Out of that 0.5D was taken by the second floor slab. Hence, loading below the second floor reshores would be 0.5D.

Load on first floor slab is 1.5D, since we are equally distributing the 1.5D load among three slabs.

Stage 8: Once the fourth floor slab is finished, workers will leave. Hence construction live load of 0.4D is removed. New load is 1.1D. This amount has to be distributed among three slabs equally.

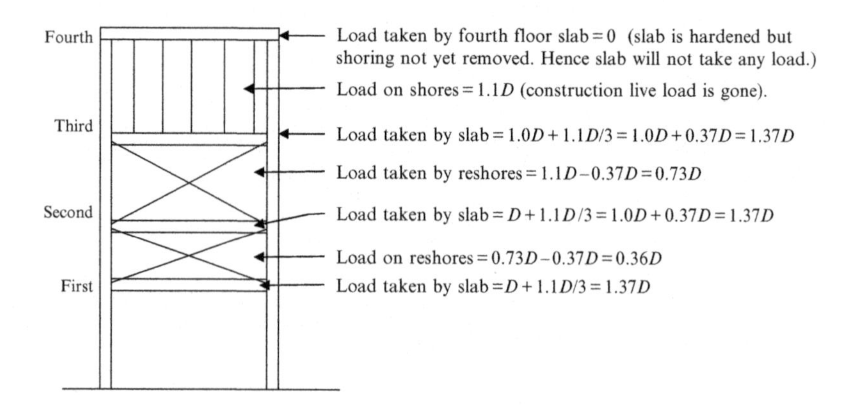

The fourth floor has a load of 1.1D including shoring weight. The construction live load is removed. This 1.1D will be taken by shoring below the fourth floor. The third floor slab will take one-third of the load that is coming from top. One-third of 1.1D is 0.37D.

Hence, the third floor slab will take 0.37D plus its own weight of 1.0D. The third floor slab will take a total of 1.37D.

Now let us see how much loading will be taken by reshores below the third floor.

The fourth floor loading was $1.1D$. Out of that $0.37D$ was already taken by the third floor slab. What is remaining is $1.1D-0.37D=0.73D$.

Hence, reshores below the third floor will take $0.73D$.

We agreed to distribute the $1.1D$, equally between the three slabs. Hence, loading taken by the second floor slab will be $1.0D+1.1D/3=1.37D$.

Now how much loading will be taken by reshores below the second floor?

The load coming from reshores above is $0.73D$. Out of that $0.37D$ was taken by the second floor slab. Hence, loading below the second floor reshores would be $0.73D-0.37D=0.36D$.

The load on first floor slab is $1.37D$, since we are equally distributing the $1.1D$ load among three slabs. (Note that there is rounding of decimals to second decimal.)

3.26 REINFORCEMENT BARS OR REBARS

Concrete is very weak in tension. Therefore rebars are installed to provide tensile strength to concrete.

The typical rebar details for various conditions are shown in Fig. 3.34.

Typical detail for corner of a wall:

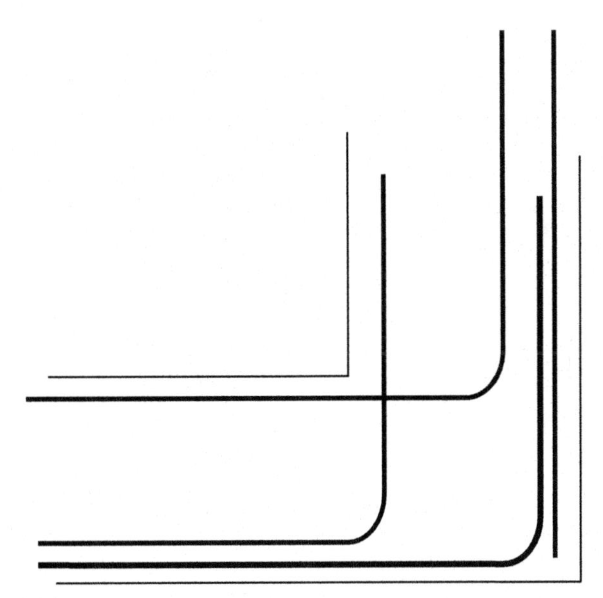

Fig. 3.34 Bent rebars.

Typically, all external bars are hooked. As per ACI, standard hook is 12 × diameter.

Practice Problem 3.6

What is the length of a standard hook for a No. 9 bar? (Fig. 3.35 and Plate 3.10)

Solution

$$\text{As per ACI 318, the standard hook is 12d.}$$

$$\text{The diameter of a No. 9 bar} = 9/8\,\text{in.}$$

$$\text{The standard hook length} = 12 \times 9/8\,\text{in.} = 13.5\,\text{in.}$$

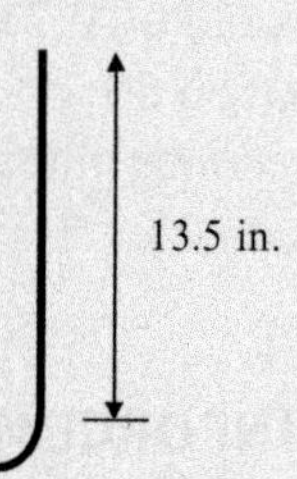

Fig. 3.35 Standard hook.

Plate 3.10 Rebars.

Plate 3.11 shows rebars are extended for a wall above (Fig. 3.36 and Plate 3.12).

Plate 3.11 In this photograph, rebars are extended for a wall and a floor slab.

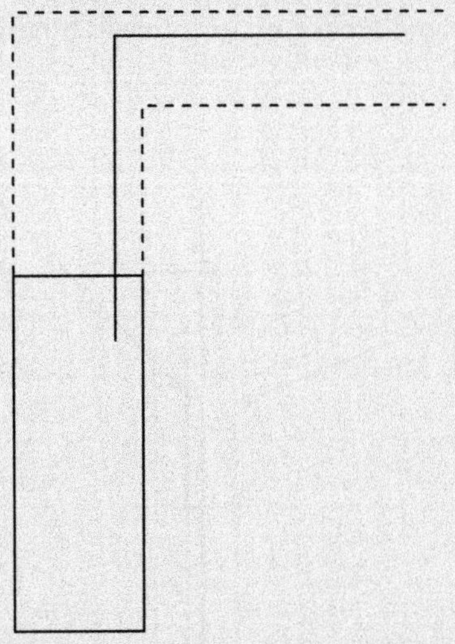

Fig. 3.36 Rebars installed for a future wall and slab. The schematic of the photograph is shown. The dotted line indicates the future construction of the wall portion and the slab. The rebars need to be extended as shown in the photograph, otherwise, they will have to be drilled in.

Continued

Plate 3.12 Rebars can be bent in the field.

3.27 DOWELS OR REBARS FOR MASONRY WALLS

Typically, dowels are provided for masonry walls by the concrete contractor (Fig. 3.37).

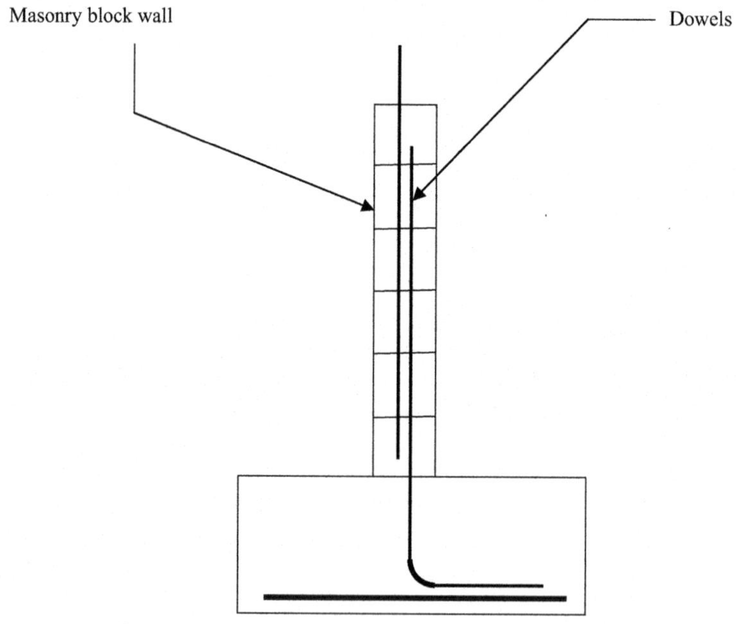

Fig. 3.37 Rebars for a masonry wall.

3.28 CONSTRUCTION JOINTS IN SLABS

It is not possible to concrete a whole slab or a wall in one concrete pour. Hence, construction joints are needed. Typically, a keyway is provided to engage the old concrete with new concrete (Fig. 3.38).

Construction joints in slabs:

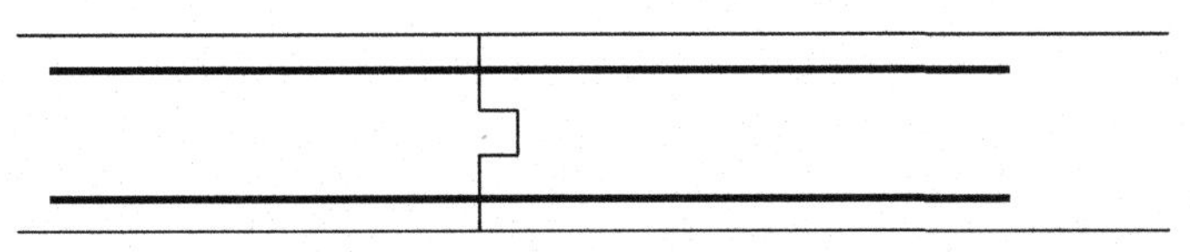

Fig. 3.38 Keyway and a joint.

Typically, rebars are extended through the joint.

3.29 CONCRETE MIX DESIGN

Concrete contains cement, sand, aggregates (gravel), and water. Fine aggregates is another name for sand.

Mix ratios: Concrete mix ratios are provided in weights.

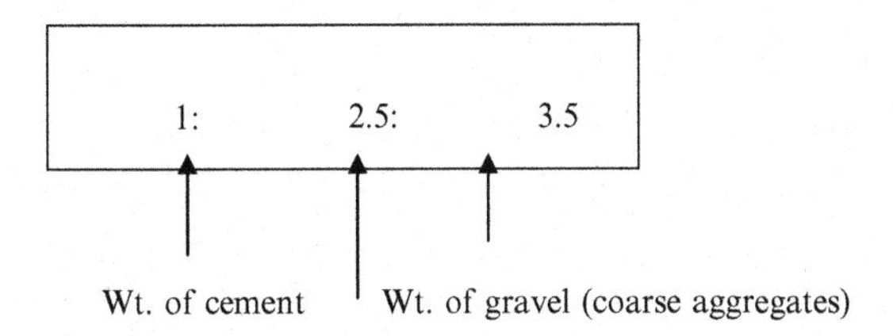

Practice Problem 3.7

Find the cement, sand, and aggregate ratio of six sacks of cement that are mixed for 1 cu. yd of concrete. The contractor mixed 1700 lbs of sand and 2200 lbs of aggregates. Assume one sack of cement is 94 lbs.

Solution

$$\text{Wt. of cement} = 6 \times 94 = 564\,\text{lbs}$$

$$\text{Wt. of sand} = 1700\,\text{lbs}$$

$$\text{Wt. of aggregates (gravel)} = 2200\,\text{lbs}$$

Continued

$$\text{Total} = 564 + 1700 + 2200 = 4464\,\text{lbs}$$

$$\text{Cement ratio} = 564/4464 = 0.126$$

$$\text{Sand ratio} = 1700/4464 = 0.381$$

$$\text{Aggregate ratio} = 2200/4464 = 0.493$$

Make the cement ratio 1.0.
This can be done by dividing the cement ratio by 0.126.

$$\text{Cement ratio} = 0.126/0.126 = 1.0$$

Divide the sand ratio and aggregate ratio by 0.126 also.

$$\text{Sand ratio} = 0.381/0.126 = 3.02$$

$$\text{Aggregate ratio} = 0.493/0.126 = 3.91$$

$$\text{Cement}: \ \text{Sand}: \ \text{Aggregates (by weight)}$$

$$1.0: \quad 3.02: \quad 3.91$$

Practice Problem 3.8

The cement, sand, and aggregate ratio of a concrete mixture is 1:2.5:3.0. Specifications require 7.2 bags of cement to be mixed for 1 cu. yd of concrete.

Following *specific gravities* are provided:
Cement: 3.1
Sand: 2.6
Aggregates: 2.65
Find the water cement ratio of the concrete. (One bag of cement weighs 94 lbs.)

Solution

$$\text{Cement, sand, aggregate ratio} = 1:2.5:3.0$$

$$\text{Weight of cement per cu. yd} = 7.2 \times 94 = 676.8\,\text{lbs}$$

$$\text{Weight of sand} = 2.5 \times 676.8 = 1692\,\text{lbs}$$

$$\text{Weight of aggregates} = 3.0 \times 676.8 = 2030\,\text{lbs}$$

$$\text{Density} = \text{Weight}/\text{Volume}$$

$$\text{Specific Gravity} = \text{Density}/\text{Density of water}$$

Cement

$$\text{Specific gravity of cement} = \text{Density of cement}/\text{Density of water} \quad (3.1)$$

$$\text{Density of cement} = \text{Weight of cement}/\text{Volume of cement}$$

$$\text{Density of cement} = 676.8/\text{Volume of cement}$$

From Eq. (3.1)

$$\text{Specific gravity of cement} = 676.8/(\text{Volume of cement} \times \text{Density of water})$$

$$3.1 = 676.8/(\text{Vol. of cement} \times 62.4)$$

$$\text{Vol. of cement} = 676.8/(3.1 \times 62.4)\ \text{cu. ft} = 3.5\ \text{cu. ft}$$

Sand

From Eq. (3.1)

$$\text{Specific gravity of sand} = 1692/(\text{Volume of sand} \times \text{Density of water})$$

$$2.6 = 1692/(\text{Vol. of sand} \times 62.4)$$

$$\text{Vol. of sand} = 1692/(2.6 \times 62.4)\ \text{cu. ft} = 10.4\ \text{cu. ft}$$

Aggregates

From Eq. (3.1)

$$\text{Specific gravity of aggregates} = 2030/(\text{Volume of aggregates} \times \text{Density of water})$$

$$2.65 = 2030/(\text{Vol. of aggregates} \times 62.4)$$

$$\text{Vol. of aggregates} = 2030/(2.65 \times 62.4)\ \text{cu. ft} = 12.3\ \text{cu. ft}$$

$$\text{Total volume of cement, sand and aggregates} = 3.5 + 10.4 + 12.3 = 26.2$$

$$\text{One cu. yd contains 27 cu. ft}$$

$$\text{Volume of water} = 27 - 26.2 = 0.8\ \text{cu. ft}$$

$$1\ \text{cu. ft} = 7.48\ \text{gallons}$$

$$\text{Volume of water} = 0.8 \times 7.48 = 5.98\ \text{gallons}$$

$$\text{Water cement ratio} = 5.98\ \text{gallons per cu. yd of concrete}$$

In this example, 7.2 bags of cement were mixed per 1 cu. yd of concrete.

$$\text{Water cement ratio } 5.98/7.2 = 0.83\ \text{gallons per bag.}$$

Concrete practice problems:

Practice Problem 3.9

A contractor is planning to pump concrete to a high-rise building. The pump mix needs to have a higher water content. How could one design a pump mix without sacrificing the strength?

(A) Add water-reducing admixtures
(B) Add fast setting admixtures
(C) Increase the water content
(D) Add mineral admixtures

Solution

Increasing the water content leads to lower strength. On the other hand, water-reducing admixtures increase the workability and pumpability without affecting the strength. Hence, the correct answer is A.

Practice Problem 3.10

What is the correct statement below:

(A) Type II cement can develop higher strength in a mass concrete than Type I cement.
(B) Development of high temperature in mass concrete increases the strength.
(C) Type I cement can develop higher strength in a mass concrete than Type II cement.
(D) None of the above

Solution

Ans (A)

In mass concretes, the heat at the core is trapped. Therefore the temperature at the core will rise. This leads to low strength concrete. Type II cement generates less heat and the problem can be avoided.

Practice Problem 3.11

Is the following statement correct?

Replacing cement with pozzolans reduces the strength of concrete

Solution

This may or may not be true. Pozzolans generally react with by-products of cement hydration. Hence, it is not necessarily a bad thing to replace cement with pozzolans.

Practice Problem 3.12

Name three highly used pozzolans.

Solution

- Fly ash
- Blast furnace slag
- Micro silica or silica fume

Practice Problem 3.13

Workability can be increased by introducing fly ash into a concrete mix. What is the reason for this?

Solution

Fly ash particles are more spherical in shape than cement particles. Thus it is believed that fly ash particles can roll instead of slide. This allows the particles to move around easily, increasing the workability.

Practice Problem 3.14

What is the length of a No. 6 standard hook?

Solution

A standard hook is 12*d*.

$$\text{Length of standard hook} = 12 \times d = 12 \times 6/8 = 9\,\text{in.}$$

Practice Problem 3.15

A quality control inspector has been monitoring the ambient temperature. Following is his finding.

Sunday: Average air temperature 32°F

Monday: Average air temperature 38°F

Tuesday: Average air temperature 32°F

Wednesday: Average air temperature 22°F

Highest air temperature for one half the day from midnight to midnight = 55°F

Should the contractor follow the cold weather plan?

Solution

ACI 306 defines cold weather as follows:

Continued

Cold weather is defined as a period when, for more than 3 consecutive days, the following conditions exist:

(1) The average daily air temperature is <40°F (5°C) and

(2) The air temperature is not >50°F (10°C) for more than one-half of any 24-h period.

The average temperature is <40°F for more than 3 consecutive days. Therefore the first condition implicates that cold weather protection should be used.

On the other hand, highest air temperature for one half the day was more than 50°F. Cold weather plan should be followed if any of the conditions are met. Hence, the contractor needs to provide cold weather protection and any other required precautions specified.

Practice Problem 3.16

Above what temperature should the contractor be obligated to provide hot weather protection?

Solution

ACI does not provide a temperature in the case of hot weather condition.

Practice Problem 3.17

What does a curing compound do?

(A) Curing compounds allow water to evaporate.

(B) Curing compounds do not allow water to evaporate.

(C) Curing compounds keep the concrete hot.

(D) None of the above

Solution

Ans B

Practice Problem 3.18

What is the maximum height that concrete can be dropped?

Solution

ACI does not provide a maximum height. However, ACI requires that there are no rebars or any other obstructions on the way when dropping concrete.

Practice Problem 3.19

New concrete has to be poured next to old concrete. What should the contractor do prior to placing the new concrete?

Solution

The bond between the new concrete and the old concrete is very weak. Hence, the contractor should apply a bonding agent to the old concrete prior to concreting.

Practice Problem 3.20

Concrete slab of a large warehouse need to be finished. What is the best equipment that can be used for this purpose?

(A) Hand trowel
(B) Bull Float
(C) Ride on trowel machine
(D) None of the above

Solution

Ans C

Practice Problem 3.21

A concrete structure needs to be built in an area that has severe winters and very hot summers. The structural engineer has recommended fly ash to be added to the concrete mix. Is this a good idea?

(a) Yes. Fly ash generates less heat and that is beneficial.
(b) Yes. Fly ash increases the air content and resists cracking due to freezing and thawing.
(c) No. Fly ash decreases the air content and promotes cracking due to freezing and thawing.
(d) None of the above

Solution

Ans C

Cracks could appear due to the freezing and thawing of concrete. High air content is recommended when concrete is subjected to freezing and thawing conditions. Air in the concrete allows concrete to expand and contract without generating cracks. Fly ash particles are generally much smaller than cement particles. Hence, fly ash tends to decrease the air content. Adding fly ash promotes cracks.

Practice Problem 3.22

Elevation of top of slab on grade is given to be 13.5 ft. The thickness of the slab is 8 in. Soil grade is at elevation 12 ft 9 in.. As per ACI 117, soil grade tolerance is 3/4 in. Is the elevation of soil is within the tolerance limit?

(A) Yes
(B) No
(C) Can't say

Solution

$$\text{Top of slab elevation} = 13.5\,\text{ft}$$

$$\text{Slab thickness} = 8\,\text{in.} = 0.6666\,\text{ft}$$

$$\text{Top of soil elevation required} = 13.5 - 0.6666 = 12.8333\,\text{ft}$$

$$\text{Measured top of soil elevation} = 12\,\text{ft}\,9\,\text{in.} = 12.75\,\text{ft}$$

$$\text{Difference} = 12.83333 - 12.75 = 0.0833\,\text{ft} = 0.996\,\text{in.}$$

This is greater than ACI recommended tolerance of 3/4 in (0.75 in.).
Ans B

CHAPTER 4

Steel Construction

Steel has many advantages over concrete and is still widely used in construction. After 9/11, many proponents of concrete stated that if the World Trade Center towers had been built using concrete, they would not have collapsed. Concrete is an extremely good fireproof material, whereas under high temperatures steel connections tend to fail. Concrete beams and columns do not need additional fireproofing. On the other hand, steel beams and columns need to be fireproofed.

The performance of steel structures during earthquakes is much better than those made of concrete. This is because steel structures can be designed with more flexibility than concrete structures. Nevertheless, some argue that by using modern design techniques even concrete structures can be made very safe under earthquake loadings.

It is generally believed that the cost of steel framed structures is much higher than concrete structures. However, some experts believe that with the advent of high strength steel combined with state-of-the-art design techniques, it is possible to design cheap steel structures. One of the main advantages of concrete is its availability. Concrete is available throughout the year in most parts of the country. On the other hand, the availability of steel can be an issue if it is not ordered ahead of time.

Concrete structures can be built faster than steel frames. But if one can get the design, detailing, and fabrication done in time, steel erection can go faster. That being said, any errors that occur during detailing or fabrication can delay the project.

Another advantage of concrete is that any shape an architect imagines can be easily achieved. The same cannot be said of steel. Although complex shapes are possible with steel, fabrication issues can delay the progress. However, one can argue that the most aesthetic buildings are steel framed structures.

4.1 STEEL CONSTRUCTION PROCESS

4.1.1 Design Drawings and Shop Drawings

Once an architect has completed the architectural drawings, structural engineers design the columns and beams. Structural engineers size up the

Construction Engineering Design Calculations and Rules of Thumb
http://dx.doi.org/10.1016/B978-0-12-809244-6.00004-4

columns and beams and provide the type of sections needed at each location. In addition, they provide the size of anchor bolts and various other structural information.

Design drawings and specifications are part of the contract. The contractor who won the contract will provide the design drawings to a steel fabricator. Steel fabricator use the design drawings and develop shop drawings. During the shop drawings stage many issues that were not considered during the design phase will be considered. Some of these issues include: What is the best method to attach a gusset plate to a beam? Should the plate be welded in the field or in the shop? Welding a steel member in a shop is always cheaper than welding in the field. On the other hand, if welded in the field, workers can make minor adjustments to the piece so that it fits into the structure. In many instances, a design engineer will delegate the design of connections to a steel fabricating shop. This is known as design delegation. But as per law, this would not release the design engineer from the responsibility. Since two parties are involved, any issues that arise due to bad connection design are the responsibility of both the fabricator and the design engineer.

Steel design drawings: Design drawings tell the erector what beams and columns need to be used. Beam elevations and column elevations are also given.

Building grid lines, beams, and column schedule are shown below.

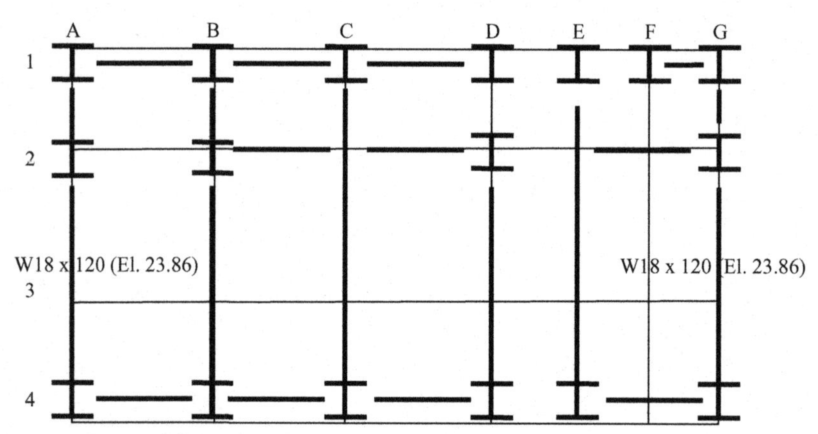

Fig. 4.1 Plan (2nd floor).

Fig. 4.1 shows the plan view of a second floor. The plan view of each floor should be provided. The size of each beam and top of the beam elevation should be given in the drawing. In Fig. 4.1, the size of two beams is shown. In addition, connection details should also be provided along with the bolt pattern and weld information.

Column schedule: Column schedule should provide column location, column size, column start and end elevations, ie, In Fig. 4.2 at C1, install a column W18 × 40 at an elevation of 81′6″. Top of column elevation is 104′6″.

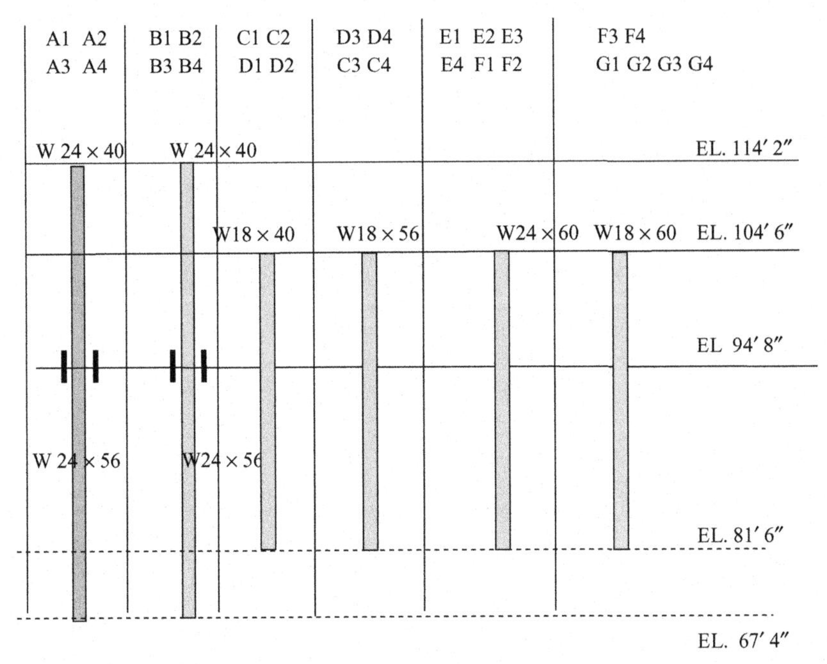

Fig. 4.2 Column schedule.

4.1.2 Erection Drawings

Once the shop drawings are approved by the design engineer, the fabricator develops erection drawings. Each steel member is given a piece number. Erection drawings would very specifically state where each steel member goes. The workers would pick the marked steel member and find out where it is to be erected. Then they would erect the pieces as shown on erection drawings. When steel members come to the site, they are sorted out according to the location where they are to be erected. This is known as shaking. Steel members would be transported to their erection location using a crane.

4.1.3 Steel Erection Process

Steel erection has different phases and different crews. Some of the steel working crews are:

- connecting crew
- bolting crew
- detailing crew
- welding crew
- decking crew
- rigging crew

The first crew to go up is the connecting crew. These workers connect beams and columns with a few bolts. Connecting is considered to be the hardest job in steel work. The connecting crew connects steel beams and columns while these steel elements are loose and dangling in air.

Following is a write-up by a steel worker:

> *As the steel pieces goes up, every time a beam is set onto a column, two pieces of steel meet in thin air. It's windy up there, and frames tend to sway without walls to stiffen them. A "connector" has to be at the top of that column, ready to pin the beam to it—and he may be 30 floors above the street. The work is simple to understand, but that doesn't make it easy. It is dirty, difficult and dangerous, and it takes a very determined man to do it. There are no gray areas. The reality of the work hits a man like a baseball bat each day. He can either do it or he can't.*

Bolting crew: Once the connecting crew has connected steel beams and columns with some bolts, the bolting crew comes in. The bolting crew attaches the rest of the bolts and tighten them. Since the steel pieces are already connected, they are not moving in the air. The bolting crew must know where each bolt should go. They must also know the proper procedure to fastening them. In addition, the bolting workers need to carry a hefty amount of bolts and various wrenches with them.

Detailing crew: After the bolting crew is finished with the bolting, the detailing crew follows up. As the name implies this crew goes through all the details to make sure that the connection is completed as per the design specifications. The detailing crew addressed any changes that the design engineer has ordered.

Welding crew: Some connections also need welding. The welding crew will conduct any welding required.

Decking crew: After all the connections are completed, the decking crew will install the deck.

Rigging crew: Riggers tie up beams and columns; they signal the crane operator to take steel pieces to steel workers. It is important that riggers know various knots and chokers.

Steel wire ropes: Steel wire ropes are made of strands of wires. There is a fiber or steel core at the center of each steel wire rope.

"6 × 24-FC" means there are six strands. Each strand has 24 wires and FC means fiber core.

Plate 4.1 Columns are erected. Columns are placed on base plates. Also some horizontal beams are installed.

Erecting columns: Plate 4.1 shows erected columns. The columns are not placed on the ground. Columns are placed on base plates.

Erecting beams: After erecting the columns, the beams are erected. Typically the connecting crew uses few bolts to install the beams. The connecting crew may use a bucket truck (Fig. 4.3).

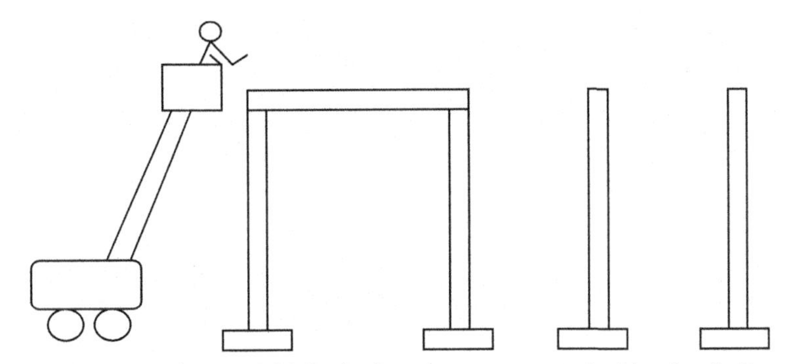

Fig. 4.3 Connecting beams. Initially the beams are connected with a few bolts.

Bolting crew: After the connecting crew has installed beams, the bolting crew installs the remaining bolts (Plates 4.2 and 4.3, Fig. 4.4).

Plate 4.2 The connecting crew installed one bolt in this case. The bolting crew will complete rest of the bolts.

Plate 4.3 Connection (welded and bolted connection).

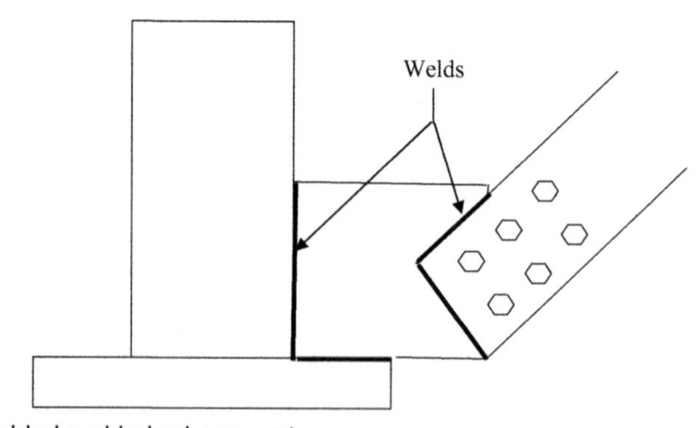

Fig. 4.4 Welded and bolted connection.

Some connections have only bolts while others have only welds. Some connections need both welds and bolts.

Welding crew: After the bolting crew has finished all the bolts, the welding crew welds where necessary.

4.2 ELEMENTS OF A STEEL BUILDING

In this sub-chapter let us look at elements of a steel building. Generally steel building has following elements:

- steel columns
- steel beams
- connections (bolted or welded)
- metal deck

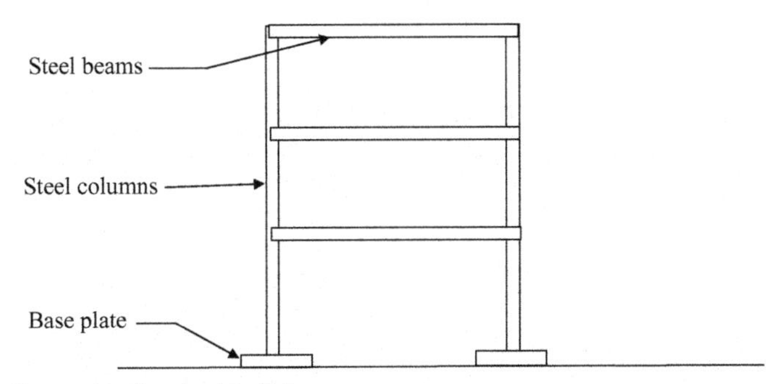

Fig. 4.5 Elements of a steel building.

The metal deck is not shown in Fig. 4.5. The metal deck spans between beams (Fig. 4.6).

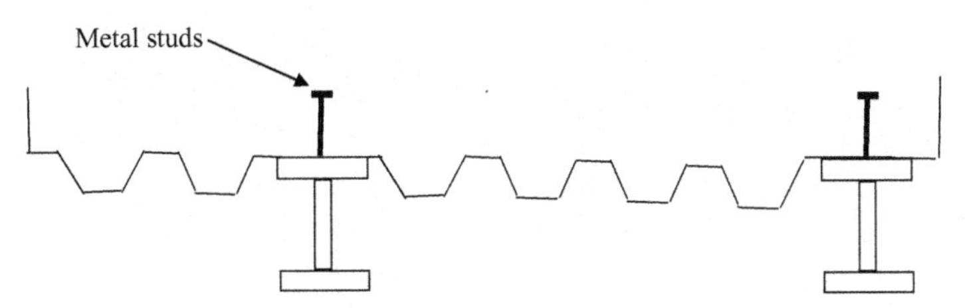

Fig. 4.6 Corrugated metal deck (the metal deck spans between beams).

4.2.1 Corrugated Metal Deck

Corrugated metal decks are typically concreted. Rebars or wiremesh is placed to increase the tensile strength. Studs are welded to the beams and they resemble metal bolts. A hole is cut through the metal deck to pass through the stud. When concreted, these studs connect the deck to the beams (Fig. 4.7).

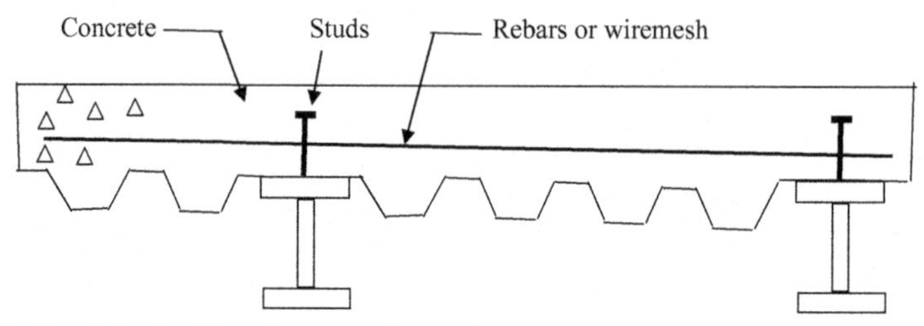

Fig. 4.7 Metal deck with rebars or wiremesh.

The studs and hardened concrete act as one unit. Since the studs are welded to the beam, the deck is attached to the beam through the studs (Plates 4.4–4.13).

Plate 4.4 A metal deck is shown in the picture. Row of studs (resembling bolts) are seeing at the bottom of the photograph. These studs are welded to the beam below the metal deck. Also a wiremesh is shown placed on top of the deck to increase the tensile strength.

Plate 4.5 Metal deck: in this photograph rebars are also installed. The deck on right and left are already concreted.

Plate 4.6 Rebars on a metal deck are shown. Rebars are placed on small metal wires known as chairs. Typically rebars are placed 1 in. above the top of the deck.

Plate 4.7 This photograph shows concreting of a metal deck. Studs and wiremesh are also seen in the photograph. The large hose is the concrete pumping hose.

Plate 4.8 There are three rows of studs in this photograph. These studs are welded to the beam underneath. When the concrete is hardened the studs would rigidly attach the hardened concrete to the beams.

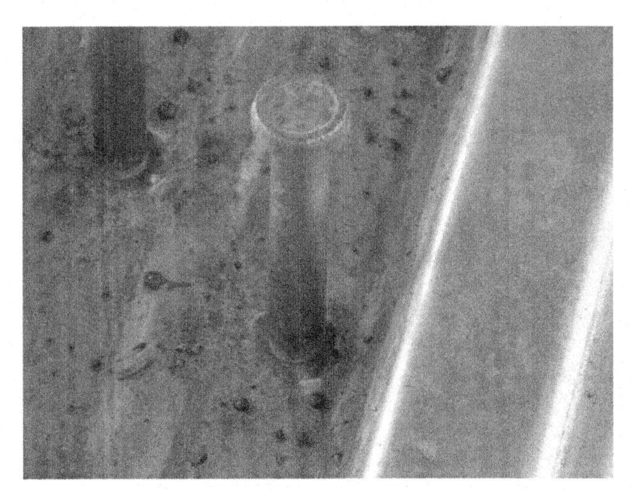

Plate 4.9 Stud of a metal deck.

Plate 4.10 The edge of the metal deck is shown in the above photograph. Typically hook bars are used at the edges to account for bending stresses.

Plate 4.11 The opening in a metal deck is shown. Openings are necessary to bring air ducts to the floor. Also openings are needed for plumbing and electrical lines. Hook bars are used at the openings as well. The beam under the metal deck is also visible.

Plate 4.12 The metal deck is seen from below. The metal deck is placed on beams. Also, note the conduits (steel tubes) going into the slab. These conduits will take electric cables and telecommunication cables.

Plate 4.13 The metal deck is concreted.

4.3 INSTALLATION OF BASE PLATES

Now let us backtrack a little. Let us look at base plates. Base plates are a very important part of a steel building. Columns are placed on base plates (Fig. 4.8).

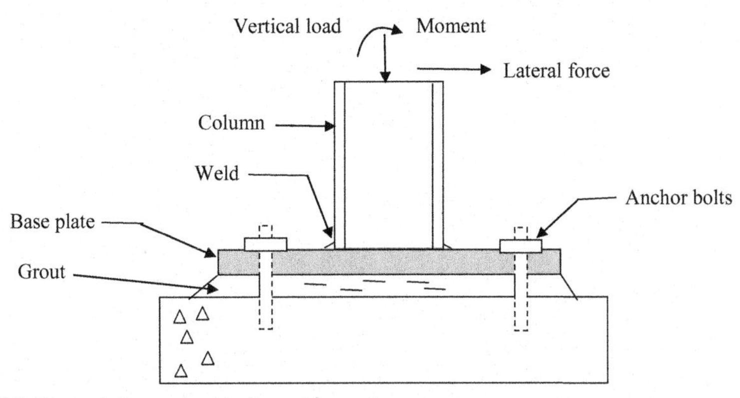

Fig. 4.8 Base plate, anchor bolts, and grout.

Columns carry vertical loads, lateral loads, and bending moments. A lateral loads can be converted into a bending moment. Bending moments occur due to wind loading and earthquake loading. Let us look at the vertical load. The columns transfers the vertical load to the base plate. The base plate transfers the load to the grout. The grout transfers the load to the concrete footing below.

$$\text{Vertical load} \rightarrow \text{Base plate} \rightarrow \text{Grout} \rightarrow \text{Concrete footing}$$

You may ask what is the purpose of anchor bolts?

Anchor bolts are needed to resist lateral forces and uplift forces due to bending (Fig. 4.9).

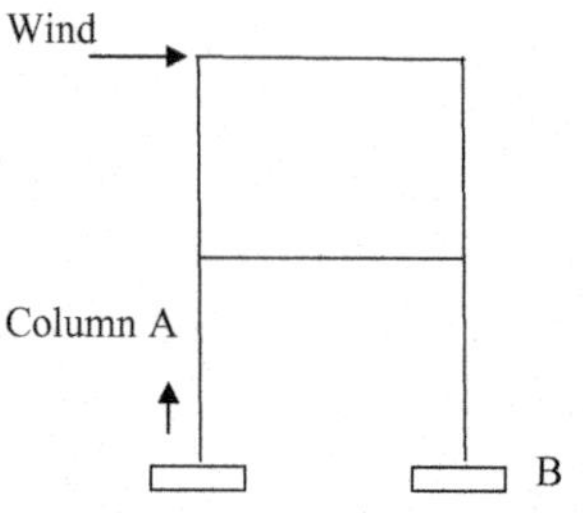

Fig. 4.9 Schematic of a building.

In this situation, column A has an uplift force due to wind forces. These uplift forces have to be resisted by anchor bolts. In addition, the anchor bolts resist any lateral loads. During construction, anchor bolts are necessary to keep the column in place until the column is grouted (Plate 4.14).

Plate 4.14 A column sitting on a base plate. The grout under the base plate and bolts are visible.

4.3.1 Construction Procedure of Base Plates

Let us look at the construction procedure of base plates.

STEP 1: Build the concrete footing (Fig. 4.10).

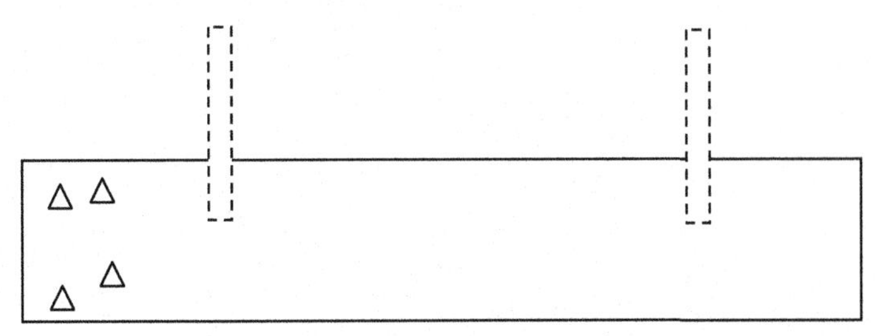

Fig. 4.10 Concrete footing.

Typically anchor rods are already installed in the footing.

STEP 2: Place metal shims to the proper height (Fig. 4.11).

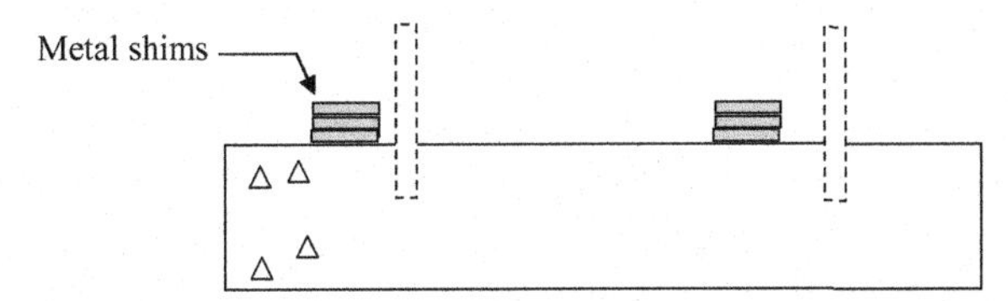

Fig. 4.11 Metal shims are placed on the concrete footing. Metal shims are steel plates.

Metal shims are placed on the footing to the proper height. Shims are small metal plates.

STEP 3: Place the base plate and the column on anchor bolts.

Typically the column and base plate are welded together in the shop. Shop welding is much cheaper than field welding. The column and the base plate are supported by shims. The shims are installed since the anchor bolts are not capable of supporting vertical loads. Anchor bolts are there to resist uplift forces and lateral forces (Fig. 4.12).

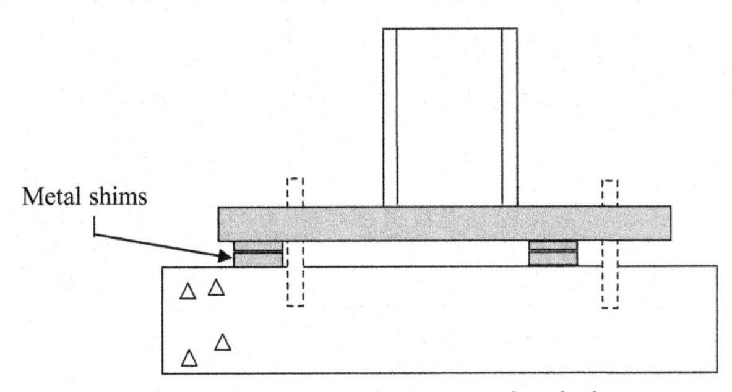

Fig. 4.12 Place the column and the base plate on anchor bolts.

STEP 4: Grout under the base plate.

The last step is to grout under the base plate. Typically, grout is pumped under the base plate using a grout pump. Then the bolt head is placed and tightened (Fig. 4.13).

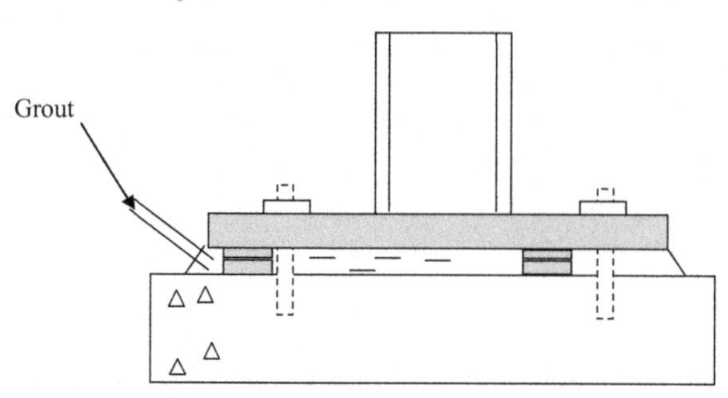

Fig. 4.13 Grout under the base plate.

Metal shims are not an integral part of the system. They are needed to support the base plate and the column until the grout is hardened.

In addition, anchor bolts do not support the vertical load. The grout base supports the vertical load. Anchor bolts are there to resist lateral forces and uplift forces (Plates 4.15 and 4.16).

Plate 4.15 Base plate viewed from above.

Plate 4.16 Base plate prior to grouting.

4.4 SUMMARY OF STEEL CONSTRUCTION

The following process is adopted in steel construction:

- Install base plates on concrete footings. Typically, the base plate is installed on shims. In addition, we learned that anchor bolts do not resist vertical loads. Anchor bolts are there to resist horizontal loads and uplift forces.
- Erect columns on top of the base plates. In some cases, the base plate and column is welded in the shop and arrive together.
- Erect beams with few bolts. This is done by the connectors or the connecting crew.
- Install all the bolts. The bolting crew does this.
- Weld connections where necessary.
- Install the metal deck.
- Provide rebars on the metal deck.
- Concrete the metal deck.

Now let us discuss welding and bolting. Steel pieces are connected using welds or bolts. In the past rivets were also used. Rivets are not capable of resisting tensile forces and are no longer used.

4.5 WELDING

Welding is done for aircrafts, ships, bridges, and pipelines. The following five joint types are commonly used. The terms "joint" and "weld" are used interchangeably in textbooks. For example, one textbook may use the term "lap joint" while another book may call it a "lap weld." Similarly, one book may use "butt joint" and another one may say "butt weld."

(1) Butt joints
(2) T-joints
(3) Lap joints
(4) Corner joints
(5) Edge joints

4.5.1 Butt Joints

Two members need to be welded are butted together. Hence, the joint is known as a butt joint. In most cases, members to be welded are provided with grooves.

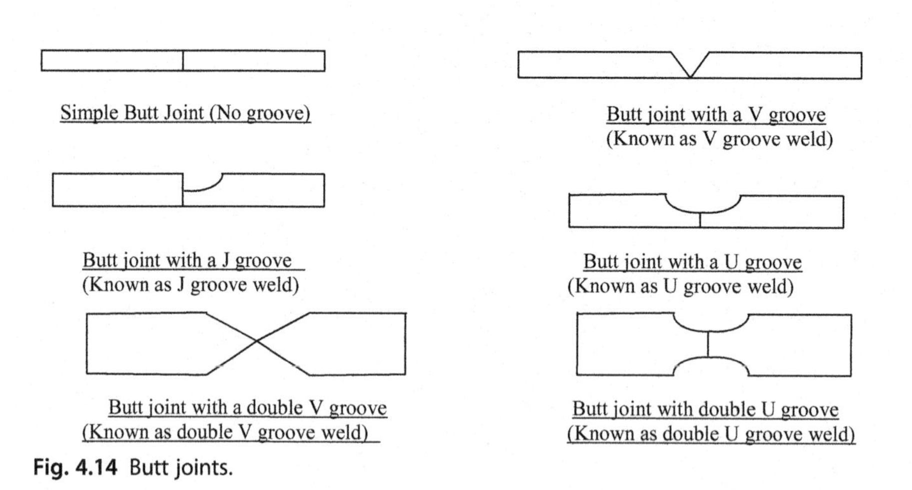

Fig. 4.14 Butt joints.

Fig. 4.14: V groove butt weld (also a backer rod is used under the plates for a better weld)

Fig. 4.14 shows a V groove weld.

4.5.2 T-Joints

T-joints are used when members are perpendicular to each other (Fig. 4.15).

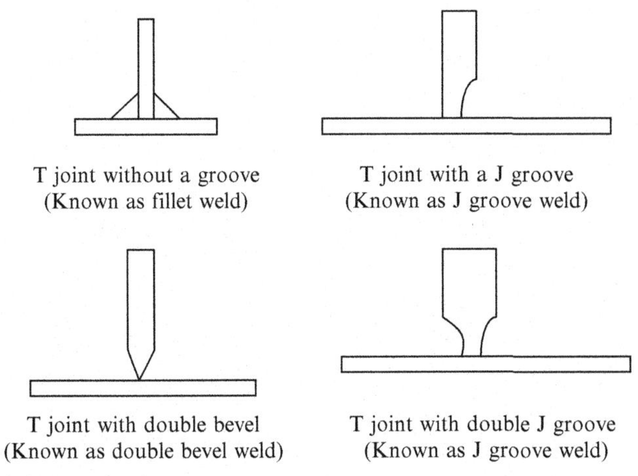

Fig. 4.15 T-joints.

4.5.3 Lap Joints

When steel members are overlapped, lap joints are used (Fig. 4.16).

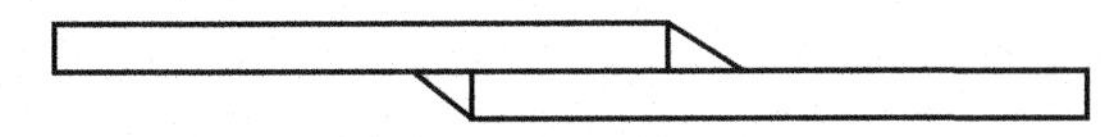

Fig. 4.16 Lap joints (also known as lap weld).

4.5.4 Corner Joints

Corner joints, as the name implies, are applied at the corners (Fig. 4.17).

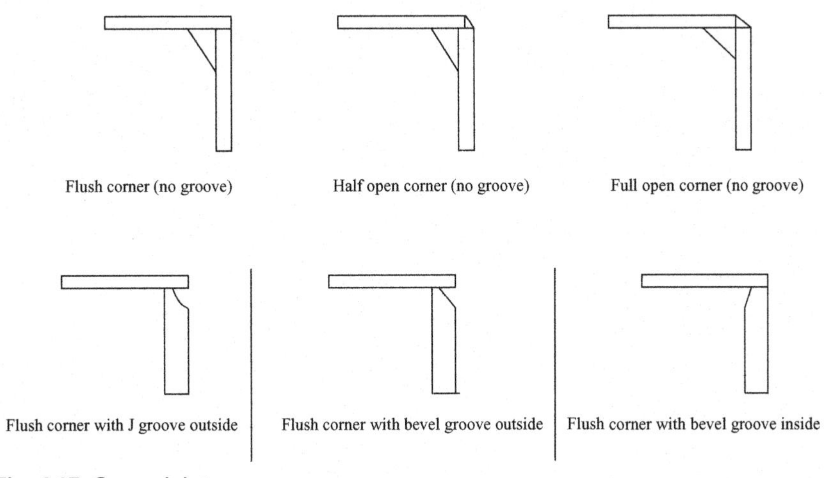

Fig. 4.17 Corner joints.

Grooves can be J grooves, bevel grooves, U grooves or V grooves. These grooves could be inside or outside as necessary.

4.5.5 Edge Joints

As the name implies, edge joints are applied at the edges (Fig. 4.18).

Some commonly used terms:

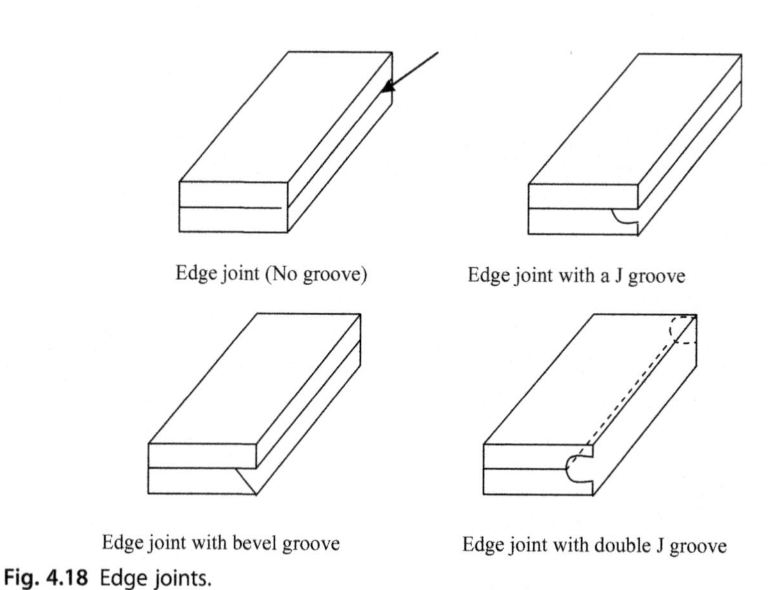

Fig. 4.18 Edge joints.

4.5.6 Fillet Weld

Whenever there is no groove, such welds are called fillet welds. When there is a groove present, it is called by its groove shape (Ex: J groove weld, U groove weld, double U groove weld, etc.).

Fillet Weld Examples (Fig. 4.19):

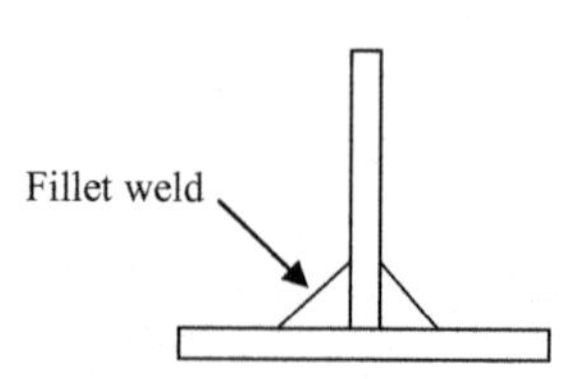

Fig. 4.19 Simple T joint with fillet welds.

4.5.7 Seam Weld

When the seam of a pipe or any other object is welded, it is called seam weld.

Welding symbols: The following welding symbols are as per AWS (American Welding Society). All the symbols are available on their website.

Fillet welds (no groove) (Fig. 4.20):

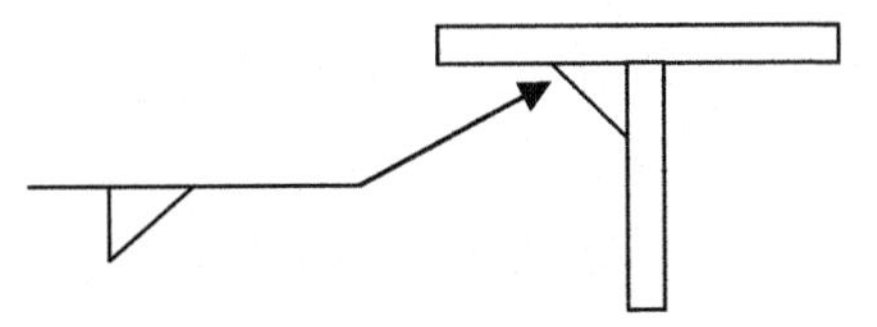

Fig. 4.20 Fillet weld (weld is on the side of the arrow).

The triangle represents the fillet weld. If the triangle is below the line it means the weld is required only on the arrow side.

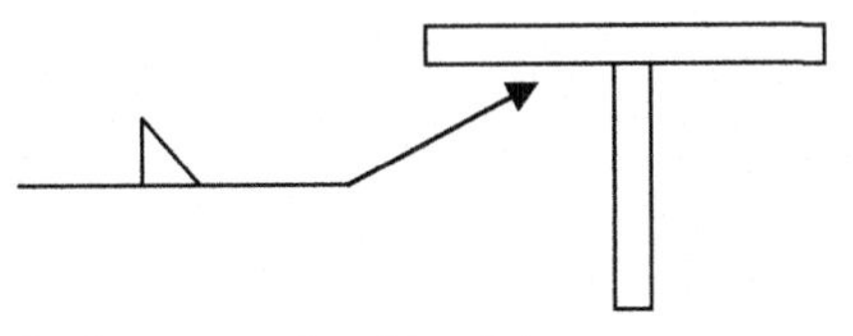

Fig. 4.21 Weld is required on opposite side of the arrow.

If the triangle is on top of the line, it means the weld is required on the opposite side of the arrow, as shown in Fig. 4.21.

If two triangles are present, above and below the arrow line, both sides have to be welded (Fig. 4.22).

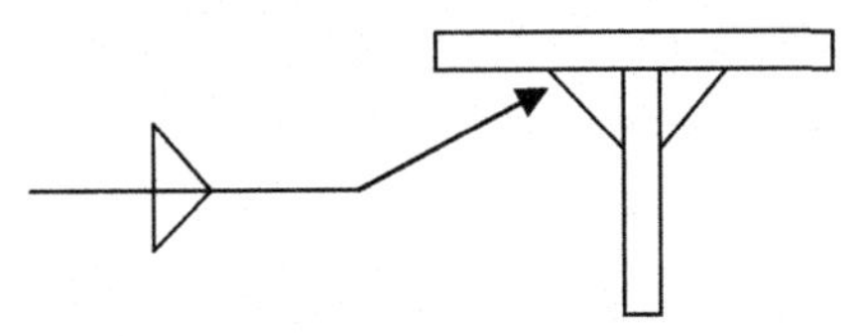

Fig. 4.22 Welds are required on both sides.

4.5.8 Size of the Weld

It is necessary to indicate the size and length of welds as well.

Since the triangle is at the bottom, the weld is on the arrow side. The number before the triangle (1/4) indicates the size of the weld. The size of the weld is the dimension on the side. Please see Fig. 4.23. If there is only one size given, then both sides have the same size as in this case. The number after the triangle (6) indicates the length of the weld. The symbol in Fig. 4.23 indicates that a fillet weld should be made on the arrow side with ¼ in. size extending to a length of 6 in.

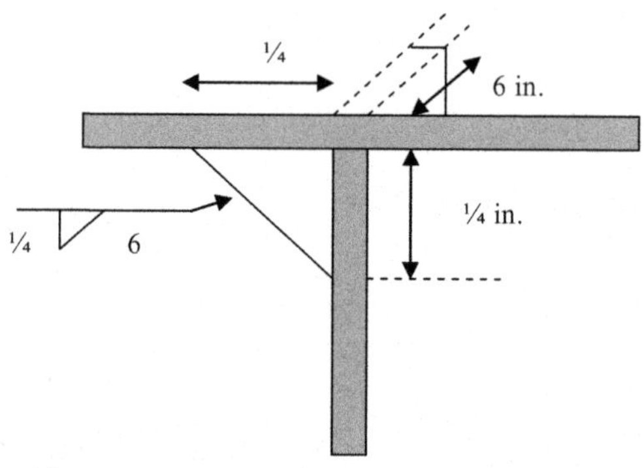

Fig. 4.23 Size of welds.

4.5.9 Size of the Weld Uneven

Side of the arrow: The symbol in Fig. 4.24 indicates that the weld should be ½ in. and ¼ in. on the side of the arrow to a length of 6 in. The question is how would the welder know which side is ½ in. and which side is ¼ in.? Typically, the drawing would be prepared with the longer side drawn longer as shown in Fig. 4.24. The vertical length of the triangle is drawn longer, so that the welder would know ½ in. weld is on the vertical side.

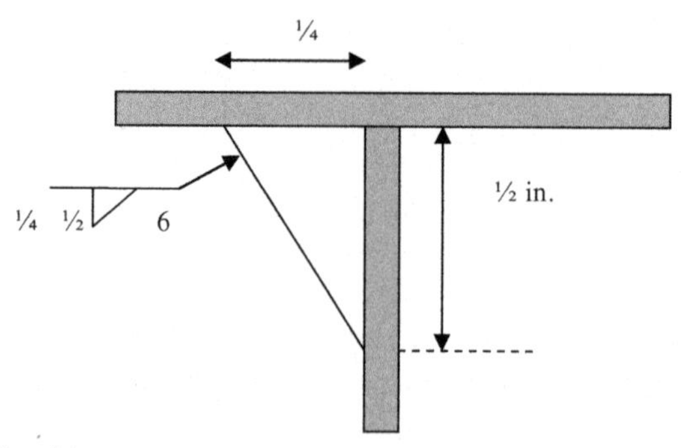

Fig. 4.24 Size of welds uneven.

Size of the weld uneven (weld on both sides) (Fig. 4.25):

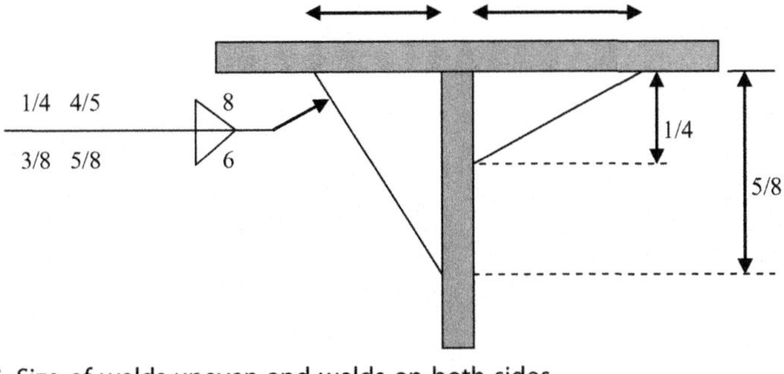

Fig. 4.25 Size of welds uneven and welds on both sides.

Arrow side: Triangles on the top and bottom indicate welds on both sides. A bottom triangle indicates the arrow side. The size of the weld on arrow side is 3/8 and 5/8. Since the vertical edge of the triangle is drawn longer 5/8 weld should be provided on the vertical leg. The length of the weld on the arrow side is 6 in.

Other side: The weld on other side of the arrow is ¼ and 4/5 in. The horizontal leg of the triangle is drawn longer. Hence, horizontal leg is 4/5 in. and vertical leg is ¼ in. The length of the weld is 8 in.

4.5.10 V Groove Weld Symbol

The V groove symbol is placed below the arrow. This indicates that the weld is on the side of the arrow. The depth of the groove is ½ in. and the angle of the groove is 100 degrees (Fig. 4.26).

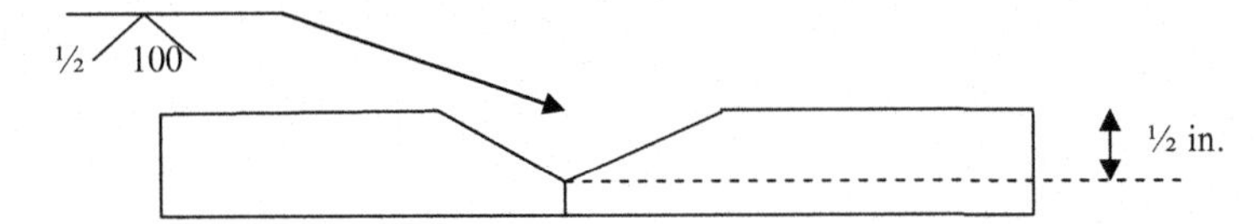

Fig. 4.26 V groove welds.

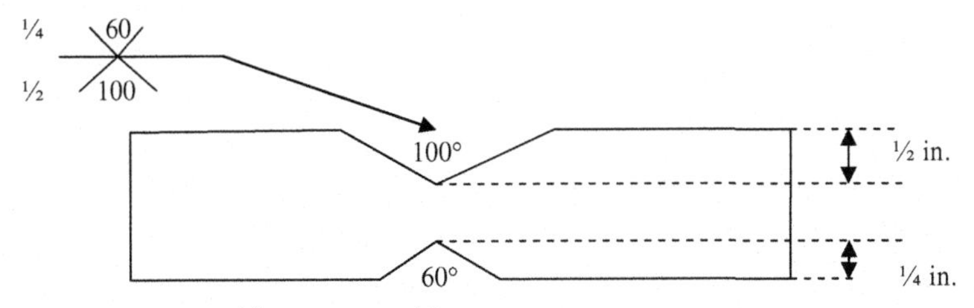

Fig. 4.27 V groove welds on top and bottom.

Arrow side: The V groove is ½ in. deep and has an angle of 100 degrees on arrow side (in this case top, Fig. 4.27).

Opposite of the arrow: The V groove is ¼ in. thick and has an angle of 60 degrees (in this case bottom, Fig. 4.27).

4.5.11 Square Weld

Above first number is the base dimension of the square. The second number is the thickness or the height of the weld (Fig. 4.28).

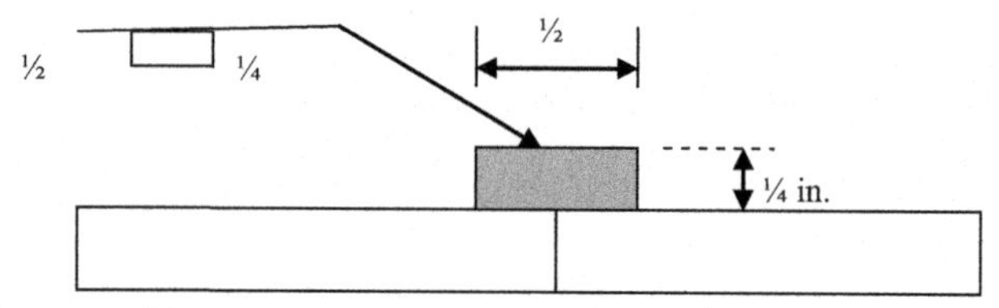

Fig. 4.28 Square weld.

Note: Complete coverage of the subject of welding is beyond this book. This book will provide you with background information to further improve your knowledge. The AWS website is a very good reference source for welding.

4.5.12 Weld Testing

Welding defects are observed by obtaining X-ray photographs. Typical welding defects are:

- incomplete penetration
- incomplete fusion
- porosity and longitudinal cracking.

4.5.13 Weld Material Weight

The cost of welding is dependent on the amount of weld material deposited. Weld material alone is not the best way to assess the cost of welding. Welding on the 10th story of a building is not the same as welding in ground floor. Also welding above one's head is more difficult than welding at hand level. Other factors such as type of weld and the complexity of the weld also affects the cost of welding.

Computation of weld material quantity is still important.

Practice Problem 4.1

Find the weight of the weld material deposited in the weld shown. The weld is 15 ft long. The weight of weld material is 0.283 lbs/cu in. (Fig. 4.29).

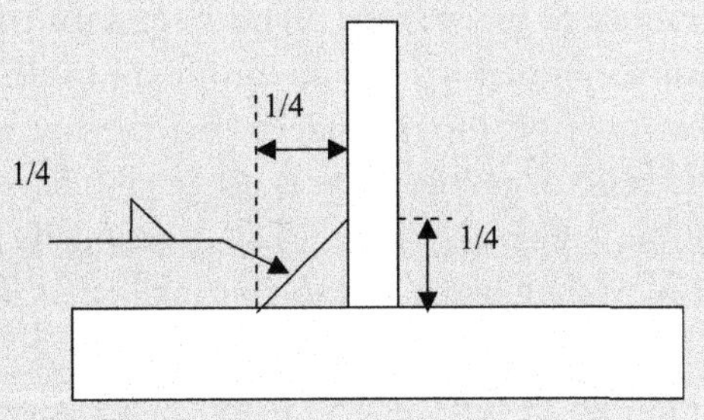

Fig. 4.29 Weight of material in a weld.

Solution

$$\text{Volume of deposited material in cu in.} = \frac{1}{2} \times \frac{1}{4} \times \frac{1}{4} \times (15 \times 12) = 5.625\,\text{cu in.}$$

$$\text{Weight of deposited material} = 0.283 \times 5.625 = 1.59\,\text{lbs}$$

4.6 BOLTS

Steel beams and columns can be connected using bolts. Let us look at parts of a bolt.

4.6.1 Parts of a Bolt

Assume two plates need to be connected using a bolt. Insert the bolt and tighten with the nut. Rotation occurs on the nut side (Fig. 4.30).

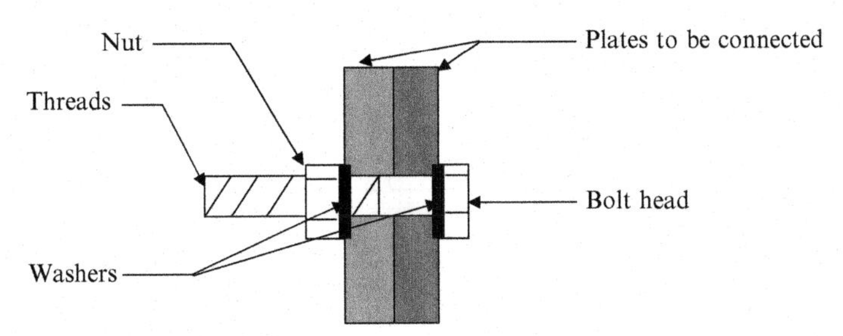

Fig. 4.30 Elements of a bolt.

4.6.2 Washers

What is the purpose of washers?

- Washers spread the load evenly on the plate.
- When the nut is tightened, the plate surface could get damaged. The washer will protect the surface of the plate.
- It is a good practice to use washers on both sides (nut side and head side). However, many use the washer only on the nut side. The argument to use the washer only on the nut side is that the nut is what is rotating. Since the head is not rotating, there is no need to have a washer on head side. In addition, some may argue that occasionally the head can also rotate. Since a washer helps spread the load, it is a good practice to use washers on both sides. In large jobs, one may be able to save some money by providing washers only on one side.

4.6.3 Joint Types

Bolts are used for attaching two elements together. There are two types of bolted connections that are used to resist shear.

(a) Slip critical joints or friction type joints
(b) Bearing type joints

4.6.3.1 Slip Critical Joints

When the two plates try to move due to shear, friction develops between the two surfaces. Friction resists the shear force. Such joints are known as slip critical joints. The holes are oversized since the bolt shaft need not butt against the metal plates. Shear resistance is achieved through friction between the plates. In this type of connection, proper tightening of bolts are important. A design engineer will provide the tension required in the bolt. Tension in bolts are proportional to the frictional resistance (Fig. 4.31).

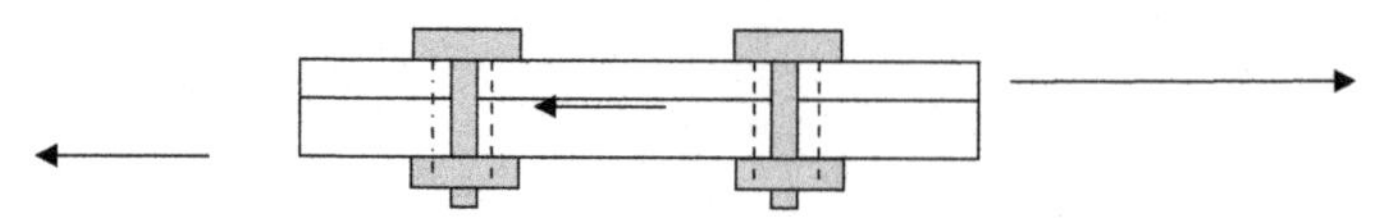

Fig. 4.31 Slip control joints.

4.6.3.2 Bearing Type Joints

Shear resistance is attained by the bolt shaft butting against the metal plates. The holes should not be oversized for bearing type bolts (Fig. 4.32).

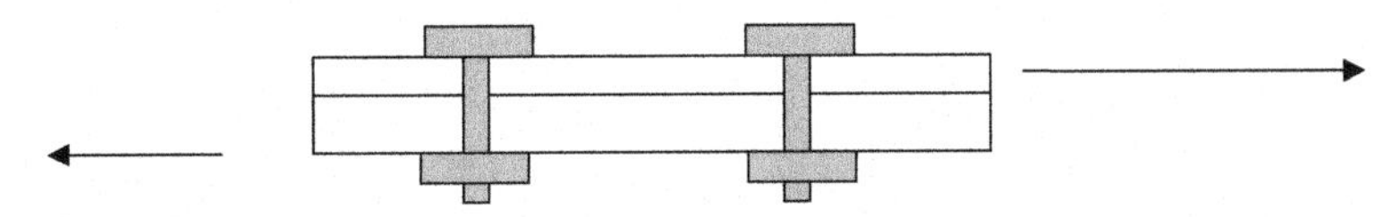

Fig. 4.32 Bearing type of bolts.

Bolts do not need to be tightened to a pre-specified tension. Bolts are typically snug fitted.

Snug fitted bolts: Snug fitting is achieved by the force of an average worker using a spud wrench. Snug fitting is done to make sure that the bolts will not fall off.

4.6.4 Bolt Tension

When the bolt is turned, the bolt body is under tension (Fig. 4.33).

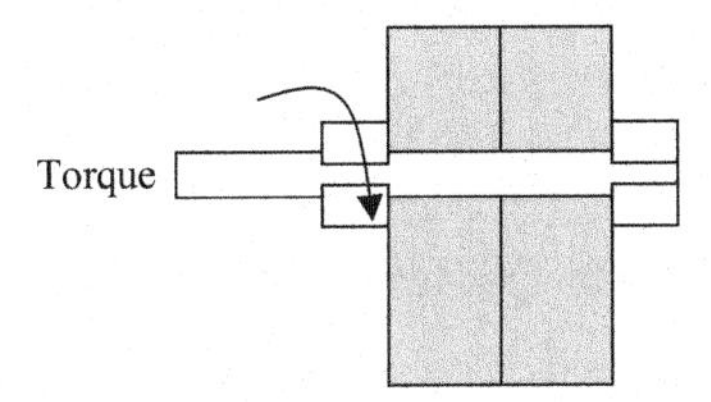

Fig. 4.33 Torque of a bolt.

When a torque is applied to the nut, the two plates compress together. The plates will be under compression and the bolt shaft will be under tension. Nevertheless, there are some reservations in obtaining the tension in bolts by measuring the torque. I will discuss that issue later in the chapter.

4.6.4.1 Bolt Tightening Methods

As mentioned earlier, tightening of bolts to the specified tension is extremely important for friction type bolts.

Turn of the nut method: In this method, the bolt is snug fitted first. This is done by an average worker tightening the bolt with a spud wrench. After that, the bolt is tightened with the rotation specified. For an instance bolts with length less than 8 in., should be rotated 1/2 turn after snug fitting. Bolts with lengths exceeding 8 in. should be rotated 2/3 turn. In this method, the worker should mark the initial position of the nut and then rotate to the correct rotation specified (Fig. 4.34).

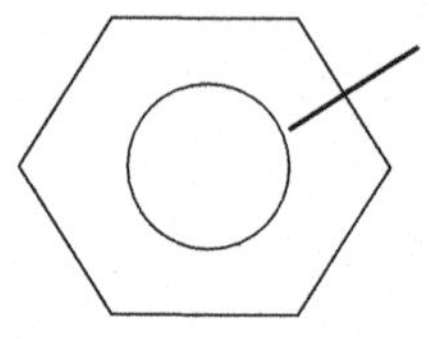

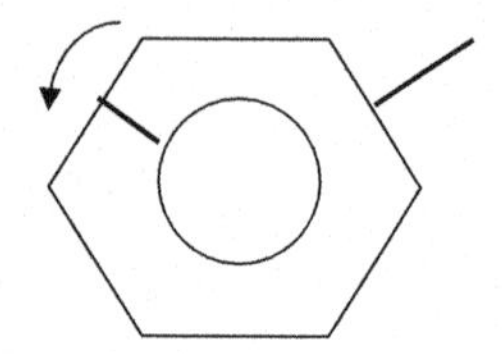

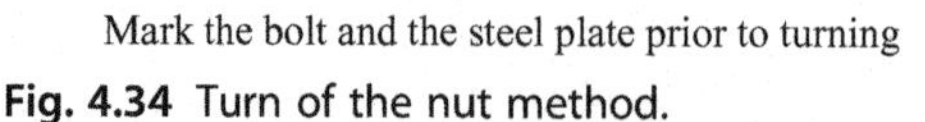

Fig. 4.34 Turn of the nut method.

Calibrated wrench tightening: Calibrated wrenches are used to obtain the proper tension in bolts.

Torque wrench: Direct tension indicator: Instead of a torque wrench, a direct tension indicator can be used. These machines show the tension in bolts. The bolt is tightened until the bolt has achieved the required tension. This method cannot be used for all the bolts.

Direct tension indicating washers (DTI washers): The whole idea is to find the tension in the bolt. Tension in the bolts will create compression in the plates. Design compression between plates needs to be attained to get the required friction.

4.6.5 Tension vs Torque Debate

What is required is the friction between plates. Achieving the correct compression is required to obtain the required friction between plates. Compression between plates is achieved by tensioning the bolt shaft. The nut is rotated to create tension in the bolt shaft. The question is can we predict the tension in the bolt shaft by measuring the torque?

It is easy to measure the torque. A torque wrench can easily be used to measure the torque. It is not easy to measure the tension in bolts. Hence, many measure the torque and extend the torque results to calculate the tension in the bolt shaft.

If the bolt is slightly rusted, a large portion of the torque will go to get the bolt rotated. In addition, if a bolt is slightly out of tolerance then that would also create additional torque. Other factors that would affect the torque is the washer type and rust in washers. Friction between parts are not the same for each bolt. Hence, torque is not a good indicator of tension in bolts.

CHAPTER 5

Construction Equipment

5.1 CRANES

5.1.1 Mobile Cranes

The most common type of mobile cranes are attached to caterpillar type wheels.

These cranes can move to the location, rotate the arm, and lift objects. The height that can be lifted is limited to the height of the boom. Modern high-rise buildings could have 20 or more stories and in such situations, these cranes may not be suitable.

Things to consider during the operation of cranes are:

(1) Make sure that the weight is within the manufacturer's specified range of the crane. Never try to lift more than what is specified by the manufacturer of the crane.

(2) Check all the cables and brake mechanisms prior to lifting.

(3) Make sure that the crane is placed on stable ground. If the crane is placed on unstable ground, it will sink.

(4) Make sure the object being lifted is not bolted, lagged, or clamped to the floor or to another surface.

(5) Make sure that the object is properly balanced (Fig. 5.1).

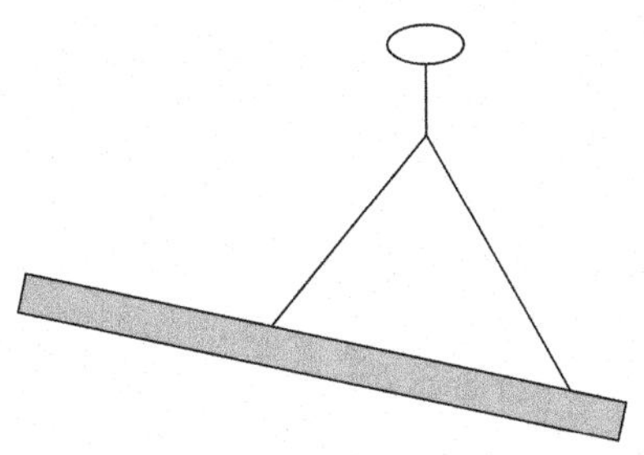

Fig. 5.1 The object is not properly balanced.

(6) Remove loose pieces from the load before lifting.

(7) Long loads, such as beams, tend to swing during lifting. Long loads can be controlled during lifting by attaching ropes to one or both ends of

Construction Engineering Design Calculations and Rules of Thumb
http://dx.doi.org/10.1016/B978-0-12-809244-6.00005-6

the load. Workers on the ground can work these ropes to help control load swinging.

(8) Workers should never ride on the load.

(9) Other workers should stand clear of the load.

(10) If the load is lifted over traffic, special permission should be obtained.

(11) Observe the chains and chain links for damages, nicks, bends, or elongation prior to lifting.

(12) Never leave the load suspended for a long period of time.

5.1.2 Lattice Boom Cranes

Lattice boom cranes are named for their lattice structure. Fig. 5.2 shows a crawler or caterpillar type lattice boom crane.

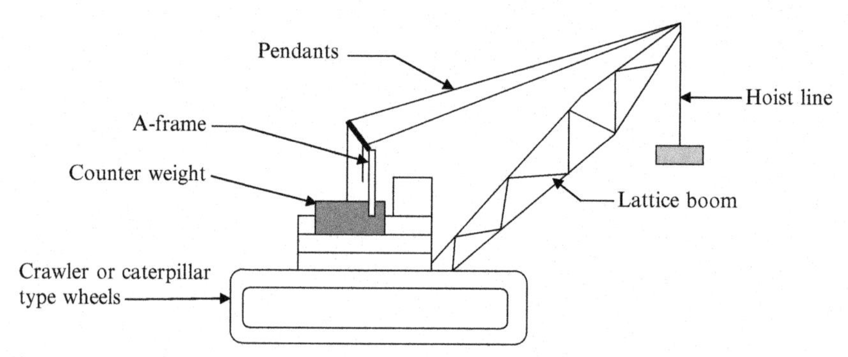

Fig. 5.2 Lattice boom crane.

The load is attached to a hook. The load is lifted using the hoist line. Two cables that are attached to the lattice boom are known as pendants. Counter weight is used to balance the load so that the crane does not tip over. The frame supporting the cables (pendants) is known as an A-frame. Heavy lattice boom cranes are typically on crawler type wheels. These wheels are more stable than tires (Plate 5.1).

Luffing jib: Lattice boom cranes can have a luffing jib. A luffing jib is a smaller lattice boom attached to extend the vertical and horizontal reach of the crane. It is important to know the angle of the boom. The operator of the crane needs to closely monitor the boom angle (Fig. 5.3).

A luffing jib is another lattice structure that is added to the crane to provide more capacity.

Plate 5.1 A lattice boom crane is shown in the figure. The boom is made of a metal lattice structure, much like a steel bridge. The A-frame can be seen behind the lattice boom. The counter weights can be seen behind the A-frame.

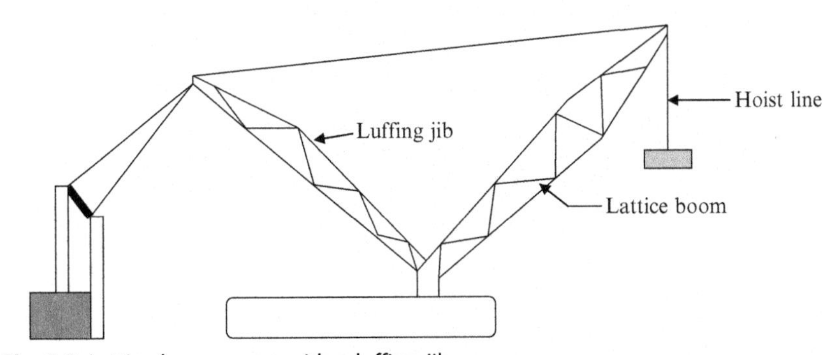

Fig. 5.3 Lattice boom crane with a luffing jib.

5.1.3 Telescopic Boom Cranes

Instead of a lattice boom, telescopic cranes have a boom similar to a telescope (Fig. 5.4).

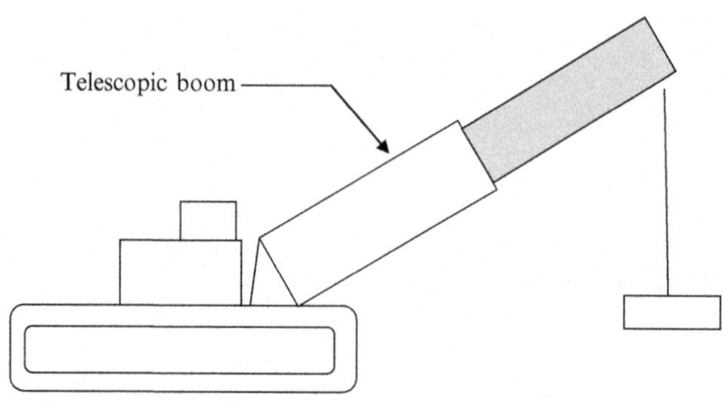

Fig. 5.4 The telescopic crane has a boom that looks like a telescope.

5.1.4 Tower Cranes

Tower cranes, as the name implies, sit on top of a tower. Tower cranes are widely used for high-rise buildings since other cranes may not be able to lift to high elevations. Tower cranes cannot be moved. However, thanks to its long rotating arm, a tower crane can move loads effectively (Plate 5.2).

Plate 5.2 *Tower cranes:* picture shows two tower cranes bringing material to the roof. Tower cranes are typically attached to the building.

Fig. 5.5 shows the basic elements of a tower crane.

- Foundation
- Tower structure
- Lattice boom (Could be a telescopic boom as well)

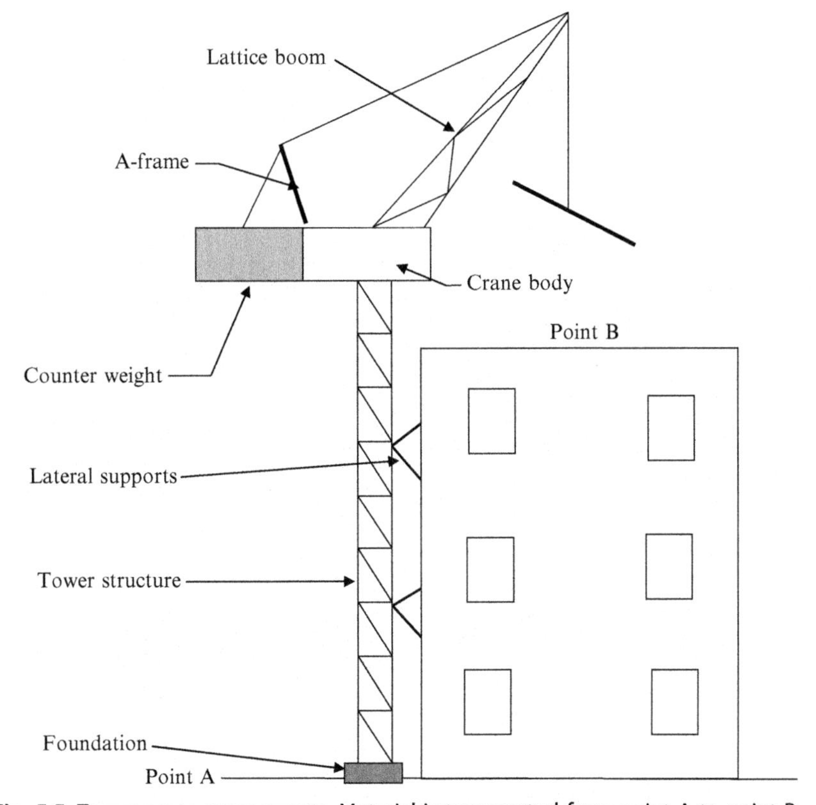

Fig. 5.5 Tower crane components. Material is transported from point A to point B.

- Counter weights
- Lateral supports (Attachments to the building)

Some booms could be horizontal.

How is the crane body, lattice boom, and counter weight lifted to the top of the tower? This is one of the questions many people contemplate when they see a tall tower crane. How did they take the counter weights, boom, and crane body to the top of the tower? Most tower cranes are self-assembling or self-erecting. In addition, some call them self-climbing cranes. All these terms mean the same process.

Tower crane self-erecting process: Self-erecting cranes have a jacking mechanism. The self-erecting mechanism is shown below.

STEP 1: Build the foundation and erect the tower crane to a moderate height using a mobile crane.

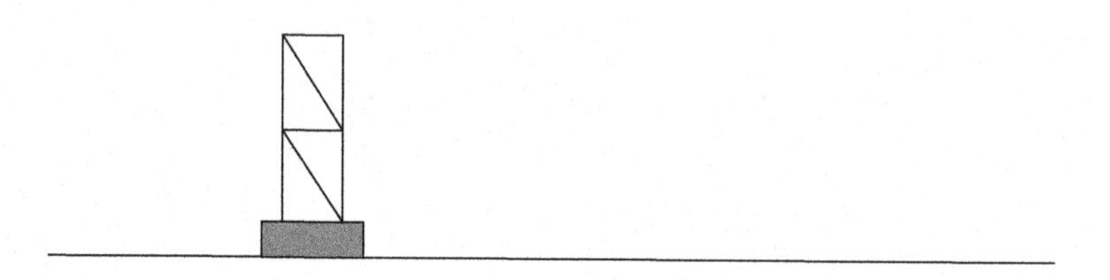

STEP 2: Place the jacking structure using a mobile crane.

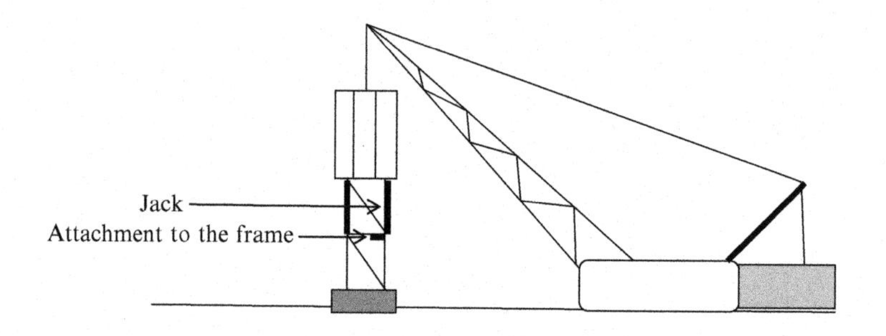

Jacking structure is a metal structure with a cylindrical jack.

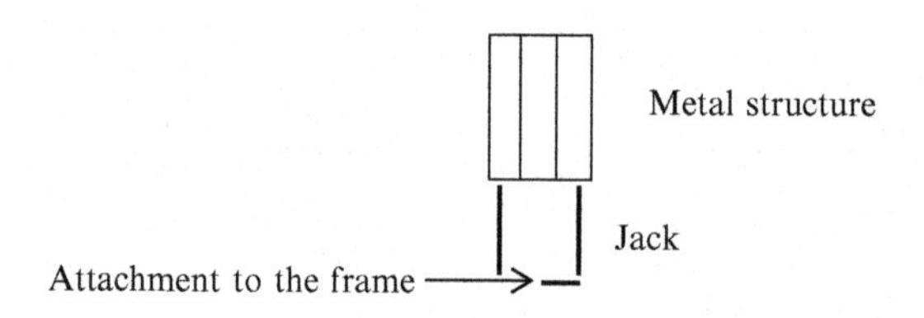

STEP 3: Place more tower crane elements on the jacking structure using the mobile crane.

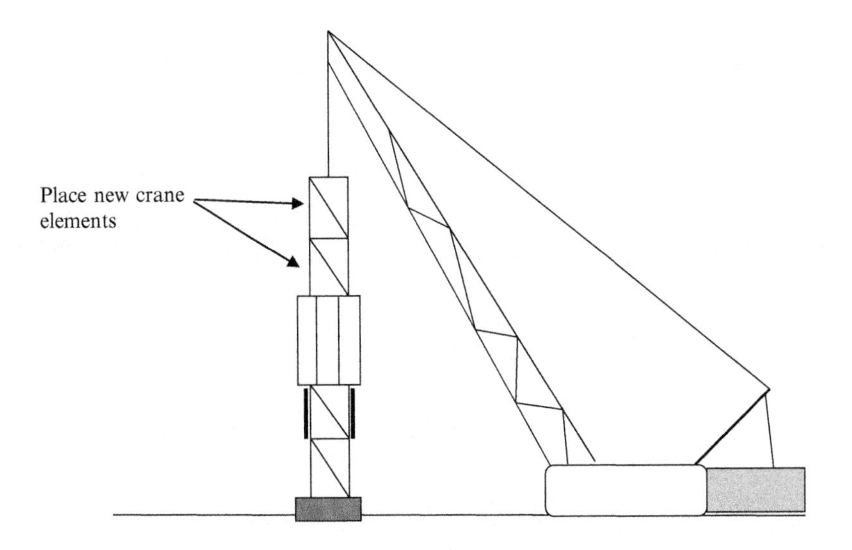

STEP 4: Place the crane body, counter weights, and the boom using the mobile crane.

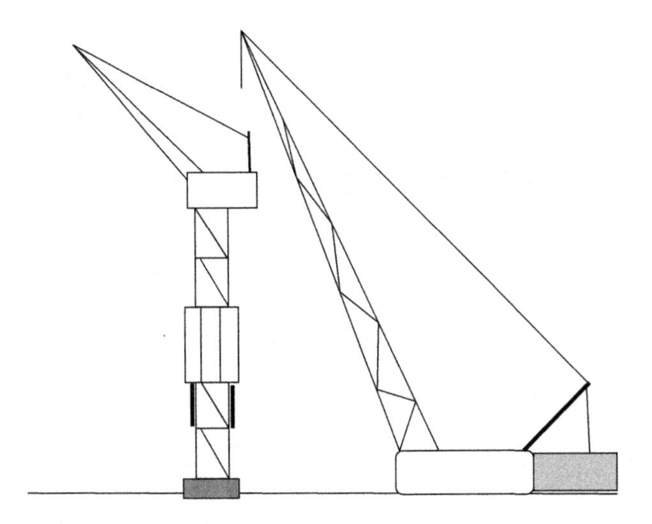

STEP 5: Raise the tower crane with the jacks.

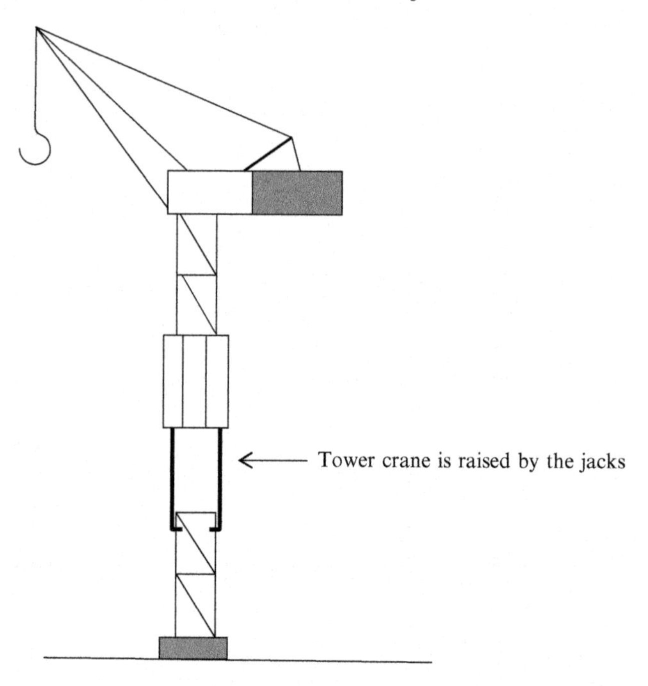

STEP 6: Bring in a new crane element.

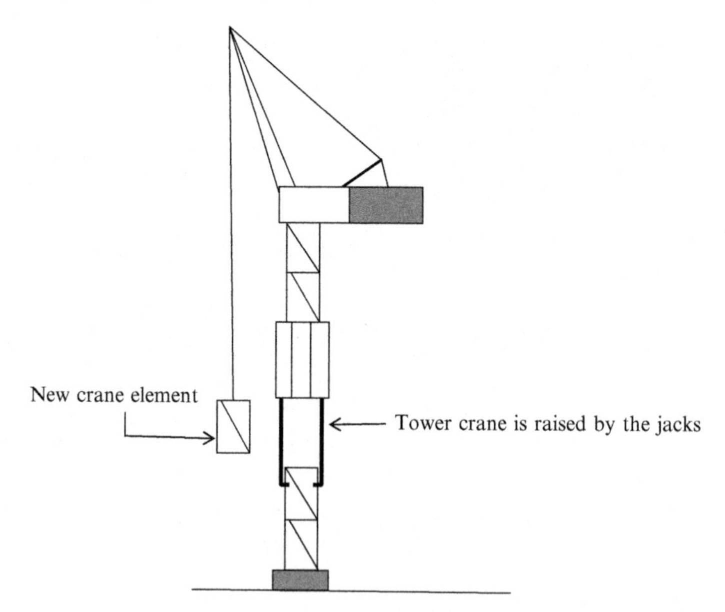

STEP 7: Insert the new crane element into the void and move the jacks up.

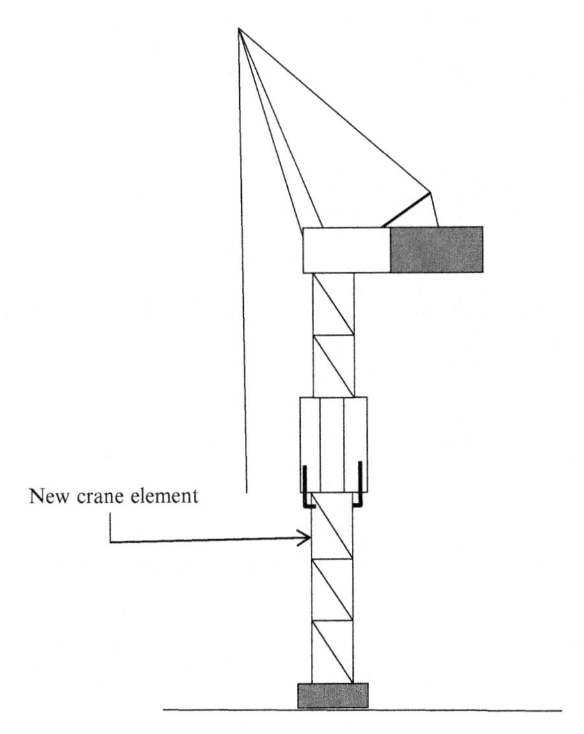

Typically, there are rails in the tower crane to roll the new crane element into place.

STEP 8: Lift the crane again using the jacks.

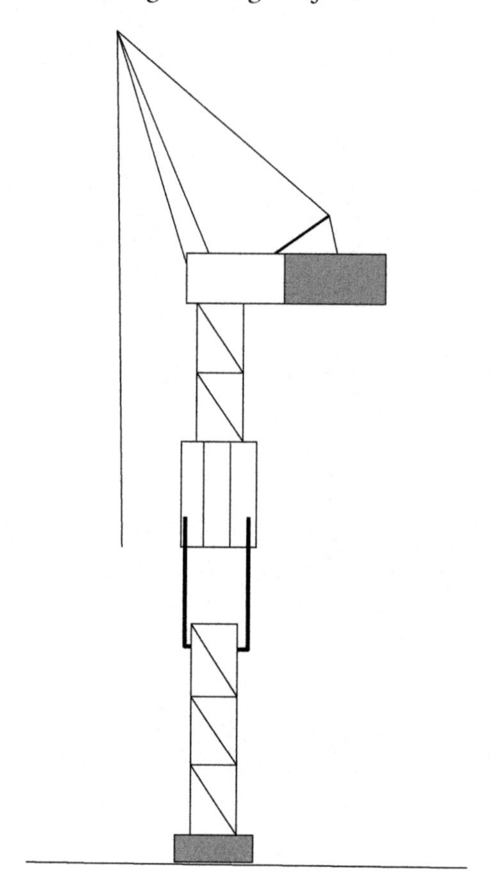

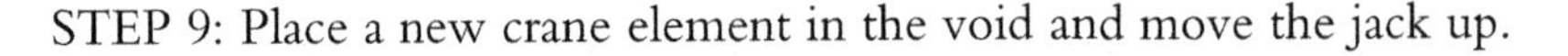

STEP 9: Place a new crane element in the void and move the jack up.

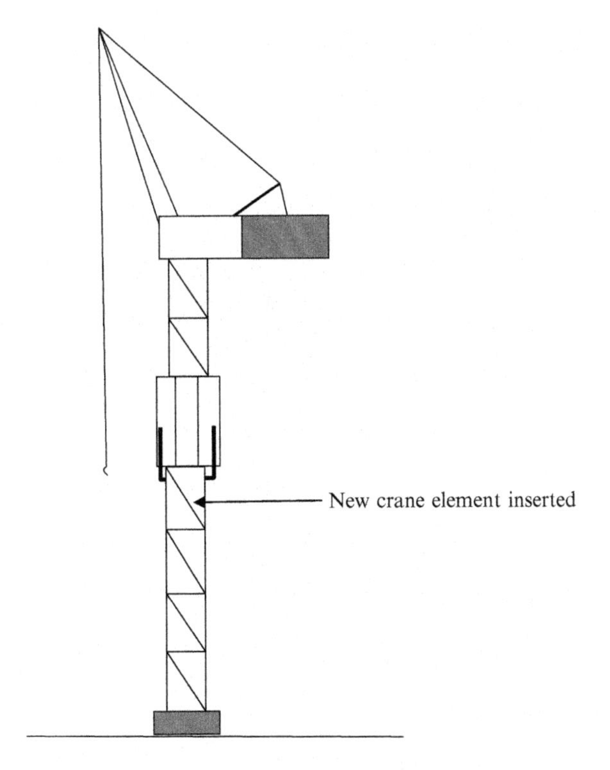

The process is repeated until the desired height is reached (Plate 5.3).

5.1.5 Gantry Cranes

Gantry cranes are placed on trolleys so that they can move. They are not suitable for high-rise buildings.

5.2 CRANE SELECTION, ERECTION, AND STABILITY

Many factors have to be considered during the selection of cranes for a particular construction site. The weight of typical loads, height needs to be lifted, stability of ground, cost implications, nearby traffic, and the safety of workers are some of the factors that need to be considered during the selection process of cranes. Tower cranes have become the most widely used crane type for high-rise buildings. Caterpillar cranes may be suitable for many construction projects as long as the height is not an issue. Gantry cranes are suitable for short buildings and warehouses.

Plate 5.3 A tower crane is shown in the photograph. The jacking structure is slightly larger than the other elements.

The parts of cranes such as chains, chain links, hooks, and moving parts constantly need to be inspected by qualified professionals. Crane failures can cause accidents and delays on the construction site, including fatal injuries.

Practice Problem 5.1

Find the tension in two slings (Fig. 5.6).

Solution

From symmetry, one can deduce that the tension in two slings to be equal.

From trigonometry

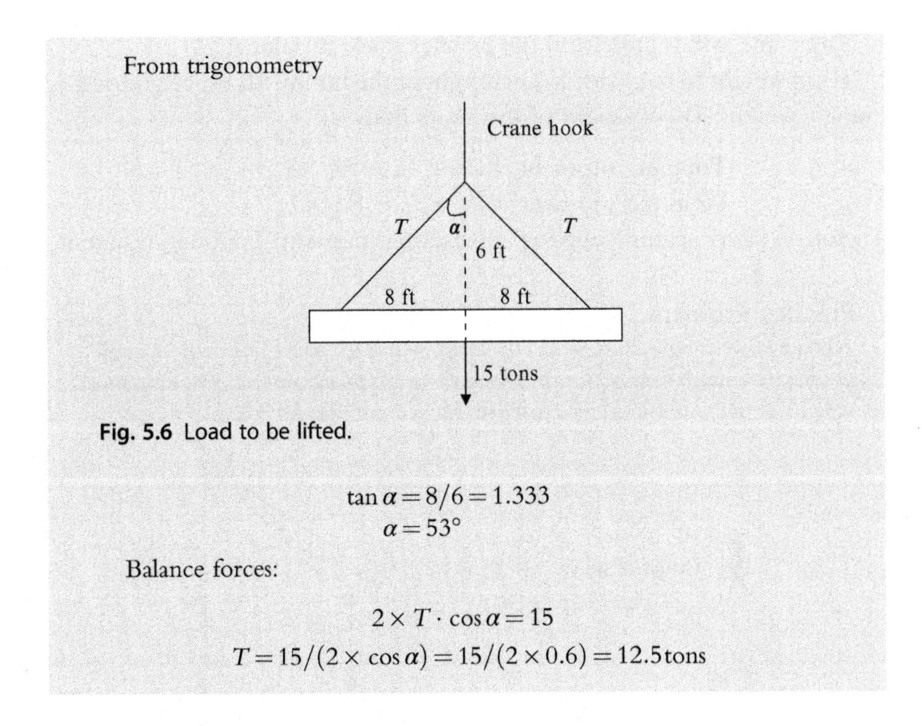

Fig. 5.6 Load to be lifted.

$$\tan\alpha = 8/6 = 1.333$$
$$\alpha = 53°$$

Balance forces:

$$2 \times T \cdot \cos\alpha = 15$$
$$T = 15/(2 \times \cos\alpha) = 15/(2 \times 0.6) = 12.5\,\text{tons}$$

5.2.1 Mobile Crane Stability

Mobile cranes have counter weights to balance the weights that need to be lifted and the weight of the boom (Fig. 5.7).

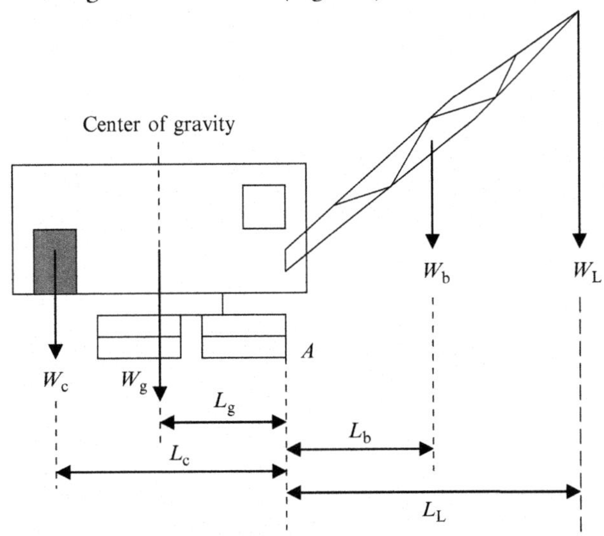

Fig. 5.7 Crane mechanism.

The crane will topple from the crawler track at point A.

W_L = weight of the load; W_b = weight of the boom; W_c = weight of the counter weight; W_g = weight of the crane body.

$$\text{Toppling moment} = W_L \times L_L + W_b \times L_b$$
$$\text{Resisting moment} = W_c \times L_c + W_g \times L_g$$
$$\text{Factor of safety against toppling} = \text{Resisting moment/Toppling moment}$$

Practice Problem 5.2

A crane body weighs 20 tons and the boom weighs 2 tons. The crane is rated to lift a maximum load of 2 tons. Distances are as shown in Fig. 5.8. Find the weight of the counter weight to have a factor of safety of 2.5.

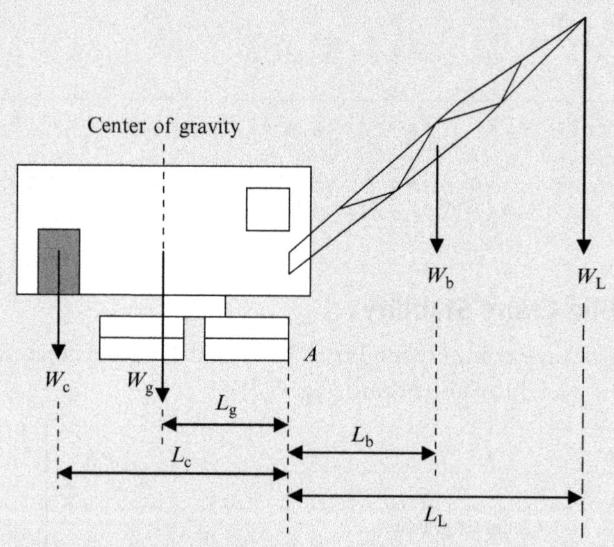

Fig. 5.8 Crane mechanism with loads.

$W_L = 2$ tons, $W_b = 2$ tons, W_c = Weight of the counter weight,

$W_g = 20$ tons $L_L = 20$ ft, $L_b = 10$ ft, $L_c = 10$ ft, $L_g = 6$ ft

$$\text{Toppling moment} = W_L \times L_L + W_b \times L_b = 2 \times 20 + 2 \times 10 = 60 \text{ ton ft}$$

$$\text{Resisting moment} = W_c \times L_c + W_g \times L_g = W_c \times 10 + 20 \times 6$$
$$= 10W_c + 120$$

Factor of safety against toppling = 2.5

$$= \text{Resisting moment/Toppling moment}$$

$$2.5 = (10W_c + 120) \times /60$$

$$150 = 10W_c + 120$$

$$W_c = 3 \text{ tons}$$

Practice Problem 5.3

A 30 ft × 4 ft beam is shown. There are two openings in the beam with diameters of 1 ft and 1.5 ft. One side of the wire rope is 25 ft. Find the length of the wire rope (y) required to properly balance the beam horizontally (Fig. 5.9).

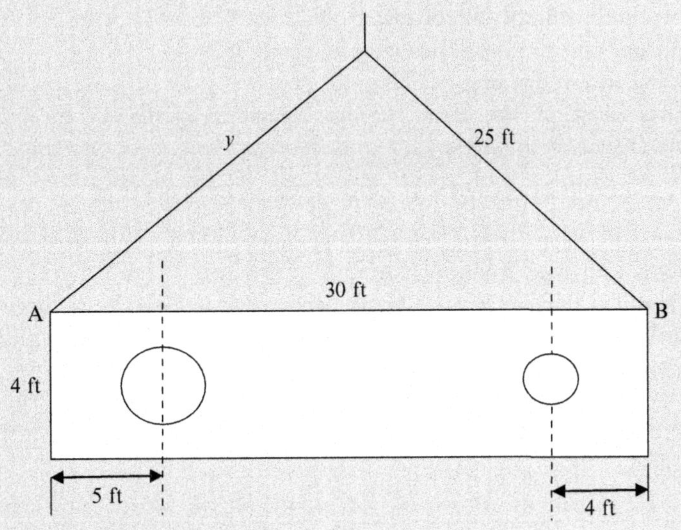

Fig. 5.9 Figure for the problem.

Solution

To properly balance the beam, the vertical string should go thru the center of gravity of the beam.

STEP 1: Find the center of gravity (C.G) of the beam (Fig. 5.10).

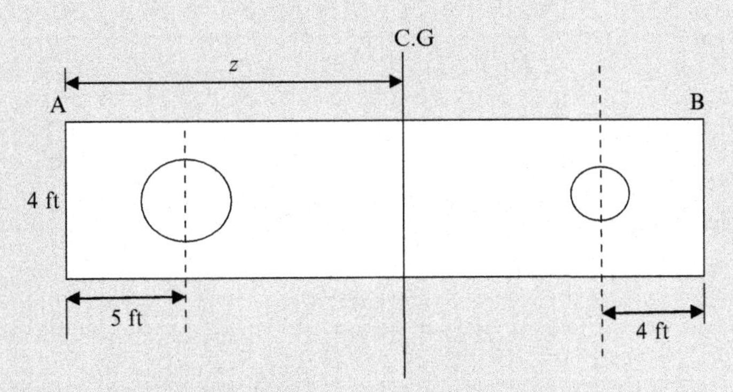

Fig. 5.10 Locate center of gravity.

Continued

Equation for finding center of gravity:

Assume the center of gravity of the beam with holes lies "z" distance from point A;

Total area × distance to center of gravity = Individual areas × distances to center of gravity of that area

Total area of the beam = (30 × 4) − area of openings

Area of opening 1 (1.5 diameter hole) = 1.767 ft^2

Area of opening 2 (1.0 diameter hole) = 0.785 ft^2

Total area of openings = 2.553

Total area of the beam (minus openings) = $(30 \times 4) - \pi \times (1.5)^2/4 - \pi \times (1.0)^2/4 = 117.45$

$$117.45 \cdot z = (30 \times 4) \times 15 - 1.767 \times 5 - 0.785 \times 26$$

Area of original beam without openings = 120 ft^2 and its center of gravity is 15 ft away from point A.

Openings reduce the area. Hence, they need to be reduced from the original beam. The first opening is 5 ft away from point A. The second opening is 26 ft away from point A (Fig. 5.11).

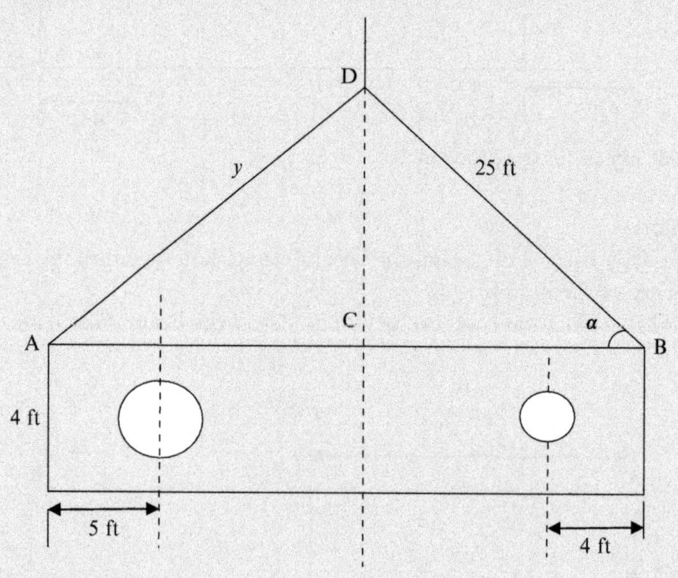

Fig. 5.11 Balanced load.

$$z = 15.077 \text{ ft}$$

$$AC = 15.077$$

$$BC = 30 - 15.077 = 14.923$$

Assume angle DBC to be α.

$$BC/DB = \cos\alpha = 14.923/25 = 0.59693$$
$$\text{Hence, } \alpha = 53.349°.$$
$$DB \cdot \sin\alpha = DC$$
$$25 \times \sin\alpha = DC$$
$$\text{Hence } DC = 20.0573\,\text{ft}.$$

AD can be found using Pythagoras theorem.

$$AD^2 = AC^2 + DC^2$$
$$AD^2 = 15.077^2 + 20.0573^2$$
$$AD = y = 25.092\,\text{ft}$$

5.3 DOZERS

Cutting of soil is usually done using dozers. Dozers are exclusively used to cut soil. They can also transport soil for short distances by pushing the soil with the blade. When the distance of soil that needs to be moved exceeds 100 yards or so, dozers become extremely inefficient in moving soil. For longer distances, scrapers can be used (Fig. 5.12).

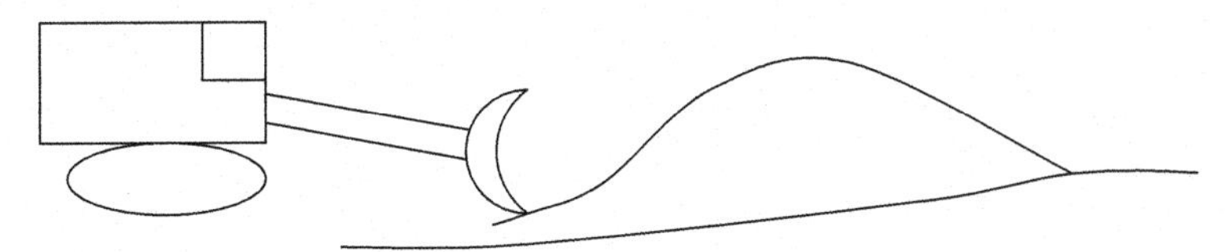

Fig. 5.12 Dozer.

5.4 SCRAPERS

Unlike dozers, scrapers have an underbelly that can store soil. A scraper scrapes soil and stores inside the underbelly. Then it can transport soil to the final location, dump it, and level it (Fig. 5.13).

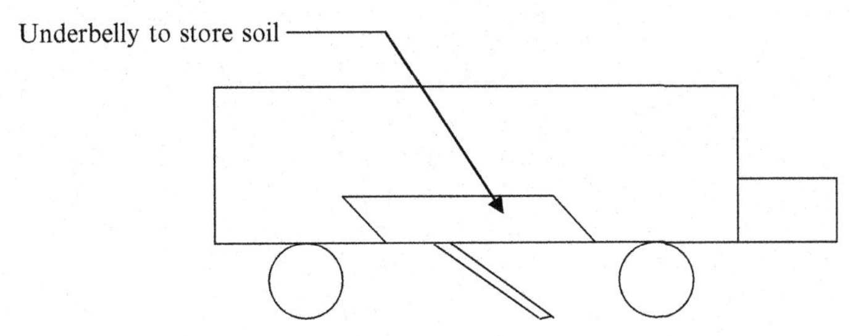

Fig. 5.13 Scraper.

There are different types of scrapers.

- *Elevating scrapers:* These scrapers can elevate the under belly after scraping the soil.
- *Towed scrapers:* This type of scrapers are towed and the soil is moved up to the belly while it is moving.

When the soil is hard, dozers are used to push the scrapers so that more soil can be scooped up quickly. It will take more time to scoop the soil if the scraper is working alone. Using a dozer to push the scraper will accelerate the project. In addition, one should remember that a dozer rental cost will add to the cost.

5.5 LOADERS

Loaders have a bucket in front that can be used to load soil into trucks. The bucket can also be used for grading the soil.

Loaders are either wheel mounted or crawler mounted. Loaders dig out material, transport it, and dump it to a truck. They are also capable of minor grading activities.

Loader capacity is measured using the size of the bucket. The size of the bucket can range from ¼ CY to 20 CY. The average loader has a bucket capacity of 8 CY.

Factor of safety against tipping: The tipping of the loader can happen due to heavy bucket loads. Typically, loaders are provided with a factor of safety of 2.5–3.5 against tipping when the bucket is loaded with material having a density of 3000 lbs per cubic yard (lbs/CY) (Plate 5.4).

Plate 5.4 Loader.

Practice Problem 5.4

Heaped capacity of a bucket of a loader is 8 CY. The manufacturer claims a factor of safety of 3.0 against tipping when loaded with material having a density of 3000 lbs/CY. What is the weight of the loader? Assume that moment arms are at equal distance from the tipping point (Fig. 5.14).

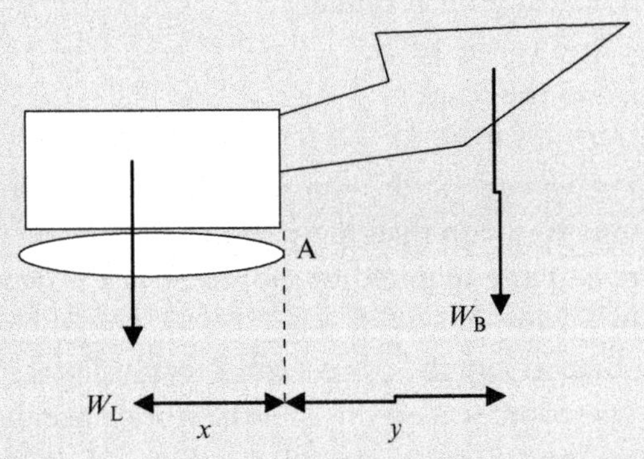

Fig. 5.14 Loader.

W_L is the weight of the loader and W_B is the weight of the bucket, arm, and weight of soil.

Solution

$$\text{Weight of the material in the bucket} = 8 \times 3000\,\text{lbs} = 24,000\,\text{lbs}$$
$$\text{Weight of the loader} = 3.0 \times 24,000 = 72,000\,\text{lbs}$$

Practice Problem 5.5

A loader is loaded with 3000 lbs/CY material. The capacity of the bucket is 6 CY. The loader has a safety factor of 3.5 against tipping. When the loader is fully loaded, its maximum speed is 15 mph. What is the horsepower of the loader?

Solution

$$\text{Weight of the bucket material when fully loaded} = 6 \times 3000 = 18,000\,\text{lbs}$$
$$\text{Weight of the loader} = 18,000 \times 3.5 = 63,000\,\text{lbs}$$
$$\text{Total weight} = 63,000 + 18,000 = 81,000\,\text{lbs}$$
$$\text{Speed} = 15\,\text{mph} = 22\,\text{ft/s}$$
$$\text{Horsepower} = \text{weight} \times \text{speed}/550 = 81,000 \times 22/550 = 3240\,\text{HP}$$

Bucket fill factor of loaders: Some soil can be scooped by the bucket easily. Soils such as moist loam have a bucket fill factor of 1.0–1.2. On the other hand, cemented material has a fill factor of 0.85.

Loader cycle time: Loader cycle time is the time taken for the loader to scoop up soil, transport it to the truck, dump it, and come back.

Breakdown of the loader cycle time:

- Loading the bucket
- Spot the truck (In some cases loader may have to wait for the truck to arrive. This is due to poor planning.)
- Transport the load to the truck
- Dump the load to the truck
- Return

Loading the bucket depends on the type of material to be loaded. Hard soils may take more time to scoop than loose soils.

Spotting the truck: Time required to spot the truck is dependent on the work location. In a large site, with many trucks doing many operations, it may take more time to spot the correct truck that is allocated for this operation. In a small operation, it is easy to spot the truck since there is only one truck for the operation.

Transporting the load to the truck depends on work conditions and distance to the truck. In a small area, where maneuvering is difficult, it might take longer to complete this task. Dumping the load is a standard operation for a given loader and a truck combination. Return time after dumping also depends on terrain, distance, and space available for loader movement.

Practice Problem 5.6

The following time intervals have been computed for a loader operation:

Loading the bucket = 0.1 min

Spot and transport to the truck = 0.15 min

Dump the load and return = 0.2 min

The loader has a bucket of 6 CY and bucket fill factor for the soil is 0.9.

Operator works 50 min/h and 8 h/day.

How many days are needed to remove 20,000 CY of material?

Solution

$$\text{Total time required for one cycle} = 0.1 + 0.15 + 0.2 = 0.45 \text{ min}$$
$$\text{Number of cycles per hour} = 50/0.45 = 111.1$$

Note that operator is only working 50 min/h.

$$\text{Number of CY soil removed in a cycle} = 6 \times \text{bucket fill factor} = 5.4\,\text{CY}$$
$$\text{Number of CY soil removed in a hour} = 5.4 \times 111.1\,\text{CY} = 599.9\,\text{CY}$$

$$\text{Number of CY soil removed in a day} = 599.9 \times 8\,\text{CY} = 4799.2\,\text{CY}$$

$$\text{Number of days required to remove } 20{,}000 \text{ CY of soil} = 20{,}000/4799.2 = 4.17 \text{ days}$$

5.6 EXCAVATORS

Excavators are also known as backhoes. Excavators could be wheel type or crawler type. Small scale backhoes are normally wheel mounted while large backhoes are crawler mounted. When the weight of the bucket is large, the arm length is made shorter. On the other hand, some backhoes have longer arms and small size buckets.

Heap volume and struck volume: A backhoe bucket has two volumes: Heap volume and struck volume (Fig. 5.15).

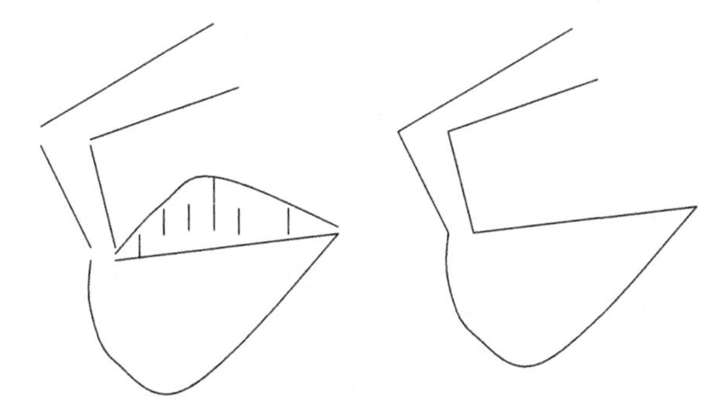

Fig. 5.15 Heap volume (left) and struck volume (right).

Soil type: Hard cemented soil is difficult to excavate compared to lose sand or loam. Hence, soil type plays a major role in determining the efficiency of a backhoe operation. Bucket efficiency is high when the operator can obtain a full bucket in one scoop. It is possible to obtain a full bucket in one scoop for soils such as loam and loose sand. This is not possible for stiff clay and weathered rock (Table 5.1).

Efficiency of backhoes: Backhoes have an optimum depth of excavation. A backhoe works at its highest efficiency at this depth. When the depth is too shallow or too deep, the efficiency decreases. The optimum depth of excavation of a given backhoe depends upon its arm lengths and bucket sizes. Backhoe manufacturers will provide the optimum depth of the backhoe.

Angle of operation: A backhoe bucket goes down into the excavation and digs soil. Then the bucket lifts up and rotate to dump the soil. The angle that needs to be rotated to dump the soil to a truck is known as angle of operation (Fig. 5.16).

Table 5.1 Bucket fill factor vs soil type

Soil type	Bucket fill factor (%)
Loose sand	100
Medium dense sand	90
Dense sand	80
Soft clay	95
Medium stiff clay	90
Stiff clay	80
Clayey gravel (stiff)	85
Shale and other rocks (weathered)	75
Shale and other rocks (medium hard)	50

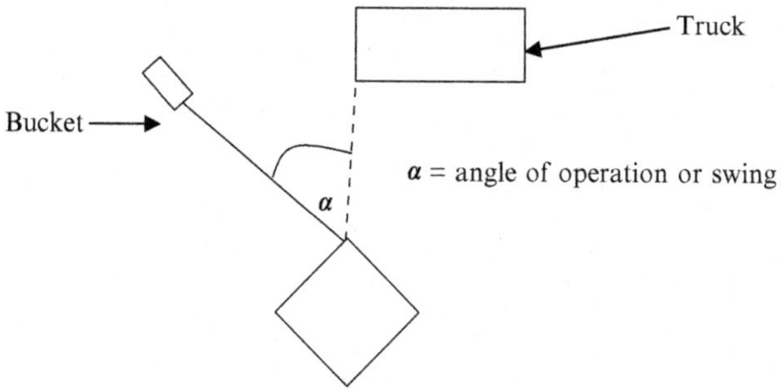

Fig. 5.16 Angle of operation.

When the angle of operation is low, faster operation can be expected. When the angle of operation is high, excavation work slows down. Hence, it is important to make sure that the truck is placed so that the angle of operation is low. In some situations, it may not be possible to maintain a low angle of operation. There could be existing structures, trees, or other obstructions that necessitate trucks being parked far away from the backhoe. Angle of operation is also known as angle of swing.

Angle of operation and depth of operation: Table 5.2 gives the productivity factor for various depths of operation and various angles of swing. Productivity factor is also known as A:D factor.

How to use Table 5.2? Let's assume a certain excavator has an 8 ft optimum depth of operation where it has the highest efficiency. Let us say it is excavating at a depth of 9.6 ft.

Table 5.2 Productivity factor (P) or A:D factor

	Angle of swing (A) degrees					
Optimum depth ratio (D/D_o)	**45**	**60**	**75**	**90**	**120**	**150**
0.40	0.93	0.89	0.85	0.80	0.72	0.59
0.60	1.1	1.03	0.96	0.91	0.81	0.66
0.80	1.22	1.12	1.04	0.98	0.86	0.69
1.00	1.26	1.16	1.07	1.00	0.88	0.71
1.20	1.2	1.11	1.03	0.97	0.86	0.70
1.40	1.12	1.04	0.97	0.91	0.81	0.66
1.60	1.03	0.96	0.90	0.85	0.75	0.62

Source: Power crane and shovel association.

$$\text{Optimum depth } (D_o) = 8\,\text{ft}$$
$$\text{Depth of operation} = 9.6\,\text{ft}$$
$$D/D_o = 9.6/8 = 1.2 = 120\%$$

If you travel along a row of 120%, you will see the productivity factor decreasing when the angle of operation increases. If possible, the best productivity can be obtained when the angle of operation is closer to 45 degrees.

On the other hand, let us assume that this backhoe is operating at an angle of 75 degrees. If you come down along the 75 degree column line you will see productivity increasing until $D/D_o = 1.00$. Then the productivity starts to drop. Productivity is highest at optimum depth. When the operating depth is shallow or too great, productivity decreases.

Look at $D/D_o = 1.00$ and angle of swing is 90 degrees. You will see productivity factor to be 1.00. This is the base. Productivity factor goes above 1.00 when the angle of swing is reduced.

Computation of excavator production: Excavator production is given by the following equation

$$q = \frac{3600B \times E \times P}{C}$$

q = volume of soil excavated and dumped in a truck by the excavator (CY/h)
B = bucket struck capacity (CY)
E = bucket efficiency factor from table A
P = productivity factor from table B
C = cycle time for 90 degree angle and optimum depth

Practice Problem 5.7

An excavator with a bucket that has a struck capacity of 1.2 CY and an optimum depth of 8 ft was used to dig a pit that is 10 ft deep. The angle of operation is found to be 120 degrees. The soil is medium stiff clay. The manufacturer of the excavator lists cycle time for 90 degree angle and optimum depth to be 18 s. Find the productivity of the excavator in CY/h.

Solution

$$q = \frac{3600B \times E \times P}{C}$$

q = volume of soil excavated and dumped in a truck by the excavator (CY/h)
B = bucket struck capacity (CY) = 1.2 CY
E = bucket efficiency factor from table A
E = 0.9 for medium stiff clay.
P = Productivity factor from table B

$$D/D_o = 10/8 = 1.25$$

$$\text{Angle of operation} = 120$$

D/D_o = 1.2 and angle of operation 120, gives a P value of 0.86.

Using interpolation P is found to be 0.85.

C = cycle time for 90 degree angle and optimum depth = 18 s (usually provided by the manufacturer).

If the cycle time is not provided by the manufacturer, this value can be ascertained in the field by practice. Have an operator work at optimum depth with an angle of operation of 90 degrees and evaluate the cycle time.

$$q = \frac{3600B \times E \times P}{C} = \frac{3600 \times 1.2 \times 0.9 \times 0.85}{18} = 180\,\text{CY/h}$$

5.7 DRAGLINES

A dragline is a combination of a crane, bucket, and two cables (Fig. 5.17).

When the upper cable (hoist cable) is tightened, the bucket moves up and to the right. When the lower cable (drag cable) is tightened, the bucket moves down and towards the crane. This way, the operator can manipulate the bucket to excavate and move soil.

Scrapers become too inefficient and expensive when soil has to be transported much longer distances. In such situations, trucks are used.

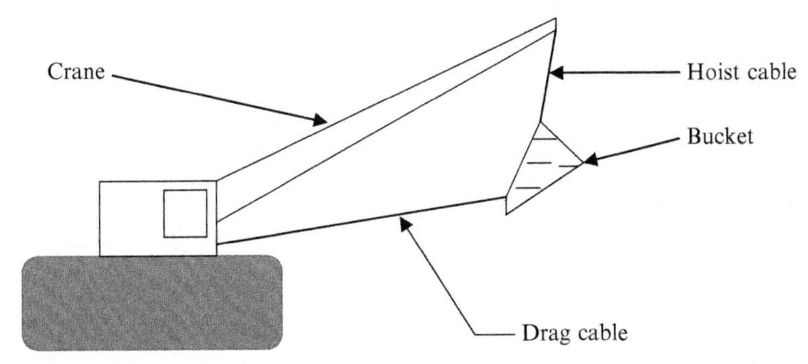

Fig. 5.17 Dragline.

Backhoes and trucks: Dozers can cut soil and make a small mound. Then a backhoe loads the trucks. Trucks take the soil to necessary location and dump there. Several dozers at the destination level and grade the soil.

5.8 GRADERS

Grading is a very important activity on any construction site. Graders are specialized equipment built for grading. Graders are equipped with a blade that is used for grading purposes. The blade of a grader is designed for minor cutting and grading. Dozers should do the main cutting and rough grading before graders are utilized. Graders cannot do the work of dozers since they are not capable of cutting deep into the soil.

Grader cycle: Graders move forward with the blade. Then it returns to its original position. Assume that the forward pass distance is d_f and return distance is d_r. In many cases $d_f = d_r$ (Fig. 5.18).

For most cases $d_f = d_r$

$$\text{Time to go forward} = T_f = d_f / v_f$$

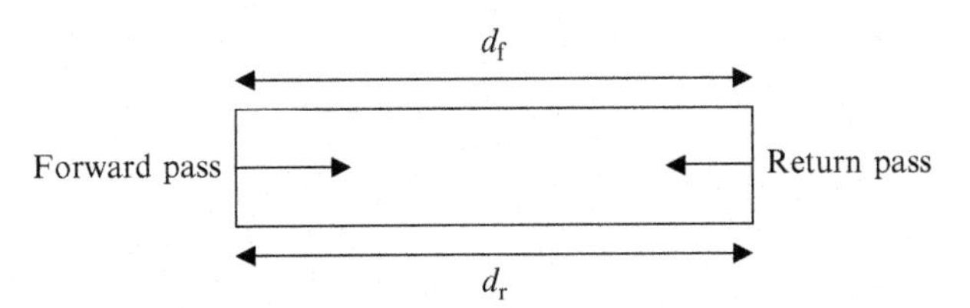

Fig. 5.18 Forward pass and return pass.

d_f = forward pass distance; v_f = velocity of the grader.

$$\text{Time to return} = T_r = d_r/v_r$$

d_r = return distance; v_r = return velocity of the grader (probably in reverse gear).

$$\text{Cycle time} = T_c = T_f + T_r = \text{Time to go forward and return}$$
$$T_c = T_f + T_r = d_f/v_f + d_r/v_r$$

If the grader has a width of W ft, a grader can grade $(W \times d_f)$ ft^2 in going forward and it can grade $(W \times d_r)$ ft^2 during return.

Normally more than one pass is necessary to grade a given area.

In many cases $d_f = d_r = d$ Then the above equation for cycle time can be simplified.

$$T_c = T_f + T_r = d/v_f + d/v_r = d(1/v_f + 1/v_r)$$

If we assume an average velocity (v_a) of the grader for forward movement and return movement, then this equation can be further simplified.

$$T_c = 2d/v_a$$

If each location has to be passed N times, then time to grade a distance of "d" = T_g

$$T_g = 2Nd/v_a$$

Note that N passes are needed to grade one point. Each pass consists of one forward pass and one return pass. Sometimes efficiency term E is introduced to account for operator efficiency, other construction traffic, grading in slopes, movement by surveyors, masons, and various other construction workers and dust control activities. Remember that no construction operation is done in complete isolation. Other issues can always cause a delay for a given operation. These other issues are dependent upon the specific site. To account for site-specific factors, efficiency term is introduced.

$$T_g = 2Nd/(Ev_a), \quad E = \text{efficiency}$$

Practice Problems 5.8

100 × 3000 ft Area need to be graded. It has been noted that grader has to pass each location twice. Each pass contains one forward pass and one backward pass. The grader has a width of 10 ft and has an average velocity (both forward and reverse) of 5 mph. The operator is new and

does not have much experience in grading. Dust control activities and surveying work in the vicinity expect to delay the grading operation. Due to these reasons, efficiency of the grading operation was found to be 0.85. The operator works 50 min/h. Find how many hours are required to grade the area.

Solution

Grader is moving along the length. Grader width is 10 ft. There are 10 strips (Fig. 5.19).

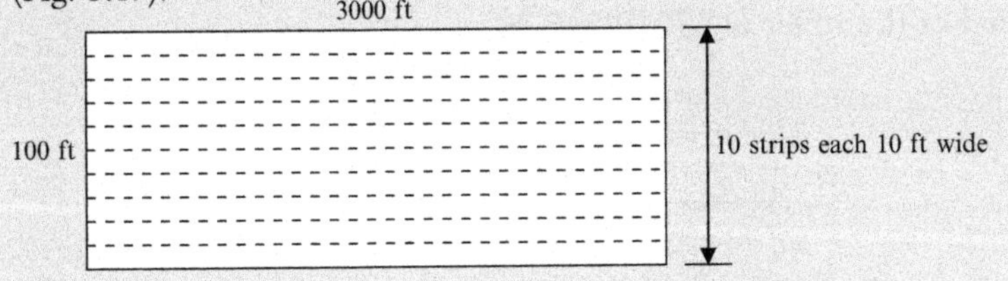

Fig. 5.19 Grader movement.

Average velocity of the grader = 5 mph = 7.33 ft/s.

Time for one pass (one forward pass and reverse to the initial position) = T_c = 6000 ft/7.33 = 818.5 s.

Number of passes needed for one strip = 2.

Time to complete one strip with 100% efficiency = 2 × 818.5 s = 1637 s.

Due to other activities, grading operation can be delayed. Hence, introduce the efficiency factor of 0.85.

Time to complete one strip with 85% efficiency = 1637/0.85 = 1926 s = 32.1 min.

Time to complete 10 strips = 32.1 × 10 = 321 min.

Operator works only 50 min/h.

Number of hours required to complete the project = 321/50 = 6.42 h.

5.9 COMPACTION EQUIPMENT

Compaction of fill material is required to make sure that no settlement follows construction. Typically, compaction is done 95% of modified proctor value. Modified proctor is a compaction test done in a laboratory. Typically, it is required to achieve at least 95% of that value. Compaction equipment has to be selected as per soil conditions.

- Sand and gravels—static rollers, vibratory rollers
- Clay soils—sheep foot rollers (vibratory)
- Trenches—small size rollers, vibratory plates, or jumping jacks

Static rollers could be steel drums or pneumatic tires. They could be made to vibrate to increase their compaction effort.

Productivity of rollers: Heavy rollers achieve the required compaction faster than a smaller roller. At the same time, smaller vibratory type roller can be more effective than a larger static roller. Suitable equipment must be selected for the soil type that needs to be compacted.

Assume thickness "B" soil layer is compacted with a roller that has a width of "W" ft. Assume that each location needs to be rolled "N" times. The speed of the roller is "S" (Fig. 5.20).

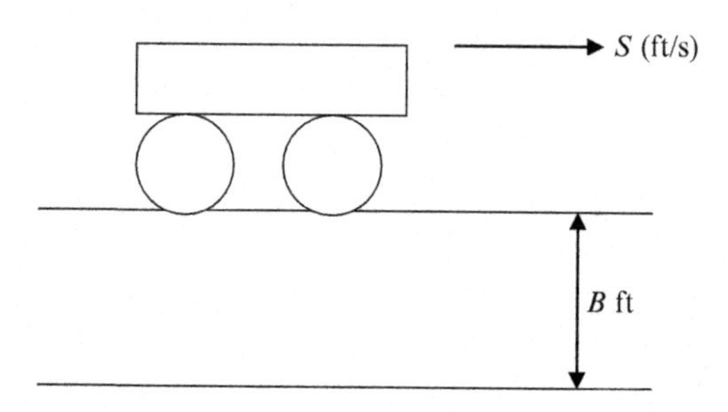

Fig. 5.20 Roller.

Width of the roller $= W$ ft, $N =$ number of passes required to compact one location, $S =$ speed (ft/s).

$$\text{Area of compaction ft}^2/\text{s in one pass} = (W \times S)/N \text{ ft}^2/\text{s}.$$

"N" passes are required to compact one location. After N passes area $(W \times S)$ will be compacted. Only portion of area $(W \times S)$ will be compacted in one pass.

Assume four passes are required to compact one location. One pass will compact only one-fourth of the total compaction.

In other words, compaction of area $(W \times S)$ in one pass $= (W \times S)/4$ ft^2/s.

$$\text{Volume of compaction } (\text{ft}^3/\text{s}) = (W \times S \times B)/N \text{ft}^3/\text{s}$$
$$\text{Volume of compaction } (\text{ft}^3/\text{h}) = 3600 \times (W \times S \times B)/(N) \text{ ft}^3/\text{h}$$
$$\text{Volume of compaction } (\text{CY}/\text{h}) = 3600 \times (W \times S \times B)/(N \times 27) \text{CY}/\text{h}$$

$$\text{Volume of compaction } (\text{CY}/\text{h}) = 3600 \times (W \times S \times B)/(N \times 27) \text{CY}/\text{h}$$
$$W(\text{ft}),\ S(\text{ft/s}),\ B(\text{ft}),\ N = \text{number of passes.}$$

Practice Problem 5.9

A soccer field (300 ft × 150 ft) has to be compacted with 18 in. lifts. Each lift has to be rolled three times to achieve the necessary compaction with a 10 ton static roller with a width of 6 ft. The speed of the roller is 5 mph. How many hours are required to compact one lift?

Solution

$$\text{Volume of compaction (CY/h)} = 3600 \times (W \times S \times B)/(N \times 27)\,\text{CY/h}$$

W in ft, S in ft/s and B in ft.

$$W = 6\text{ft},\quad S = 7.33\text{ft/s},\quad B = 1.5\text{ft}$$

$$\text{Volume of compaction (CY/h)} = 3600 \times (6 \times 7.33 \times 1.5) \times /(3 \times 27)\ \text{CY/h}$$

$$= 2932\,\text{CY/h}$$

$$\text{Total cubic yards per lift} = (300 \times 150 \times 1.5) \times /27 = 2500\,\text{CY}$$

$$\text{Hours required to compact } 2500\,\text{CY} = 2500/2932 = 0.85\,\text{h}$$

Jumping jacks and vibratory plates: Jumping jacks have a plate that moves up and down. In the case of vibratory plates, as the name indicates, compaction is done by a vibrating plate.

5.10 MACHINE POWER

Machine power is measured by horsepower and Watts (SI). When early engines were manufactured, it was important to compare the power of the engines with horses. It was known at that time, the average horse could pull 550 lbs at a rate of 1 ft/s. If a certain machine can pull or move a vehicle and create a frictional force of 550 lbs at a rate of 1 ft/s then we can say that particular machine has a horsepower of 1.0.

Average horse can pull 550lbs at a rate of 1 ft/s (1HP)

Note that pull force is not equal to the weight. Pull force of a vehicle is equal to the frictional resistance of wheels. This is also known as rolling resistance (Fig. 5.21).

When $P = 550$ and speed $= 1$ ft/s, the power of the engine is 1 HP.

1 HP = 550 lb ft/s.

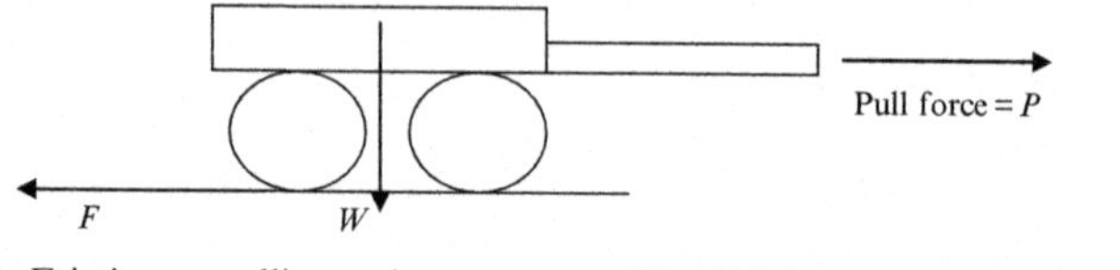

F = Friction or rolling resistance W = Weight P (pull force) = F.

Fig. 5.21 Horse power.

Practice Problem 5.10

A steam engine can move a train that weighs 25 tons at a speed of 30 mph. The friction coefficient between rail and wheels is 0.13. Find the horsepower of the engine.

Solution

$$\text{Pull force} = \text{Friction}$$
$$\text{Friction} = 0.13 \times 25\,\text{tons} = 0.13 \times 25 \times 2000 = 6500\,\text{lbs}$$
$$\text{Speed} = 30\,\text{mph} = 44\,\text{ft/s}$$
$$\text{Work done per second} = 6500 \times 44 = 286,000\,\text{lbs ft/s}$$
$$\text{Work done per second in HP} = 286,000/550 = 520\,\text{HP}$$

Power in a machine: Engines generate power. The power of the engine is then transferred to a flywheel. Energy loss occurs during this process. Power is then transferred to a gear mechanism and then to the axle.

Engine power and flywheel power: Engine horsepower is much larger than flywheel horsepower. Hence, it is important to know the flywhcel HP of an engine rather than the engine HP.

5.11 RIM PULL

Rim pull is the pull needed to overcome friction. Typically at lower gears of a vehicle (gears 1 and 2), the very high rim pulls can be generated. However, at low gears, one has to maintain a lower speed. At high gears of a vehicle, high speeds can be achieved but the rim pull will be less (Fig. 5.22).

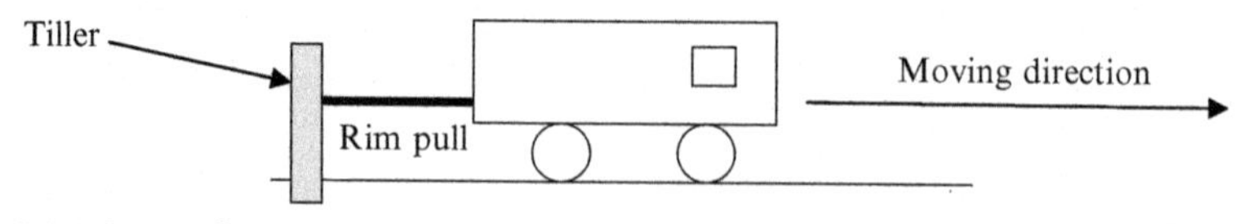

Fig. 5.22 Rim pull.

Practice Problem 5.11

A tractor needs to pull a tiller. The horsepower of the tractor is 70 HP. The tiller exerts a pull of 2000 lbs. What is the maximum speed that the tractor can travel?

Solution

$$70\,\text{HP} = 70 \times 550\,\text{lbs ft/s} = 38{,}500\,\text{lbs ft/s}$$

$$\text{Horsepower} = \text{Work done per second} = \text{Force} \times \text{speed} = 2000 \times \text{speed}$$

$$70\,\text{HP} = 38{,}500 = 2000 \times \text{speed}$$

$$\text{Maximum speed} = 38{,}500/2000 = 19.25\,\text{ft/s}$$

5.12 DRAWBAR PULL

Drawbar pull is the same as rim pull except that this term is used for vehicles that do not have a rim but a crawler type tracks. It does not make a difference whether it is pull or push (Fig. 5.23).

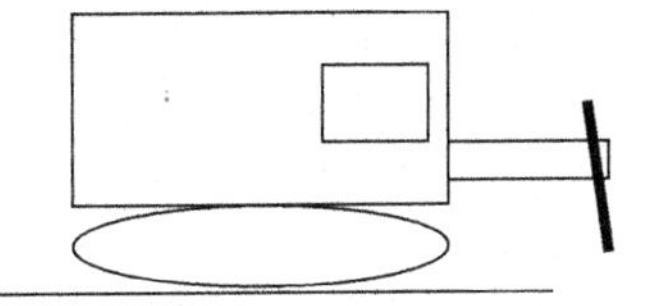

Fig. 5.23 Crawler type tracks.

Practice Problem 5.12

A dozer needs to cut through hard soil. The force exerted on the soil is computed to be 900 lbs (Fig. 5.24). The dozer has to move at a speed of 10 mph to maintain the schedule. The contractor is planning to use a 30 HP dozer. Can this machine do the work required?

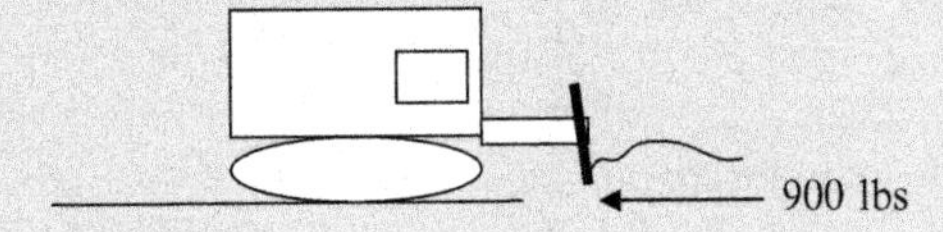

Fig. 5.24 Dozer cutting soil.

Solution

$$10\,\text{mph} = 14.67\,\text{ft/s}$$

$$\text{Power needed} = 900 \times 14.67\,\text{lbs ft/s} = 13{,}200\,\text{lbs ft/s} = 24\,\text{HP}$$

The machine is capable of performing the task as specified.

5.13 ROLLING RESISTANCE

There is a resistance to rolling of wheels. Rolling resistance depends on air pressure in tires, the wear and tear of tires, and road roughness (Fig. 5.25).

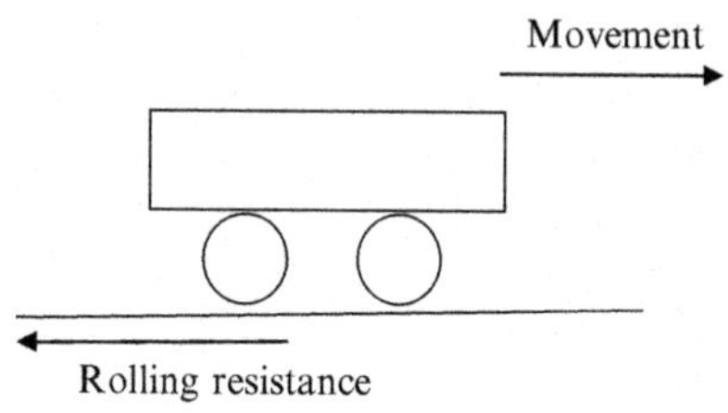

Fig. 5.25 Movement and friction.

Rolling resistance typically given in lbs/ton.

Practice Problem 5.13

Rolling resistance of a truck is given to be 30 lbs/ton. What is the rolling resistance of a truck that weighs 12 tons?

Solution

Rolling resistance of the truck $= 12 \times 30$ lbs $= 360$ lbs

Grade: the grade of a road affects the work done by machines (Fig. 5.26).

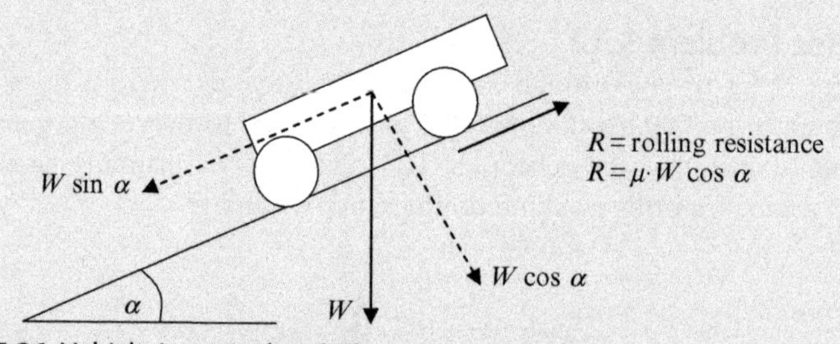

Fig. 5.26 Vehicle in an inclined plane.

Component of weight W acts along the slope ($W \sin \alpha$)
Second component acts perpendicular to the slope ($W \cos \alpha$)
Rolling resistance $= \mu \cdot W \cos \alpha$ ($\mu =$ rolling resistance coefficient)
For slope angles less than 10 degrees, $\sin \alpha$ is approximated with $\tan \alpha$
$\tan \alpha = V/H$ or the grade.

Practice Problem 5.14

A 5-ton loaded truck is moving uphill. The grade is 3.5%. What is the rim pull of the truck?

Solution

The force that needs to be overcome is $W \sin \alpha$. For small angles $\sin \alpha$ is approximated to $\tan \alpha$.

$$\text{Hence } W \sin \alpha = W \tan \alpha$$

$$\tan \alpha = \text{Grade} = 0.035$$

$$\text{Rim pull} = W \times 0.035 = (5 \times 2000) \times 0.035\,\text{lbs} = 350\,\text{lbs}$$

5.14 TRACTION

A truck may have a super powered engine but it may not be able to pull or haul anything if the road is slippery. Usable force of a truck, dozer, scraper, or any other construction machine depends on the coefficient of traction.

$$\boxed{\text{Usable force} = \text{Coefficient of traction} \times \text{Weight on driving wheels}}$$

If the truck tries to pull or haul any load greater than usable force, slippage of wheels can occur. A high coefficient of traction allows a truck to haul large loads. A truck can haul larger loads in a gravel road than in a smooth concrete road (Fig. 5.27).

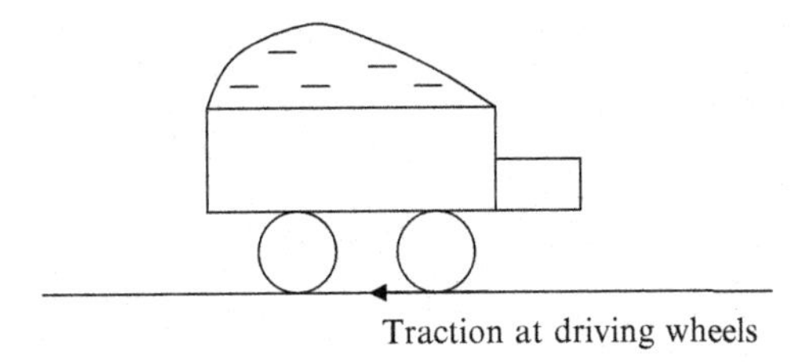

Fig. 5.27 Traction.

Practice Problem 5.15

Find the usable force of a truck if the fully loaded truck is 15,000 lbs and 55% of the weight is transferred to the driving wheels. The coefficient of traction between the wheels and the road is 0.65. Find the usable force.

Solution

$$\text{Weight transferred to driving wheels} = 15,000 \times 0.55\,\text{lbs} = 8250\,\text{lbs}$$

$$\text{Usable force} = \text{Coefficient of traction} \times \text{Weight on driving wheels}$$

$$\text{Usable force} = 0.65 \times 8250 = 5362\,\text{lbs}$$

Rim pull and usable force: In some situations, maximum rim pull may not be usable due to slippage.

Practice Problem 5.16

An empty truck weighs 10,000 lbs. The rim pull of the truck is 9000 lbs. 62% of the weight is transferred to driving wheels. The coefficient of traction between the wheels and the road is 0.65. What is the maximum volume of soil measured in CY that can be hauled with the truck when using the maximum rim pull? (1 CY of soil weighs 3000 lbs).

Solution

Assume y lbs of soil are loaded.

$$\text{Total weight} = y + 10,000$$

$$\text{Weight on driving wheels} = 0.62 \times (y + 10,000)$$

$$\text{Usable force} = \text{Coefficient of traction} \times \text{Weight on driving wheels}$$

$$\text{Usable force} = 0.65 \times 0.62 \times (y + 10,000)$$

At maximum rim pull,

$$9000 = 0.65 \times 0.62 \times (y + 10,000)$$

$$y = 12,332\,\text{lbs}$$

$$y = 12,332/3000 = 4.11\,\text{CY}$$

5.15 RIM PULL OR DRAWBAR PULL VS SPEED

The horsepower of a vehicle depends on its engine power. An engine has a maximum capacity. When the vehicle is moving fast, its rim pull is less. When the machine is moving at a lower speed, it has a higher rim pull capacity and can haul heavier loads.

Rim pull or drawbar pull (lbs): At first gear, the rim pull can be as high as 12,000 lbs. The maximum speed that can be attained in first gear is 40 mph. If the gear is shifted to the second gear, a higher speed can be achieved but the maximum rim pull or drawbar pull is 9000 lbs (Fig. 5.28).

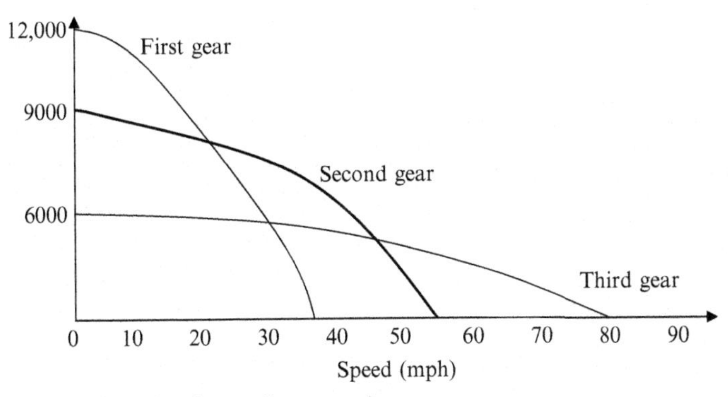

Fig. 5.28 Rim pull or drawbar pull vs speed.

Practice Problem 5.17

A dozer has to cut soil and the push force is estimated to be 7000 lbs. The contractor wants the machine to operate at 10 mph. What gear the operator should use? (use Fig. 5.28).

Solution

At 10 mph, the maximum drawbar pull for second gear is approximately 8500 lbs. Hence, it can spare 7000 lbs at second gear. At 10 mph, maximum drawbar gear is 11,000 lbs at first gear.

The operator can use either the first gear or the second gear. On the other hand, at third gear with a speed of 10 mph, maximum drawbar pull is only 6000 lbs. The dozer may not be able to exert 7000 lbs drawbar pull at third gear.

5.16 EQUIPMENT FUEL COST

Fuel consumption of a machine depends on the horsepower of the machine and the type of work it does. An excavator burns more fuel when digging in hard soil than in soft soil. A dozer burns more fuel when grading uphill. Older machines use more fuel than new machines.

Practice Problem 5.18

A 350 HP dozer consumes fuel at a maximum rate (0.02 gal/HP/h) when grading 4% grade uphill. The dozer uses 50% of the maximum rate when grading downhill, and it uses 30% of the maximum rate when moving from one location to another location. It has been noted that 40% of the time the dozer is grading uphill, 35% of the time it is grading downhill, and 25% of the time it is moving from location to location. What is the fuel consumption per hour if the operator works only 50 min/h?

Solution

$$\text{Maximum fuel consumption} = 0.02 \times 350 = 7\,\text{gal/h} = 0.117\,\text{gal/min}$$

Operator works only 50 min/h.

$$\text{Number of minutes grading uphill} = 0.4 \times 50\,\text{min} = 20\,\text{min}$$
$$\text{Fuel consumption grading uphill} = 20 \times 0.117 = 2.34\,\text{gal}$$
$$\text{Number of minutes grading downhill} = 0.35 \times 50\,\text{min} = 17.5\,\text{min}$$
$$\text{Fuel consumption grading uphill} = 17.5 \times (0.117 \times 0.5) = 1.02\,\text{gal}$$

Note that dozer uses only 50% of the maximum rate when grading downhill.

$$\text{Number of minutes moving from one location to another} = 0.25 \times 50\,\text{min} = 12.5\,\text{min}$$
$$\text{Fuel consumption during moving} = 12.5 \times (0.117 \times 0.3) = 0.44\,\text{gal}$$
$$\text{Total fuel consumption per hour} = 2.34 + 1.02 + 0.44 = 3.8\,\text{gal/h}$$

5.17 EQUIPMENT PRODUCTION

The productivity or the effectiveness of labor and equipment are of paramount importance to project success. There are situations where a contractor could decide to utilize cheaper equipment compared to high capacity equipment that cost more. The contractor needs to study the productivity of the equipment and assess which equipment are the most cost effective.

Factors that affect the effectiveness of equipment for a given project:

- Suitability of the equipment for the specific project
- Production rate
- Safety
- Repair cost
- Fuel consumption

- Cost of operators
- Need for specialized operators (difficulty of replacement in the event one leaves)
- Cost of parts if any parts are broken
- Lead time to obtain parts
- Cost of transportation
- Security of equipment from vandalism

Practice Problem 5.19

An excavation contractor has the choice to use one large backhoe or two small size backhoes for a specific project. The backhoes are used to excavate and load soil into trucks for transportation. The contractor needs to excavate and remove 1500 CY of soil. The project duration is 30 workdays.

Which is the cheaper alternative?

Following information is known of the backhoes.

Large backhoe:

Rental cost = \$700 per day

Production rate = can excavate and load one truck in 45 min. One truck has a capacity of 10 CY.

Labor required = 1 operator + 1 laborer

Operator cost = \$70 per hour with benefits

Laborer cost = \$50 per hour with benefits

Company overhead allocated to this project = \$15 per hour

Average maintenance cost per day (oiling, repairs, parts) = \$30

Fuel consumption = 1 gallon of diesel per hour (Diesel cost \$3 per gallon)

Small backhoe:

Rental cost = \$300 per day

Production rate = Can excavate and load one truck (10 CY) in 1 h and 15 min

Labor required = 1 operator + 1 laborer

Operator cost = \$60 per hour with benefits

Laborer cost = \$50 per hour with benefits

Company overhead allocated to this project = \$15 per hour

Average maintenance cost per day for one small backhoe (oiling, repairs, parts) = \$20

Fuel consumption = 0.6 gallons of diesel per hour (diesel cost \$3 per gallon)

Solution

Find the cost of the project if the large backhoe is used.

Continued

To complete the project, the contractor has to load and remove 150 trucks (1500/10).

If the large backhoe is used, the project can be completed in $150 \times 0.75 = 112.5$ h (45 min = 0.75 h).

Assuming 8-h workday, the project can be completed in 14.1 days (112.5/8).

Hence, the project can be successfully completed within the schedule if the large backhoe is used since the project contract duration is 30 working days.

Cost of the large backhoe (convert all costs to dollars per hour):

Rental cost = \$700 per day = \$87.5 per hour

Operator cost = \$70 per hour with benefits

Laborer cost = \$50 per hour with benefits

Company overhead allocated to this project = \$ 15 per hour

Average maintenance cost per day (oiling, repairs, parts) = \$30 per day = \$3.75 per hour

Fuel cost = \$3 per hour

Total cost per hour = $87.5 + 70 + 50 + 15 + 3.75 + 3 = 229.25$ per hour

Total number of hours needed = 112.5

Total cost = 112.5×229.25 = \$25,790

Option of using two small backhoes:

First, check whether the project can be completed with two small backhoes.

To complete the project, the contractor has to load and remove 150 trucks (1500/10).

Production rate per one small backhoe = 1.25 h to load one truck.

Production rate for two small backhoes = 0.625 h to load one truck

If two small backhoes are used, the project can be completed in $150 \times 0.625 = 93.8$ h = 11.7 working days

Available days = 30

The project can be successfully completed with two small backhoes.

Cost of two small backhoes (convert all costs to dollars per hour):

Rental cost = \$600 per day = \$75 per hour

Operator cost = \$120 per hour with benefits for two operators

Laborer cost = \$100 per hour with benefits for two laborers

Company overhead allocated to this project = \$ 15 per hour

Average maintenance cost per day for two backhoes (oiling, repairs, parts) = \$40 per day = \$5 per hour

Fuel consumption = 1.2 gallons per hour

Fuel cost = \$3.6 per hour

Total cost per hour = 75 + 120 + 100 + 15 + 5 + 3.6 = 318.6 per hour

Total number of hours needed = 93.8 h

Total cost = 93.8 × 318.6 = \$29,884

Using the large backhoe is cost effective in this case.

There are other factors that cannot be quantified.

Safety factor: If the workspace is too tight, using two backhoes may be relatively unsafe compared to using one large backhoe.

Cost of parts if any parts are broken: Equipment can break down. If new parts are needed it is important to know the possible cost for parts and lead time to obtain them.

Specialized operators: Unlike small backhoes, large backhoes may require operators with high skill. If the present operator leaves the company, it is important to know the availability of large backhoe operators in the market.

5.18 PRODUCTIVITY ANALYSIS AND IMPROVEMENT

The contractor needs to analyze the equipment usage with respect to suitability, productivity, safety, reliability, and cost. For example, the contractor may find concreting a high-rise building is easier using a concrete lift mechanism compared to pumping. Good contractors are constantly in the process of investigating new methods and equipment. Productivity analysis of a construction operation needs to be considered with respect to other alternatives. One method may be faster but costly. On the other hand, another method may require specialized equipment where spare parts are not easy to find.

In many cases, it is not easy to quantify the advantages and disadvantages of different alternatives.

CHAPTER 6

Earthwork Construction and Layout

6.1 EXCAVATION AND EMBANKMENT (CUT AND FILL)

Roads are constructed over uneven ground. During the construction of a road, some locations are required to be cut and other locations have to be filled. Consider the terrain shown below (Fig. 6.1). Point A to B has to be *cut* and point B to C has to be *filled.*

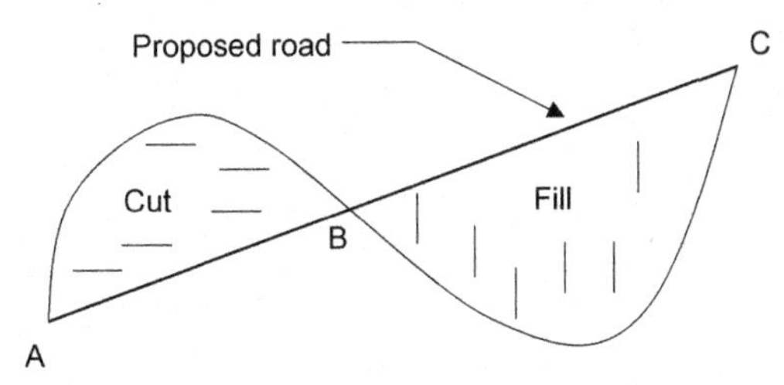

Fig. 6.1 Cut and fill.

Cut and Fill: In some situations, the soil that is removed during a cut can be used for fill. If the soil removed is not suitable, then suitable soil has to be imported to the site for fill purposes.

6.2 BORROW PIT VOLUME PROBLEMS

It is important to have a good knowledge of phase relationships in soil in order to solve borrow pit problems.

6.2.1 Soil Phase Relationships

Soil consists of solids, air, and water. Solids are soil particles. A soil matrix can be schematically represented as shown below.

Volume		**Mass**
V_a	Air	$M_a = 0$ (Usually mass of air is taken to be zero)
V_w	Water	M_w (Mass of water)
V_s	Solid	M_s (Mass of solids)

Soil phase diagram

http://dx.doi.org/10.1016/B978-0-12-809244-6.00006-8

M_a = Mass of air = 0 (usually taken to be zero)
V_a = Volume of air (volume of air is *not* zero)
M_w = Mass of water
V_w = Volume of water
M_s = Mass of solids
V_s = Volume of solids
M = Total mass of soil = $M_s + M_w$ (mass of air ignored)
V = Total volume of soil = $V_s + V_w + V_a$
V_v = Volume of voids = $V_a + V_w$

Density of water (γ_w) = The density of water can be expressed in many units.

$$\gamma_w = M_w / V_w$$

SI Units: $\gamma_w = 1\,g/cm^3 = 1000\,g/L = 1000\,kg/m^3 = 9.81\,kN/m^3$.
fps Units: $\gamma_w = 62.42$ pounds per cu. feet (pcf).

Total density of soil γ_t, γ_{wet} *or* γ: Some books use γ_{wet} and some other books uses γ_t or simply γ to denote total density of soil. Total density (also known as wet density) is simply the mass of soil (including water) divided by the volume of soil.

$$\gamma_{wet} = M/V$$
$$M = M_w + M_s \text{ and } V = V_w + V_a + V_s$$

{M = Total mass of soil including water. Mass of air ignored.}
{V = Total volume of soil including soil, water, and air. Volume of air is *not* ignored.}

Dry density of soil (γ_d): $\gamma_d = M_s / V$
M_s = Mass of solid only.
V = Total volume of soil including soil, water, and air.

$$V = V_w + V_a + V_s$$

Density of solids = M_s / V_s.

Specific gravity (G_s): Specific gravity is defined as the density of solids divided by the density of water. The density of solids is represented by G_s or by simply G.

$$\text{Specific gravity}(G_s) = M_s / (V_s \cdot \gamma_w)$$

Void ratio (e): Void ratio (e) is defined as the ratio of volume of voids to volume of solids.

$$e = V_v / V_s$$

V_v = Volume of voids (volume of water + Volume of air) = $V_a + V_w$.

V_s = Volume of solids.

Moisture Content (w): Moisture content $(w) = M_W / M_s$

M_W = Mass of water.

M_s = Mass of solids.

Porosity (n):

$$n = V_v / V$$

V_v = Volume of voids.

V = Total volume = $V_s + V_W + V_a$.

What does porosity means? Look at the top term V_v. It tells us how much voids are there in the soil. The ratio between voids and total volume is given by porosity. In other words, porosity gives us an indication of pores in a soil. Soil with high porosity has more pores than soil with low porosity. It is reasonable to assume that soils with high porosity has a higher permeability.

Degree of saturation (S): $S = V_W / V_v$

V_W = Volume of water.

V_v = Total of volume of voids.

When the total volume of voids is filled with water $S = 100\%$.

The degree of saturation tells us how much water is in the voids.

Some Relationships to Remember:

Relationship 1

$$\boxed{\gamma_d = \gamma_{wet} / (1 + w)}$$

This relationship appears in the soil compaction section as well.

It can be proven as follows:

$$\gamma_{wet} = M / V, \text{hence } V = M / \gamma_{wet}$$

$$\gamma_d = M_s / V$$

Replace V with M/γ_{wet}

$$\gamma_d = M_s / (M / \gamma_{wet}) = M_s \times \gamma_{wet} / M$$
$$M = M_s + M_W \text{ (mass of air is ignored)}$$
$$\gamma_d = M_s \times \gamma_{wet} / (M_s + M_W)$$

Divide both top and bottom by M_s.

$$\gamma_d = \gamma_{wet} / (1 + w)$$

Relationship 2

$$\boxed{S \cdot e = G_s \cdot w}$$

This relationship can be shown to be true as below:

$$S = V_W/V_V; \quad e = V_v/V_s$$

$$S \cdot e = V_W/V_s \tag{6.1}$$

$$G_s = M_s/(V_s \cdot \gamma_W); \quad \gamma_w = M_W/V_w$$

Replace γ_w in the equation.

$$G_s = M_s/(V_s) \times (V_W/M_W)$$

Hence $G_s = (M_s \times V_W)/(V_s \cdot M_W)$;

$$w = M_W/M_s$$

$$G_s \cdot w = [(M_s \times V_W)/(V_s \cdot M_W)] \times (M_W/M_s)$$

By simplification;

$$G_s \cdot w = V_W/V_s \tag{6.2}$$

Eqs. (6.1) and (6.2) are equal. Hence $S \cdot e = G_s \cdot w$.

Relationship 3

$$\boxed{n = e/(1 + e)}$$

Proof: Replace "e" with V_v/V_s in the above equation.

$$n = e/(1 + e) = (V_v/V_s)/[1 + V_v/V_s]$$

Multiply top and bottom by V_s.

$$e/(1 + e) = (V_v)/[V_s + V_v]$$
$$V_s + V_V = V$$
$$e/(1 + e) = V_v/V = n \quad (V_v/V \text{isporosity})$$

Relationship 4

$$\boxed{e = n/(1 - n)}$$

Proof: From relationship 3:

$$n = e/(1 + e)$$
$$n + ne = e$$
$$n = e - ne$$
$$n = e(1 - n)$$
$$e = n/(1 - n)$$

Relationship 5

$$\boxed{\gamma_d = \gamma_W \cdot G_s/[1 + (w/S)G_s]}$$

Proof:

$$\gamma_d = \frac{\gamma_W \cdot G_s}{[1 + (w/S)G_s]}$$

$$\gamma_d = \frac{\gamma_W \cdot M_s/(V_s \cdot \gamma_W)}{[1 + (M_W/M_s/(V_W/V_v) \times M_s/(V_s \cdot \gamma_W)]}$$

$$\gamma_d = \frac{\gamma_W \cdot M_s/(V_s \cdot \gamma_W)}{[1 + (M_W/M_s/(V_W/V_v) \times M_s/(V_s \cdot \gamma_W)]}$$

$$\gamma_d = \frac{M_s/V_s}{[1 + (M_W/(V_W/V_v) \times 1/(V_s \cdot \gamma_W)]}$$

$$\gamma_d = \frac{M_s/V_s}{[1 + (\gamma_W \cdot V_v) \times 1/(V_s \cdot \gamma_W)]}$$

$$\gamma_d = \frac{M_s/V_s}{[1 + (V_v) \times 1/(V_s)]}$$

Multiply top and bottom by V_s

$$\gamma_d = \frac{M_s}{[V_s + V_v]}$$

$$V_s + V_v = V$$

$$\gamma_d = \frac{M_s}{V}$$

Relationship 6

$$\boxed{\gamma_{wet} = \frac{\gamma_W \cdot G_s \times (1 + w)}{[1 + e]}}$$

Proof:

Write down relationship 5.

$$\gamma_d = \frac{\gamma_W \cdot G_s}{[1 + (w/S)G_s]}$$

Substitute for γ_d and S.

$$\gamma_d = \gamma_{wet}/(1 + w) \quad \text{and} \quad S \cdot e = G_s \cdot w$$

Hence $S = G_s \cdot w/e$

$$\gamma_{wet}/(1 + w) = \frac{\gamma_W \cdot G_s}{[1 + (w \cdot e/G_s \cdot w)G_s]}$$

$$\gamma_{wet} = \frac{\gamma_W \cdot G_s \times (1 + w)}{[1 + e]}$$

Relationship 7

$$\gamma_d = \frac{\gamma_w \cdot G_s}{[1+e]}$$

Proof:

$$\gamma_d = M_s/V; \quad G_s = M_s/V_s/\gamma_w; \quad e = V_v/V_s$$

Apply the values in the equation:

$$M_s/V = \gamma_w \cdot (M_s/V_s/\gamma_w)/(1+V_v/V_s)$$

γ_w cancels out.

$$M_s/V = (M_s/V_s)/(1+V_v/V_s)$$

M_s cancels out.

$$1/V = (1/V_s)/(1+V_v/V_s)$$

$$V_s/V = 1/(1+V_v/V_s)$$

$$= 1/[(V_s+V_v)/V_s] = V_s/[(V_s+V_v)] = V_s/V$$

The left-hand side and right-hand side are equal.

Practice Problem 6.1

The specific gravity of a soil sample is given to be 2.65. The moisture content and degree of saturation are 0.6 and 0.7, respectively. Find the void ratio.

Solution

$$S \cdot e = G \cdot w$$

$$0.7 \times e = 2.65 \times 0.6$$

$$e = 2.27$$

Practice Problem 6.2

The total density of a soil sample was found to be 110 pcf and the moisture content to be 60%. What is the dry density of the soil sample?

Solution

$$\gamma_d = \gamma_{wet}/(1+w) = 110/(1+0.6) = 68.75\,\text{pcf}$$

Practice Problem 6.3

A soil sample obtained from the ground was measured and weighed. The weight was measured to be 1 lb and the total soil volume was 0.01 cu. ft. The soil sample is then put in the oven and dried. The dried soil sample was weighed to be 0.7 lbs. The specific gravity of the soil is known to be 2.6.

Find the following:

(a) Total density or wet density
(b) Dry density
(c) Porosity
(d) Void ratio
(e) Degree of saturation

Solution

(a) Total density $= M/V = 1/0.01 = 100$ lbs/ft^3.
(b) Dry density $= \gamma_d = \gamma/(1+w)$.
Weight of dry soil (M_s) $= 0.7$ lbs.
Weight of water in the soil sample (M_w) $= 1-0.7 = 0.3$ lbs.
Water content (w) $= M_w/M_s = 0.3/0.7 = 0.428$.
Dry density $= \gamma_d = \gamma/(1+w) = 100/(1+0.428) = 69.9$ lbs/ft^3.
(c) Porosity (n) $= V_v/V$.
$V = 0.01$ cu. ft. This is the total volume of the soil sample.
Specific gravity (G) of the soil is given to be 2.6.
Find V_s*:*

$$G = 2.6 = M_s/(V_s \cdot \gamma_w) = 0.7/(V_s \times 62.4)$$

Since $\gamma_w = 62.4$ lbs/ft^3.
Hence $V_s = 0.0043$ ft^3.

$V = V_v + V_s$ (Total volume = Volume of voids + Volume of solids)

$$0.01 = V_v + 0.0043$$

$$V_v = 0.01 - 0.0043 = 0.0057$$

$$\text{Porosity } (n) = V_v/V = 0.0057/0.01 = 0.57$$

(d) Void ratio (e) can be found using the following equation:

$$e = n/(1-n)$$

$$e = 0.57/(1-0.57) = 1.326$$

(e) $S \cdot e = G \cdot w$

$$S = 2.6 \times 0.428/(1.326) = 0.839$$

Practice Problem 6.4

A soil sample obtained from the ground was measured and weighed. The soil sample has a diameter of 4 in. and a height of 6 in. The weight of the soil sample was measured to be 4.8 lbs. The soil sample was oven dried and weighed again. The dry weight of the soil sample was found to be 3.9 lbs. The specific gravity of the soil sample is known to be 2.65.

Find the following:

(a) Total density
(b) Water content
(c) Dry density
(d) Porosity
(e) Void ratio
(f) Degree of saturation

Solution

Total density:

$$\text{Volume of the soil sample}(V) = \pi \times d^2/4 \times h = \pi \times (4/12)^2/4 \times (6/12) = 0.044\,\text{cu. ft}$$

$$\text{Wet weight of the soil sample} = 4.8\,\text{lbs}$$

$$\text{Total density(or wet density)} = M/V = 4.8/0.044 = 109.1\,\text{lbs/cu. ft}$$

$$\text{Moisture content}(w) = M_w/M_s$$

$$M_s = 3.9\text{lbs};\quad M_w = 4.8 - 3.9\text{lbs} = 0.9\text{lbs}$$

$$w = 0.9/3.9 = 0.23$$

$$\gamma_d = \gamma/(1+w) = 109.1/(1+0.23) = 88.7\text{lbs/cu. ft}$$

Find V_s:

$$G = 2.65 = M_s/(V_s \cdot \gamma_w) = 3.9/(V_s \times 62.4),\quad \text{Since } \gamma_w = 62.4\text{lbs/ft}^3.$$

Hence $V_s = 0.024\text{ft}^3$.

$$V = V_v + V_s \;\; (\text{Total volume} = \text{Volume of voids} + \text{Volume of solids})$$

$$0.044 = V_v + 0.024$$

$$V_v = 0.02$$

Porosity $(n) = V_v/V = 0.02/0.044 = 0.45$

$$\text{Void ratio}(e) = n/(1-n) = 0.45/(1-0.45) = 0.82$$

Degree of saturation (S): $S \cdot e = G \cdot w$

$$S = 2.65 \times 0.23/0.82 = 0.74$$

Practice Problem 6.5

The degree of saturation, water content, and specific gravity are, respectively, 75%, 42%, and 2.68.

Find:

(1) Total density (γ_{wet})
(2) Void ratio (e)
(3) Porosity (n)

Solution

STEP 1: Use relationship 5:

$$\gamma_d = \frac{\gamma_w \cdot G_s}{[1 + (w/S)G_s]}$$

$$\gamma_d = \frac{62.4 \times 2.68}{[1 + (0.42/0.75) \times 2.68]} = 66.9\text{pcf}$$

From relationship 1, $\gamma_d = \gamma_t/(1 + w)$

$$\gamma_{wet} = \gamma_d \times (1 + w) = 66.9 \times (1 + 0.42) = 95\text{pcf}$$

From relationship 2, $S \cdot e = G \cdot w$

$$e = 2.68 \times 0.42/0.75 = 1.5$$

From relationship 3, $n = e/(1 + e) = 1.5/(1.5 + 1) = 0.6$

6.2.2 Borrow Pit Problems

Note: The student should master the previous chapter on soil relationships thoroughly in order to understand borrow pit problem.

Fill material for civil engineering work is obtained from borrow pits. The question is: How much soil should be removed from the borrow pit for a given project?

Usually, the final product is the controlled fill or the compacted soil. Total density, optimum moisture content, and dry density of the compacted soil will be available. This information can be used to obtain the mass of solids required from the borrow pit. If the soil in borrow pit is too dry, water can always be added in the site. If the water content is too high, then soil can be dried prior to use. This could take some time in the field since one has to wait for few sunny days to get rid of water.

Water can be added or removed from soil.

What cannot be changed is the mass of *solids*. The mass of solids is the link between borrow pit soil and soil that has been transported.

Procedure:

Find the mass of solids required for the compacted fill.

Excavate and transport the same mass of solids from the borrow pit.

Practice Problem 6.6

Road construction project needs compacted soil to construct a road 10 ft wide, 500 ft long. The project needs a 2 ft layer of soil. Soil density after compaction was found to be 112.1 pcf at optimum moisture content at 10.5%.

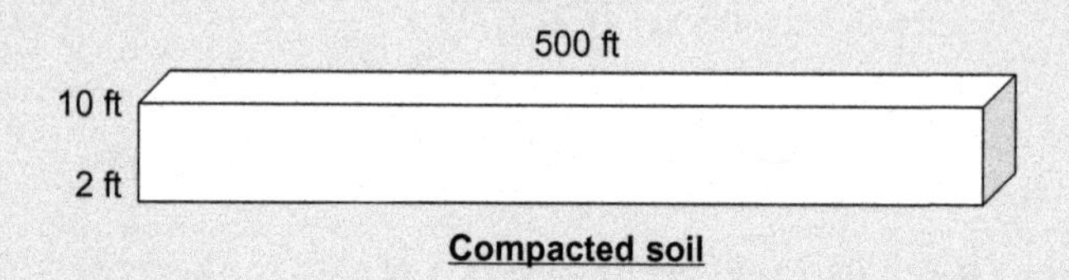

Compacted soil

The soil in the borrow pit has following properties:

Total density of the borrow pit soil = 105 pcf.

Moisture content of borrow pit soil = 8.5%.

Find the total volume of soil that needs to be hauled from the borrow pit:

$$\text{Total density} = (M_w + M_s)/V$$
$$\text{Dry density} = M_s/V$$

(See below for definitions of all terms).

Solution

STEP 1: Find the mass of solids (M_s) required for the controlled fill:

Volume of compacted soil required = 500 × 10 × 2 = 10,000 cu. ft.

Soil density after compaction (dry density) = 112.1 pcf.

Moisture content required = 10.5%.

Draw the phase diagram for the controlled fill:

Volume		**Mass**
V_a	Air	$M_a = 0$ (Usually mass of air is taken to be zero)
V_w	Water	M_w (Mass of water)
V_s	Solid	M_s (Mass of solids)

Soil phase diagram

$$V = \text{Total volume} = V_s + V_w + V_a$$

$$M = \text{Total mass} = M_w + M_s \text{ (Note that mass of air is taken to be zero)}$$

$$V_a = \text{Volume of air (Volume of air is not zero)}$$

$$M_w = \text{Mass of water}$$

$$V_w = \text{Volume of water}$$

$$M_s = \text{Mass of solids}$$

$$V_s = \text{Volume of solids}$$

$$M = \text{Total mass of soil} = M_s + M_w$$

$$V = \text{Total volume of soil} = V_s + V_w + V_a$$

$$\text{Total density} = (M_w + M_s)/V$$

$$\text{Dry density} = M_s/V$$

Soil in the site after compaction has a dry density of 112.1 pcf and moisture content of 10.5%.

$$\text{Dry density} = M_s/V = 112.1\,\text{pcf}$$

Note that total density is $(M_w + M_s)/V$

$$\text{Moisture content} = M_w/M_s = 10.5\% = 0.105$$

The road needed a 2 ft layer of soil at a width of 10 ft and length of 500 ft. Hence the total volume of soil = 2 × 10 × 500 = 10,000 cu. ft.

$$V = \text{Total volume} = 10,000\,\text{cu. ft}$$

Since M_s/V = Dry density

$$M_s/10,000 = 112.1$$
$$M_s = 1,121,000\,\text{lbs}$$

M_s is the mass of solids. This mass of solids should be hauled in from the borrow pit.

STEP 2: Find the mass of water in compacted soil:

$$\text{Moisture content in the compacted soil} = M_w/M_s = 10.5\% = 0.105$$
$$M_w = 0.105 \times 1,121,000\,\text{lbs} = 117,705\,\text{lbs}$$

STEP 3: Find the total volume of soil that needs to be hauled from the borrow pit:

The contractor needs to obtain 1,121,000 lbs of solids from the borrow pit. The contractor can add water to the soil in the field if needed.

$$\text{Mass of solids needed } (M_s) = 1,121,000\,\text{lbs.}$$

The density and moisture content of borrow pit soil is known.

Continued

$$\text{Total density of borrow pit soil} = M/V = 105\,\text{pcf}$$

$$\text{Moisture content of borrow pit soil} = M_w/M_s = 8.5\% = 0.085$$

$$\text{Since } M_s = 1,121,000\,\text{lbs} \ \ (M_s \text{ is the mass of solids required})$$

$$M_w/M_s = 0.085$$

$$M_w = 0.085 \times 1,121,000\,\text{lbs} = 95,285\,\text{lbs}$$

Solid mass of 1,121,000 lbs of soil in the borrow pit contains 95,285 lbs of water.

$$\text{Total mass of borrow pit soil} = 1,121,000 + 95,285 = 1,216,285\,\text{lbs}$$

Total density of borrow pit soil is known to be 105 pcf.

$$\text{Total density of borrow pit soil} = M/V = (M_w + M_s)/V = 105\,\text{pcf}$$

Insert known values for M_s and M_w.

$$M/V = (M_w + M_s)/V = (95,285 + 1,121,000)/V = 105\,\text{pcf}$$

$$\text{Hence } V = 11,583.7\,\text{cu. ft}$$

The contractor needs to extract 11,583.7 cu. ft of soil from the borrow pit.
The borrow pit soil comes with 95,285 lbs of water.
Compacted soil should have 117,705 lbs of water (see above STEP 2).
Hence water needs to be added to the borrow pit soil.

$$\text{Amount of water needs to be added to the borrow pit soil}$$

$$= 117,705 - 95,285 = 22,420\,\text{lbs}$$

Weight of water is usually converted to gallons. One gallon is equal to 8.34 lbs.

Amount of water needs to be added = 2688 gallons.

Summary

STEP 1: Obtain all the requirements for compacted soil.
STEP 2: Find M_s or the mass of solids in the compacted soil.
This is the mass of solids that needs to be obtained from the borrow pit.
STEP 3: Find the information about the borrow pit. Usually the moisture content in the borrow pit and total density of the borrow pit can be easily obtained.
STEP 4: The contractor needs to obtain M_s of soil from the borrow pit.
STEP 5: Find total volume of soil that needs to be removed in order to obtain M_s mass of solids.
STEP 6: Find M_w of the borrow pit. (mass of water that comes along with soil).
STEP 7: Find M_w (mass of water in compacted soil).
STEP 8: The difference in above two masses is the amount of water needs to be added.

CHAPTER 7

Site Layout and Control

7.1 INTRODUCTION

It is important to have knowledge of surveying in order to do construction work. Therefore you will be tested on surveying topics in the exam.

Magnetic North and Geographic North: See Fig. 7.1.

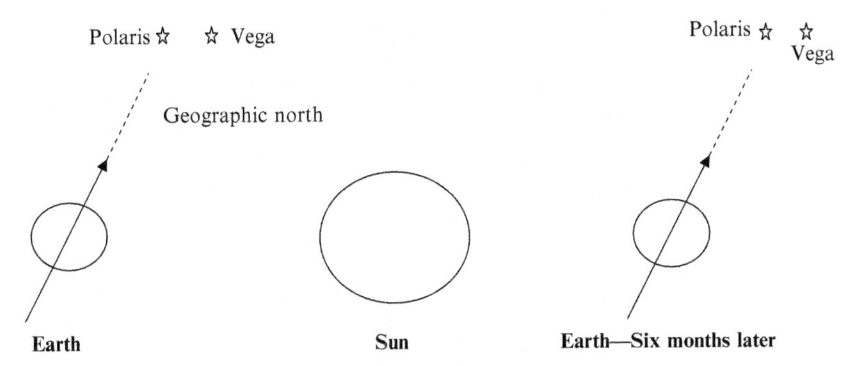

Fig. 7.1 Magnetic north and geographic north.

Geographic north is the direction of the axis of rotation; it is pointed to the polar star. However, the axis of rotation is not a constant. The axis of rotation of the earth is changing every year by a very small amount. It has been calculated that 25,000 years from now, the axis of rotation of Earth will point to a different star, known as Vega.

Magnetic North: Geographic north depends on the rotation of the Earth, whereas magnetic north depends on the magnetic field of the Earth. The fact that Earth's magnetic north is very close to the earth rotational axis is a coincidence. For instance the difference between Neptune's rotational axis and its magnetic north is 40 degrees (Fig. 7.2).

In the case of Earth, the difference between geographic north and magnetic north is very small.

Meridian: Meridian is any longitude. The meridian at a location is the longitude of that location.

Construction Engineering Design Calculations and Rules of Thumb
http://dx.doi.org/10.1016/B978-0-12-809244-6.00007-X

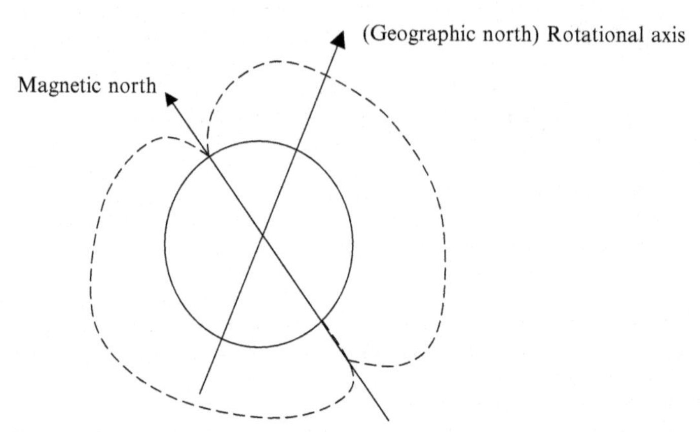

Fig. 7.2 Neptune's rotational axis and magnetic north are 40 degrees apart.

Azimuth: Azimuth is the horizontal angle made with respect to the *geographic north* (Fig. 7.3).

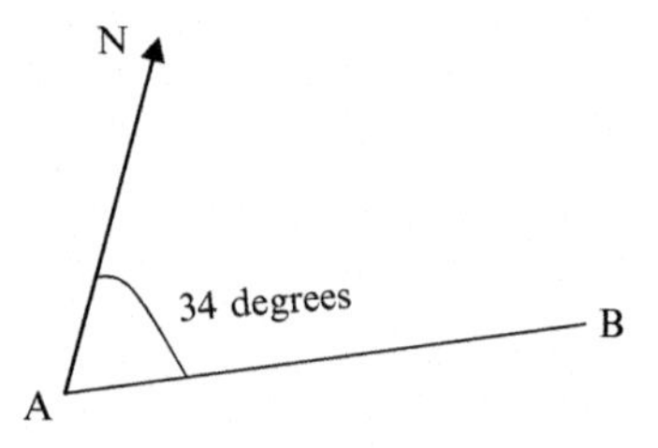

Fig. 7.3 Azimuth of line AB is 34 degrees in this case.

Zenith and Nadir: A line drawn vertically to the sky is known as Zenith. A line drawn directly to the center of the earth is known as Nadir (Fig. 7.4).

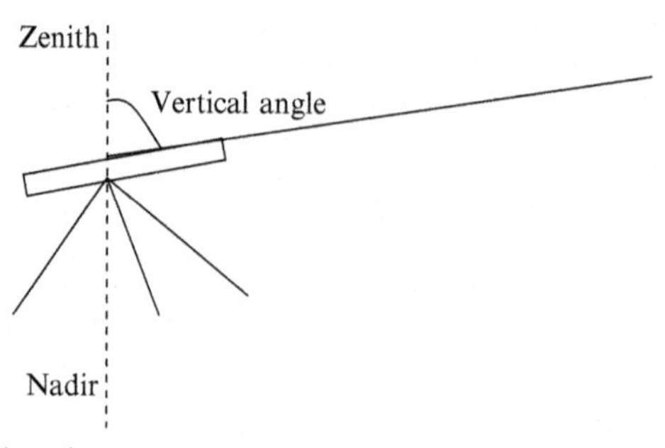

Fig. 7.4 Zenith and Nadir.

7.2 SURVEYING INSTRUMENTS

Level: Levels are used to measure the vertical lengths. Some levels are equipped with a crosshair so that they can be used to measure horizontal distances. Crosshairs are located in such a manner, the length is computed by multiplying the distance between crosshairs by 100 (Fig. 7.5).

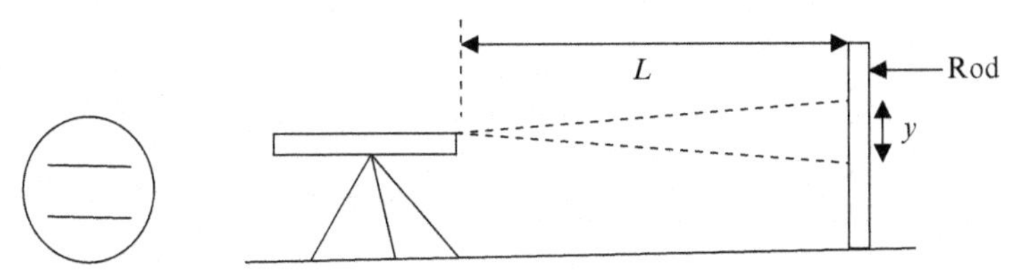

Fig. 7.5 Leveling.

The instruments are designed in a such a manner so that $L = 100 \times y$.

If the reading y is 1.2 ft, then the distance (L) is 120 ft.

Levels cannot be used for the measurement of angles.

Theodolites: Theodolites are designed to measure horizontal and vertical angles.

EDM: Electronic distance measurement or EDM can be used to measure horizontal and vertical distances.

Total stations: Total stations are electronic instruments equipped with computers that can be used to measure angles and distances.

7.3 BEARING

Bearing is the angle measured from north or south (Fig. 7.6).

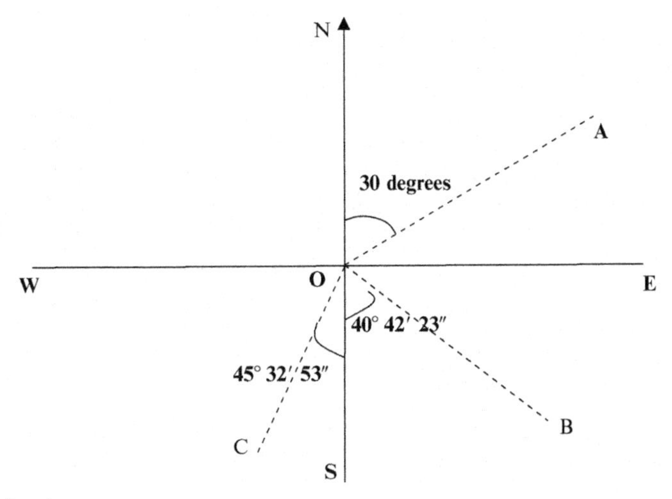

Fig. 7.6 Bearing.

Bearing of OA = N 30 degrees E
This means OA line is 30 degrees measured from N towards east.
Similarly, bearings of other lines are given below:
OB = S $40° 42' 23''$ E
OC = S $45° 32' 53''$ W

Practice Problem 7.1
Provide the bearing for above line OC from north.

Solution
Angle between OC and north is $180 + 45° 32' 53'' = 225° 32' 53''$.

7.4 TRAVERSE

Traverse is conducted by starting from a known point. Angles and distance measurements are taken to new points from the initial point.

Practice Problem 7.2
Several new houses were built in a remote area. Surveyors were called upon to locate the new houses with better accuracy relative to a known benchmark in the vicinity.

Solution
STEP 1: Select a point B, which is closer to existing houses. Find the angle of AB relative to geographic north.
STEP 2: Measure the distance AB. Now point B can be located in a map.
STEP 3: Measure distance Ap.
STEP 4: Measure pp′ perpendicular to AB. Now point p′ can be located in a map. If the house has to be exactly located in a map, more than one measurement is needed to its edges.
STEP 5: Select another convenient point C, closer to houses that need to be measured. Find the interior angle ABC and the distance BC. Now point C can be established in a map.
STEP 6: Measure Br and rr′ distances.
This way all houses can be located in a map (Fig. 7.7).

The accuracy of such a procedure depends heavily on the interior angle measurements at points A, B, C, D, and E.

Construction workers need to know the building footprint, column locations, and wall lines. Locations of these structural elements are provided by surveyors. Nevertheless, construction engineers need to have a good understanding of the process.

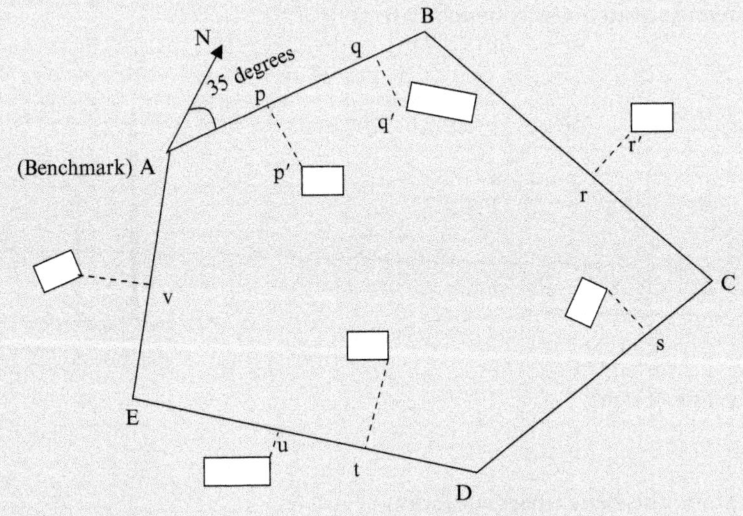

Fig. 7.7 Traverse.

Building line: Surveyors usually provide a string line to indicate the building footprint. The string line may indicate the edge of the footing (Fig. 7.8).

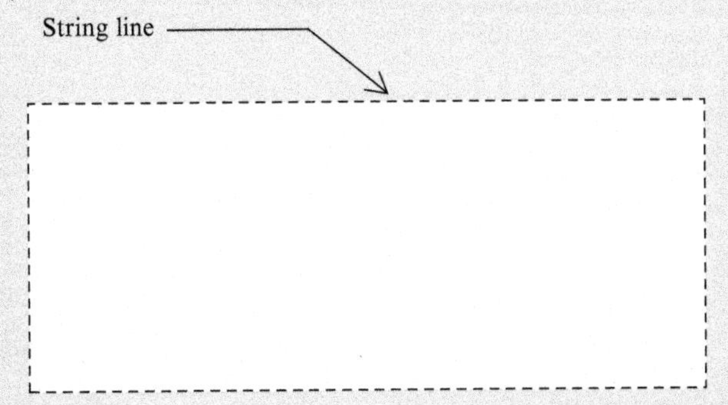

Fig. 7.8 String line.

Using the string line, the contractor can excavate for footings.

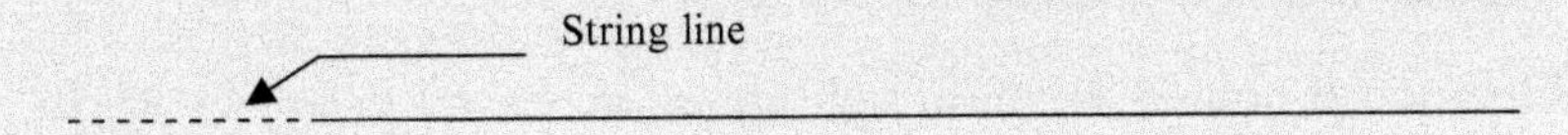

7.5 ELEVATIONS

It is important to obtain elevations for construction. Elevations are obtained with reference to a given benchmark (Fig. 7.9).

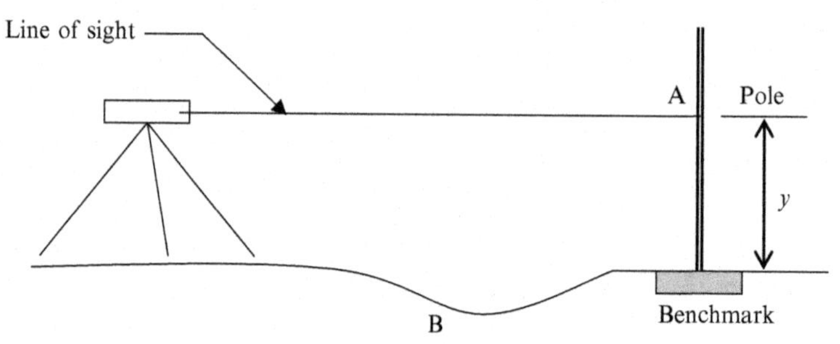

Fig. 7.9 Line of sight.

Assume the benchmark elevation is 100 ft.
Assume pole reading is 5.5 ft.

$$\text{Elevation of line of sight} = 100 + 5.5 = 105.5$$

Now place the pole in point B (Fig. 7.10).

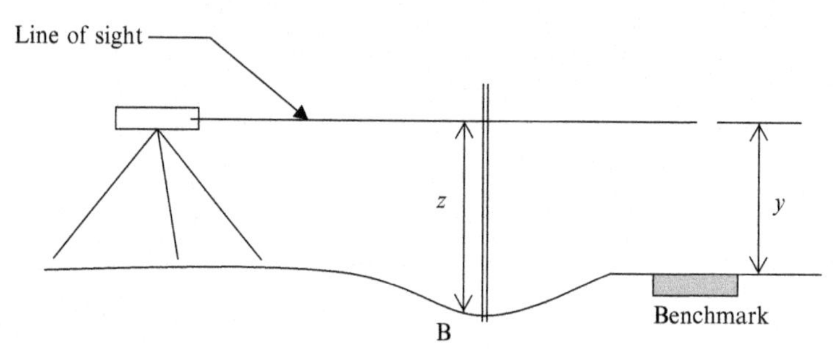

Fig. 7.10 Instrument set up.

Obtain the new reading "z."
Assume that new reading "z" is 7 ft.

$$\text{Elevation of point B} = \text{Line of sight elevation} - z = 105.5 - 7 = 98.5\,\text{ft}$$

Practice Problem 7.3

A surveyor finds a US geological benchmark in the site. The benchmark elevation was found to be 98.7 ft above mean sea level. (MSL). The surveyor placed a pole on top of the benchmark and obtained a reading of 5.7 ft. Then surveyor obtained a reading of 3.2 for point A, 6.7 for point B, and 5.0 for point C. Find the elevations of point A, B, and C.

Solution

$$\text{Benchmark elevation} = 98.7$$

$$\text{Pole reading on top of the benchmark} = 5.7\,\text{ft}$$

$$\text{Elevation of line of sight} = 98.7 + 5.7 = 104.4\,\text{ft}$$

$$\text{Reading of point A} = 3.2$$

$$\text{Elevation of point A} = 104.4 - 3.2 = 101.2\,\text{ft}$$

$$\text{Reading of point B} = 6.7$$

$$\text{Elevation of point B} = 104.4 - 6.7 = 97.7\,\text{ft}$$

$$\text{Reading of point C} = 5.0$$

$$\text{Elevation of point C} = 104.4 - 5.0 = 99.4\,\text{ft}$$

Practice Problem 7.4

A portion of surveyor's logbook is shown below. The elevation of a known benchmark is given to be 101.23 ft. Find the elevation of point C.

Location	Back sight	Fore sight	Elevation of line of sight	Elevation
BM	5.23			101.23
Point A		6.12		
Point B		7.23		
Point C		2.45		

Solution

It is advisable to draw a level and line of sights until you are familiar with backsight and foresight readings (Fig. 7.11).

Continued

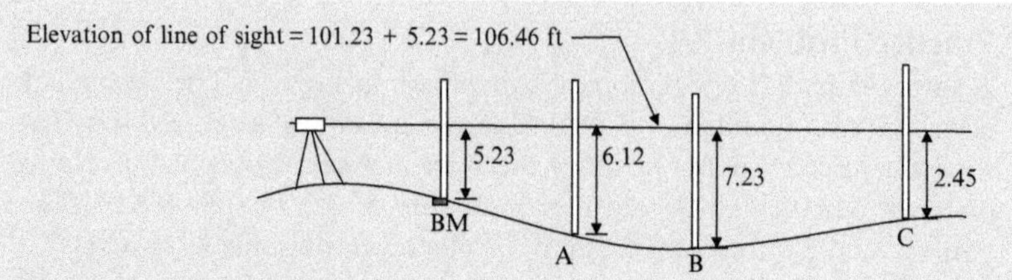

Fig. 7.11 Leveling survey.

The elevation of line of sight can be computed by adding the reading at benchmark to the elevation of benchmark.

$$\text{Elevation of line of sight} = 101.23 + 5.23 = 106.46\,\text{ft}$$

$$\text{Elevation of point A} = \text{Elevation of line of sight} - \text{Reading at point A}$$
$$= 106.46 - 6.12 = 100.34$$

$$\text{Elevation of point B} = \text{Elevation of line of sight} - \text{Reading at point B}$$
$$= 106.46 - 7.23 = 99.23$$

$$\text{Elevation of point C} = \text{Elevation of line of sight} - \text{Reading at point C}$$
$$= 106.46 - 2.45 = 104.01$$

Hence, elevation of point C is 104.01.

You do not need to find elevations of points A and B to find the elevation of point C.

Though it is not necessary to fill the table in the exam, it may be useful to learn how to fill the table.

Location	Back sight	Fore sight	Elevation of line of sight	Elevation
BM	5.23		106.46	101.23
Point A		6.12		100.34
Point B		7.23		99.23
Point C		2.45		104.01

7.6 DISTANCE MEASUREMENT

Distances are measured with tapes, Theodalites, and EDM.

Practice Problem 7.5

A surveyor had to measure the distance between points A and B. The surveyor locates a third point C and obtains angle measurements as shown. Find the distance AB (Fig. 7.12).

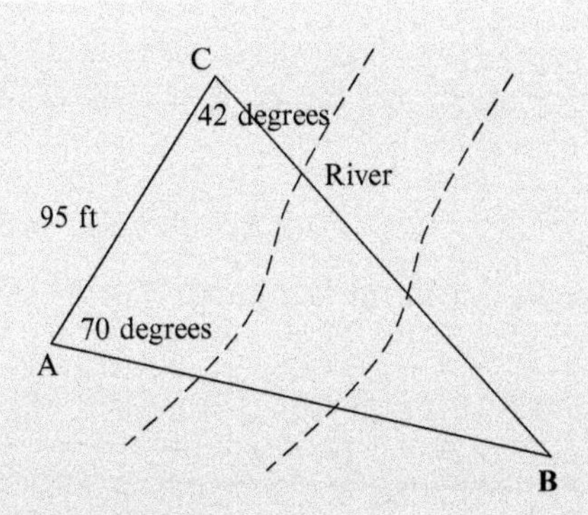

Fig. 7.12 Distance measurement.

Solution

This problem can be easily solved using the Sine law.

$$\text{Sine law}: AC/\sin B = AB/\sin C = BC/\sin A$$

$$\text{Angle}\,B = 180 - (42 + 70) = 68$$

$$AC/\sin B = AB/\sin C$$

$$95/\sin 68 = AB/\sin 42$$

$$AB = 68.6\,\text{ft}$$

CHAPTER 8

Highway Curves (Horizontal and Vertical Curves)

Surveyors are required to layout highway curves. There are two types of highway curves.

(1) Horizontal curves

(2) Vertical curves

Horizontal curves: It is important to review certain concepts of trigonometry prior to venturing into horizontal curves.

Trigonometry refresher: Knowledge of trigonometry is essential for the civil PE exam.

It is highly unlikely that you will require any more trigonometry than sin, cos, and tan functions (Fig. 8.1).

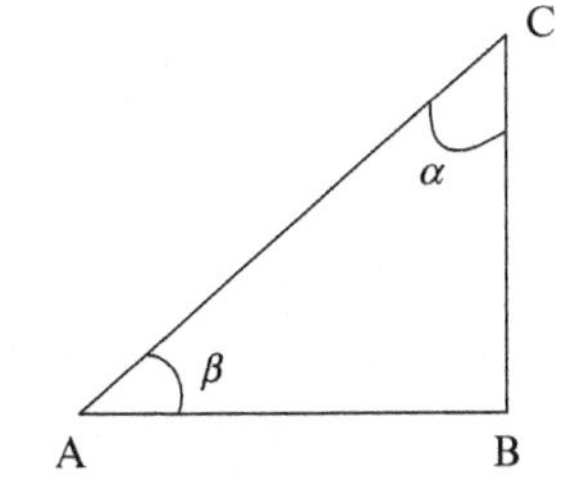

Fig. 8.1 Trigonometry.

$$\tan\alpha = AB/BC$$
$$\tan\beta = BC/AB$$
$$\alpha = 90 - \beta$$
$$BC = AC\sin\beta$$
$$BC = AB\tan\beta$$
$$AB = AC\sin\alpha$$
$$AB = BC\tan\alpha$$

Construction Engineering Design Calculations and Rules of Thumb
http://dx.doi.org/10.1016/B978-0-12-809244-6.00008-1

$$AC = BC / \cos\alpha$$
$$AC = AB / \cos\beta$$
$$AC = BC / \sin\beta$$
$$AC = AB / \sin\alpha$$
$$BC = AB / \tan\alpha$$
$$AB = BC / \tan\beta$$

Practice Problem 8.1

Find α, β, γ, AD, DC, and BC.

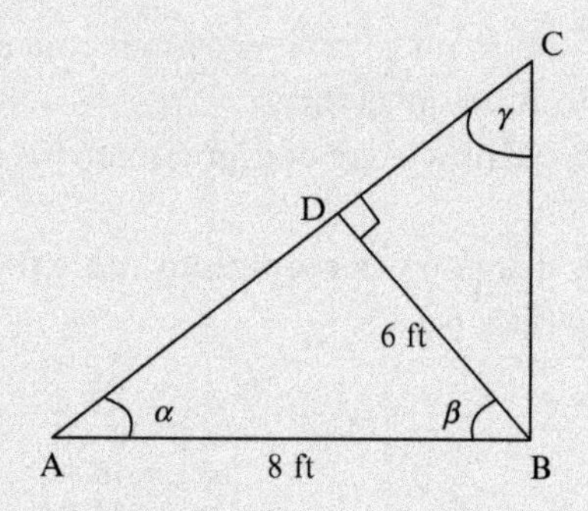

Solution

$\sin\alpha = 6/8; \alpha = 48.6°$.

Hence $\beta = 90 - \alpha = 41.4°$.

$\gamma = 90 - \alpha = 41.4$.

Note that $\beta = \gamma$ in this case.

$AD = 8\cos\alpha = 8\cos(48.6) = 5.29$.

$DC = AC - AD$.

$AC = 8/\cos\alpha = 8/\text{Cos}(48.6) = 12.1$ ft.

Hence $DC = 12.1 - 5.29 = 6.81$.

$BC = 8\tan\alpha = 8\tan(48.6) = 9.1$.

Radians: Angles can be measured with radians as well.

$360^0 = 2\pi$ radians.

$2\pi r$ = circumference of a circle.

The length of an arc is calculated by multiplying the angle measured in radians by the radius.

$$\boxed{\text{Length of an arc} = \text{Angle measured in radians} \times \text{radius}}$$

Practice Problem 8.2

Find the length of an arc that projects an angle of 45 degrees at the center of the circle. The radius of the circle is 2.5 m.

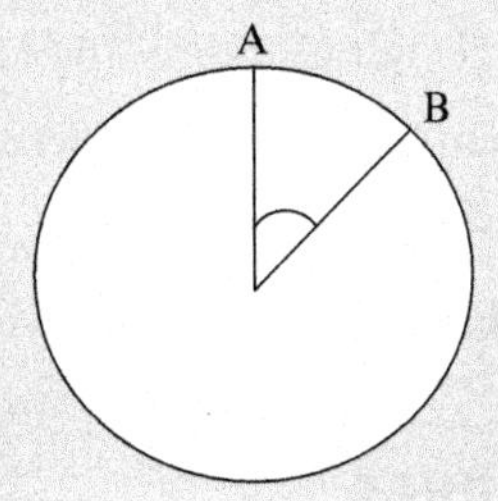

Solution

$360° = 2\pi$ radians.
$1° = (2\pi/360°)$ radians.
$45° = (2\pi/360 \times 45)$ radians $= 0.785$ radians.
Length of an arc = Angle measured in radians × radius.
Length of arc AB $= 0.785 \times 2.5 = 1.96$ m.

Practice Problem 8.3

Find the length AB of the figure shown. O is the center of the circle and radius of the circle is 50 m.

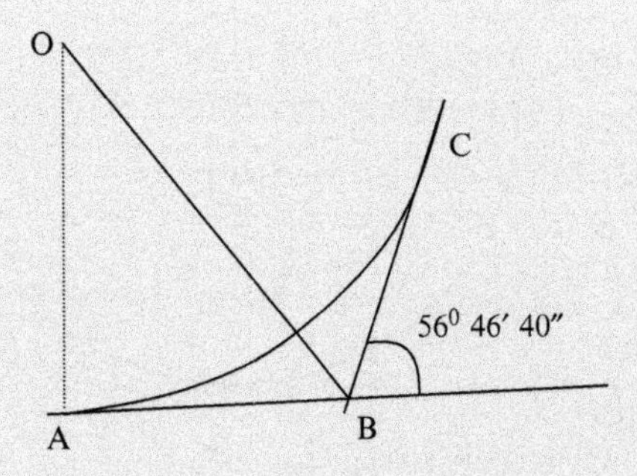

Solution

$$\text{Angle ABC} = 180° - 56°46'40''$$

It is easy to work with decimals rather than minutes and seconds.

Continued

$$56°46'40'' = 56 + 46/60 + 40/3600 = 56.78°$$

$$\text{Angle ABC} = 180 - 56.78 = 123.22°$$

$$\text{Angle OBA} = 123.22/2 = 61.61°$$

$$\tan(\text{angle OBA}) = \text{OA}/\text{AB}$$

$$\tan(61.61°) = \text{radius}/\text{AB}$$

Since radius is given to be 50 m,

$$\text{AB} = 50/\tan(61.61°) = 50/1.85 = 27.03\text{m}$$

8.1 HORIZONTAL CURVES

A straight road starts to curve at the point known as "*Point of curvature*" or PC. The curve ends and a straight road begins again at a point known as PT or point of tangent. Tangents are drawn at each point intersects at PI or point of intersection (Fig. 8.2).

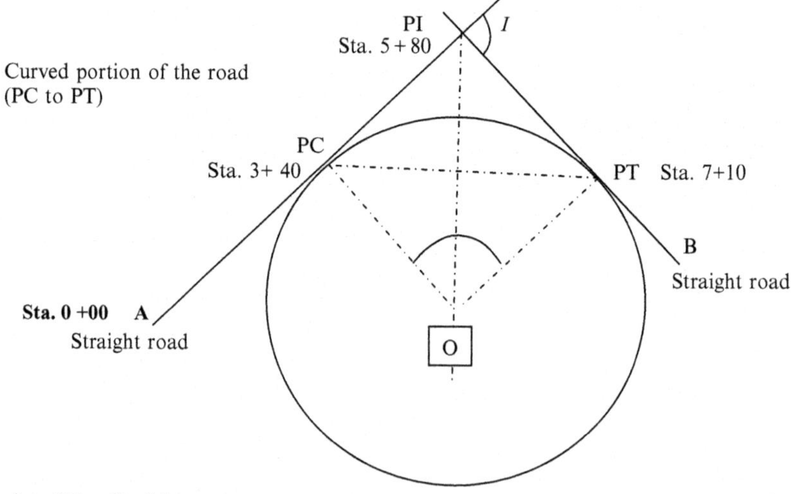

Fig. 8.2 Horizontal curve.

Station markings: Station at PC is 3 + 40 (or 340 ft from the origin of stations). Station at PI is 5 + 80. The length between PC and PI = 580 − 340 = 240 ft.

Station at PT = 7 + 10.

This station is measured along the curved portion of the road from PC to PT.

Length of the curved portion of the road from PC to PT = 710 − 340 = 370 ft.

Angle at the center of the circle: The curved length of the road from PC to PT is 370 ft. If the radius is known, the angle at the center of the circle can be calculated.

Assume the radius is 850 ft.

Whole circle is 360 degrees. The whole circle generates the perimeter of the circle.

$$360 \text{ degrees} \rightarrow 2\pi R = 2 \times 3.14 \times 850 = 5338 \text{ ft}$$

Or

5338 ft generates an angle of 360 degrees at the center.

One (1) ft generates an angle of 360/5338 degrees at the center.

370 ft arc generates an angle of $360/5338 \times 370 = 24.9$ degrees.

Practice Problem 8.4

What is the length of an arc generated by an angle of 23.5 degrees where the radius of the circle is 575 ft?

Solution

$$360 \text{ degrees} \rightarrow 2\pi R = 2 \times 3.14 \times 575 = 3613 \text{ ft}$$

$$1 \text{ degree} \rightarrow 3613/360 = 10.03 \text{ ft}$$

1-degree arc generates a curve of 10.03 ft.

23.5 degrees generates a curve of 10.03×23.5 ft = 235.7 ft.

Angle computation:

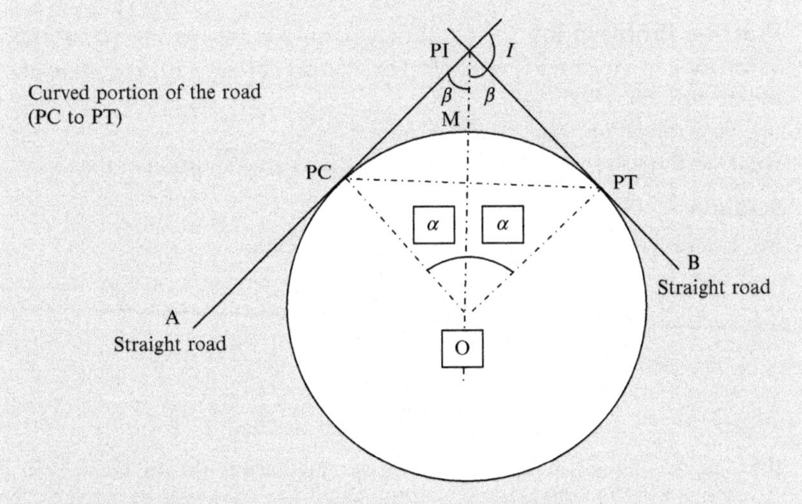

A to PC—Straight road

PC to PT—Curved portion of the road

PT to B—Straight road again

Continued

The following properties are easily determined:
Intersection angle at point PI is known as I.
Consider two legs, PCPI and PTPI.

Angle PC PI PT $= 180° - I$
(Angle at point PI, between two legs PCPI and PIPT)

From symmetry, angle PC PI O $= \beta = (180 - I)/2 = 90 - I/2$
Hence $\alpha = 90 - \beta = I/2$
If the radius of the circle is R and I are known, lengths PIPC, PCO, and PIO can be deduced.
(1) PIPC $=$ PCO tan $\alpha = R$ tan $\alpha = R$ tan $I/2$.
(2) PCO $=$ PIO cos α.
PCO $=$ R.
Hence PIO $=$ PCO/cos $\alpha = R$/cos $\alpha = R$/cos $(I/2)$.
(3) PIM $=$ PIO $- R = R$/cos $\alpha - R = R$/cos $(I/2) - R$.
(4) Length of the curved portion PC to PI can be found as follows.
Total perimeter of the circle $= 2\pi R$.
Total perimeter of the circle is due to 360 degrees.
Hence, 360 degrees extends a curve of $2\pi R$.
Hence 1 degree extends a curve of $2\pi R/360$.
α degree extends a curve of $2\pi R/360 \times \alpha$.
2α degree extends a curve of $2\pi R/360 \times 2\alpha$.
The length of the curved portion of the road from PC to PT $= 2\pi R/360 \times 2\alpha$.

Practice Problem 8.5

The station of a road at PC is 5 + 30. The station at PI is 8 + 20. The angle of intersection (I) is 40°.
(a) Find the radius (refer to above figure).
(b) Find the station at PT (measured along the curved portion of the road).

Solution

(a) Length between PC and PI is $820 - 530 = 290$ ft

$$\text{PCPI} = 290 = R \tan \alpha$$

$$\alpha = I/2 = 20°$$

$$290 = R \tan 20°$$

$$R = 290/\tan 20° = 796.8 \text{ft}$$

(b) Length of the curved portion of the road from PC to PT $= 2\pi R/360 \times 2\alpha = 2\pi \times 796/360 \times 2 \times 20 = 555.7$ ft

$$\text{Station at PC} = 5 + 30 = 530$$
$$\text{Station at PT} = 530 + 555.7 = 1085.7$$
$$\text{Station at PT} = 10 + 85.7$$

Note: Stations can be marked along a curved road or a straight line.

8.2 VERTICAL CURVES

8.2.1 Elevations and Grades

It is important to understand elevations and grades prior to dealing with vertical curves.

Practice Problem 8.6

Find the elevation of point B, if the grade is 5%. Grade = Vertical/Horizontal

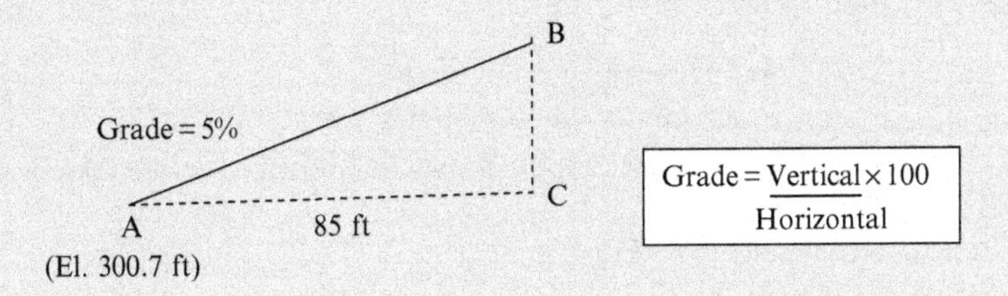

$$\text{Elevation at point A} = 300.7\ \text{ft};\quad \text{Grade} = 5\% = BC/AC$$
$$BC = 5\% \times 85 = 0.05 \times 85 = 4.25\,\text{ft}$$
$$\text{Elevation at B} = 300.7 + 4.25 = 304.95\ \text{ft}$$

Grade change: Grade changes are common in vertical curves.

Practice Problem 8.7

Find the average grade between point A and point B.

Find the change of grade per foot between points A and B

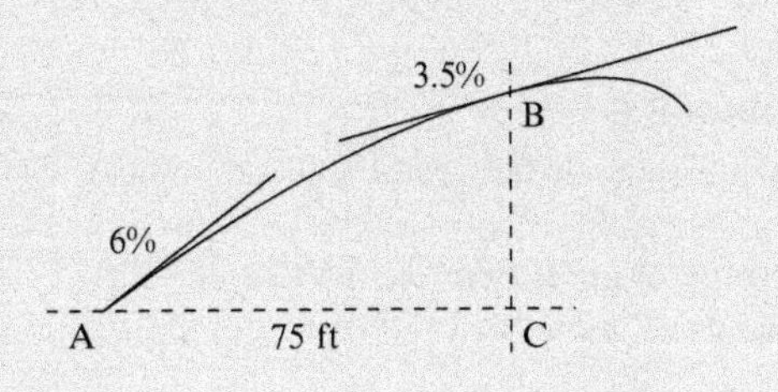

Continued

Solution

$$\text{Grade at point A} = 6\%$$

$$\text{Grade at point B} = 3.5\%$$

$$\text{Average grade between two points} = (6 + 3.5)/2 = 4.75\%$$

Change of grade between points A and B:

Grade at point A is 6%. The grade has gone down to 3.5% at point B.

Change of grade $= 6 - 3.5 = 2.5\%$.

Change of grade per foot $= 2.5/75 = 0.0333\%$ per foot.

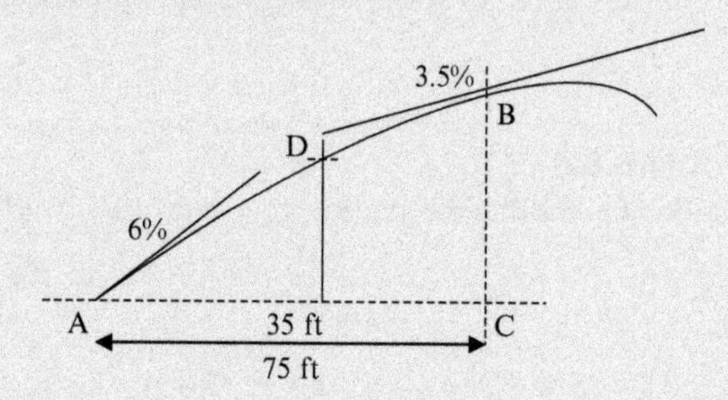

What is the grade at point D if the horizontal distance between points A and D is 35 ft.

Change of grade $= 0.0333$ per foot.

Change of grade for 30 ft $= 0.0333 \times 35 = 1.167\%$.

Grade at point D $= 6\% - 1.167\% = 4.833\%$.

8.2.2 PVC, PVT and PVI

PVC (Point of Vertical Curvature): In a vertical curve where a straight road starts to curve is known as PVC.

PVT (point of Vertical Tangent): The point where a curved road becomes a straight road again is known as PVT.

PVI (Point of Vertical Intersection): The intersection of two tangents crossing PVC and PVT is known as PVI.

Practice Problem 8.8

(a) Find the change of grade between PVC and point A. Grade at point A is 3%.

(b) Find the change of grade between PVC and PVT.

(c) Find the change of grade between PVC and highest point in the curve.

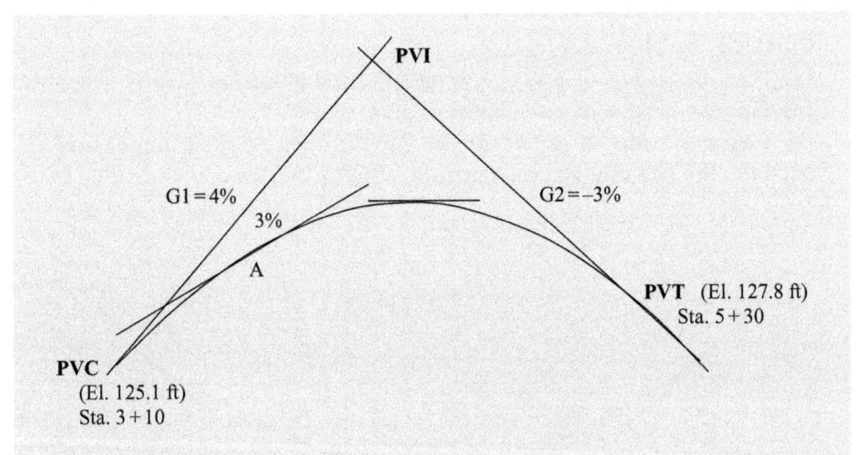

Solution

PVC = Point of vertical curvature. A straight road starts to curve at PVC. In this case, PVC is at an elevation of 125.1 and at station 3 + 10.

G1 = Grade of the tangent drawn at PVC = 4%.

G2 = Grade of the tangent drawn at PVT = −3%.

PVT = point of vertical tangent. The curved portion of the road becomes straight. In this case, PVT is at elevation 127.8 and station 5 + 30.

Point A = Point A is an arbitrary point in the curve. Grade at point A is 3%.

(a) Change of grade between PVC and point A = 4 − 3 = 1%.
Since the change is a reduction from 4 to 3, the change of grade is −1%. The minus sign indicates that the change is a reduction. In other words going from PVC to A, the grade is reduced by 1%.

(b) Change of grade between PVC and PVT = 4 − (−3) = 7%.
Again, the change is a reduction. Hence change of grade is −7%.

(c) Find the change of grade between PVC and highest point in the curve = 4 − 0 = 4%.
Change of grade is −4%

Note that in a crest curve, the highest point has a zero grade. Similarly, in a sag curve, the lowest point has zero grade.

Practice Problem 8.9

(a) Find the change of grade between PVC and point A.
(b) Find the change of grade between PVC and PVT.
(c) Find the change of grade between PVC and lowest point in the curve.
(d) Find the rate of grade change from PVC to PVT.

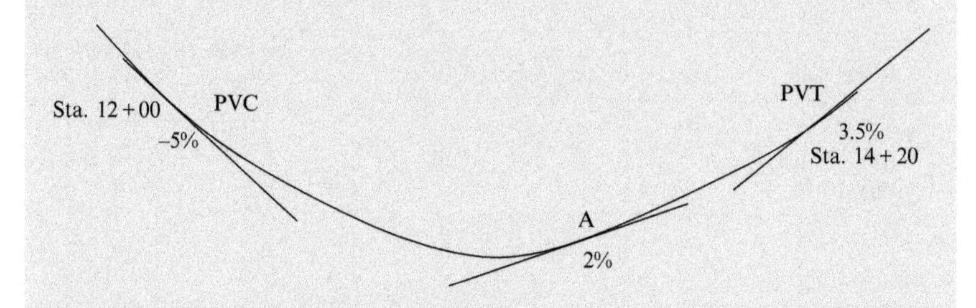

Solution

(a) *The change of grade between PVC and point A:*
The grade at PVC is −5%. This goes up to 2% when reaching point A.
The grade change is from −5% to +2%. The grade change is a positive one. Total grade change is +7%.

(b) *The change of grade between PVC and PVT:*
The grade at PVC is −5%. The grade at PVT is +3.5%. The grade is going up from PVC to PVT.
The grade change from PVC to PVT is positive. It is +8.5%.

(c) *The change of grade between PVC and lowest point in the curve:*
The grade at the lowest point of curve is zero. The grade at PVC is −5%. Hence, the grade change from PVC to the lowest point of the curve is positive. It is +5%.

(d) *The rate of grade change from PVC to PVT:*
The grade change from PVC to PVT = +8.5%.
The horizontal distance between PVC and PVT = 1420 − 1200 = 220 ft
Note that stations are measured horizontally.
The rate of grade change per foot = 8.5/220 = 0.038% per foot.
In other words, the grade increases by 0.038% per every foot.

Practice Problem 8.10

Find the horizontal distance required to have a grade change of 1% in the curve shown.

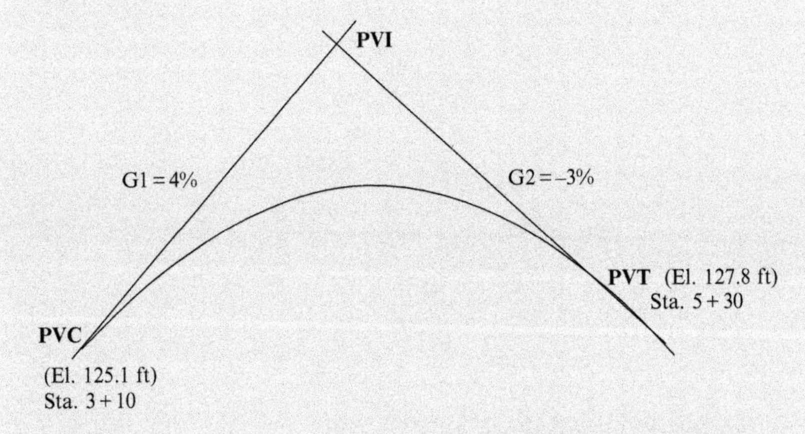

Solution

We need to find the horizontal distance for a grade change of 1%.

The grade change from PVC to PVT is negative. +4% at PVC becomes −3% at PVT.

Hence the grade change from PVC to PVT is −7%.

Horizontal distance from PVC to PVT = 530 − 310 = 220 ft (Station 3 +10 to 5 + 30).

Horizontal distance for 7% grade change is 220 ft.

Horizontal distance for 1% grade change is 220/7 = 31.4 ft.

Crest curves and sag curves: Vertical curves could be crest curves or sag curves (Fig. 8.3).

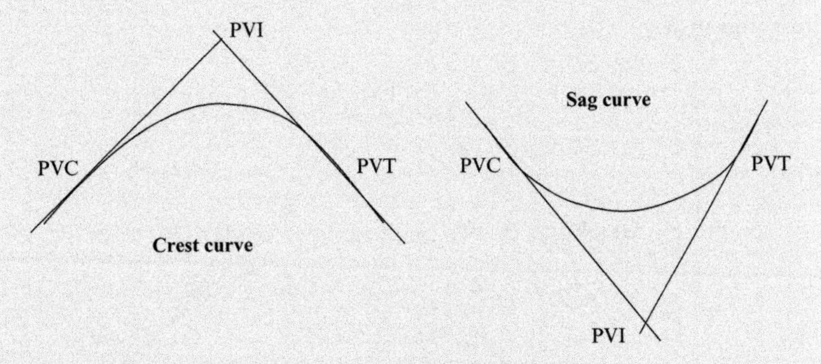

Fig. 8.3 Two type of vertical curves (crest curves and sag curves).

Practice Problem 8.11

(a) Find the grade change from PVC to PVT for the crest curve shown.
(b) Find the change of grade per foot from PVC to PVT.

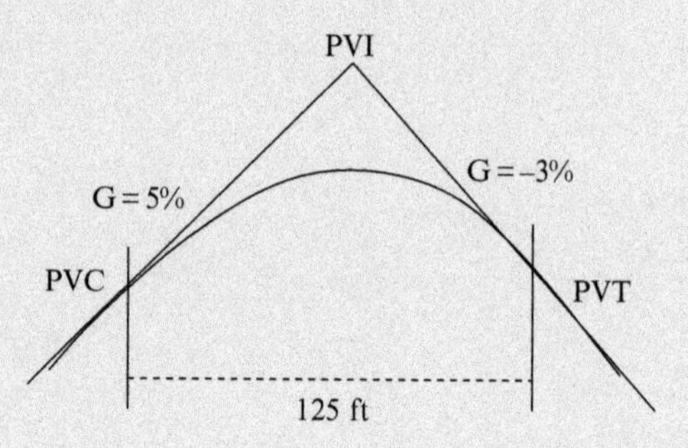

Grade at PVC = 5% Grade at PVT = −3%.

Total grade change = −3 − 5 = −8%.

5% at PVC goes down to −3 at PVT. Hence, the change of grade is negative.

The grade change per foot = −8/125 = −0.064% per foot. The grade *goes down* by 0.064% per every foot.

8.2.3 General Equation for Vertical Curves

See Fig. 8.4.

The elevation to point M from PVC is "*y*."

Horizontal distance to point M from PVC is "*x*." Both "*x*" and "*y*" are measured in ft.

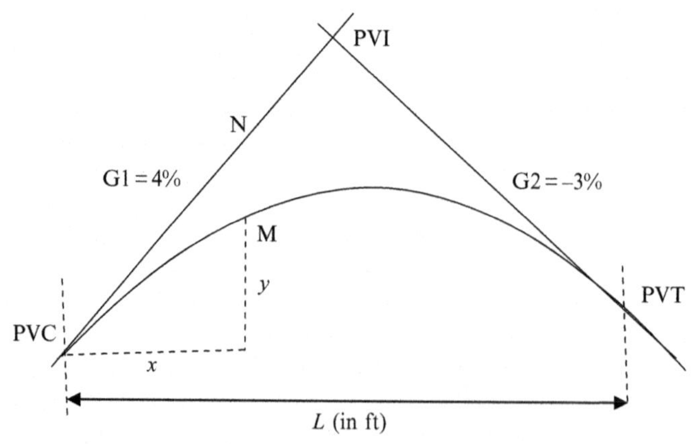

Fig. 8.4 Vertical curve (crest curve).

General equation for vertical curves shown below.

$$y = \frac{(G2 - G1)}{200L} \cdot x^2 + \frac{G1 \cdot x}{100}$$

y = Vertical height to point in the curve from BVC in ft.
x = Horizontal distance between two points in ft.
G1 and G2 are gradients in percent.
Upward gradient is taken to be positive (+) ve.
Downward gradient is taken to be (−) ve.
L = Horizontal distance between BVC and EVC in ft.

Practice Problem 8.12

The sag vertical curve of a roadway is shown in the figure. The roadway goes under an overpass. The station of BVC = 115 + 45.

Elevation of BVC = 134.56 ft.

Station of PVI = 120 + 34.

Station of overpass = 123 + 13.

Elevation of overpass = 143.25 ft.

What is the maximum height of trucks that can be allowed on the roadway assuming 1 ft clearance between the roof of trucks and the overpass.

Note that the vertical curves are constructed in a manner so that PVI station is at the center of BVC and EVC.

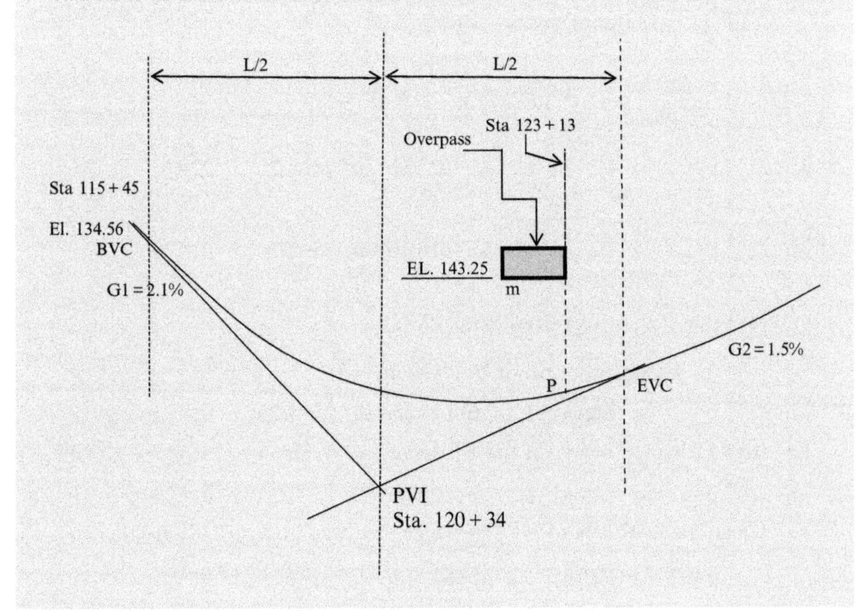

Continued

Solution

Vertical Curve Equation

$$y = \frac{(\mathrm{G2} - \mathrm{G1})}{200L} \cdot x^2 + \frac{\mathrm{G1} \cdot x}{100}$$

y = Vertical height to point in the curve from BVC in ft.
x = Horizontal distance between two points in ft.
G1 and G2 are gradients in percent.
Upward gradient is taken to be positive (+) ve.
Downward gradient is taken to be (−) ve.
L = Horizontal distance between BVC and EVC in ft.
Let's write down the given parameters.
BVC station = 115 + 45.
PVI station = 120 + 34.

The horizontal distance between BVC and PVI = 12,034 − 11,545 = 489 ft.

Note that PVI station is located midway between BVC and EVC.
Hence $L = 2 \times 489 = 978$ ft.
G1 = −2.1 (Downward gradient is considered to be negative).
G2 = 1.5 (Upward gradient is considered to be positive).

We need to find the horizontal distance between the overpass (point P) and BVC.

Horizontal distance to the overpass measured from BVC (x) = 12,313 − 11,545 = 768 ft.

STEP 1: Find the elevation at point P:

$$y = \frac{(\mathrm{G2} - \mathrm{G1})}{200L} \cdot x^2 + \frac{\mathrm{G1} \cdot x}{100}$$

$$y = \frac{(1.5 - -2.1)}{200 \times 978} \cdot 768^2 + \frac{(-2.1 \times 768)}{100}$$

$$y = 10.855 - 16.128 = -5.273$$

Note that "y" is measured from BVC.

Elevation at point P = 134.56 − 5.273 = 129.28

Elevation of the overpass = 143.25

The total clearance between the roadway and overpass = 143.25 − 129.28 = 13.97 ft

Clearance of 1 ft is needed between overpass and roof of trucks

Hence maximum height of trucks = 12.97 ft (Ans B)

CHAPTER 9

Trench Excavations

9.1 STRING LINES

Trench excavations are mostly done by using a string line. The procedure is easily explained with an example.

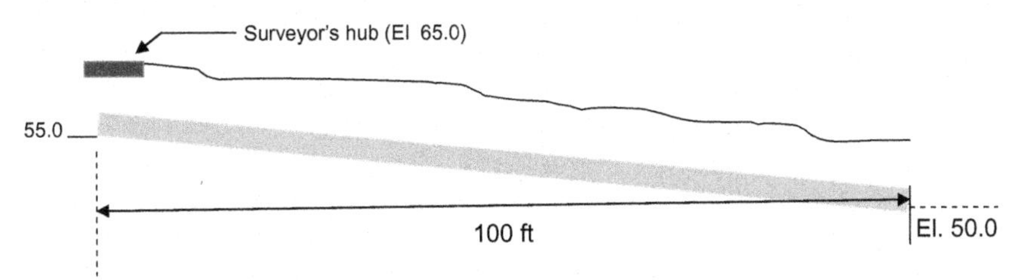

Fig. 9.1 Trench excavations.

In Fig. 9.1, the elevation of the surveyor's hub is known. The contractor will establish a string line, probably 3 ft above the surveyor's hub as shown below. The elevation of the string line would be 68.0. A machine operator will excavate 13 ft measured from the string line $(68 - 55 = 13)$. The depth to the bottom of the trench measured from ground level would vary depending upon the ground elevation (Fig. 9.2).

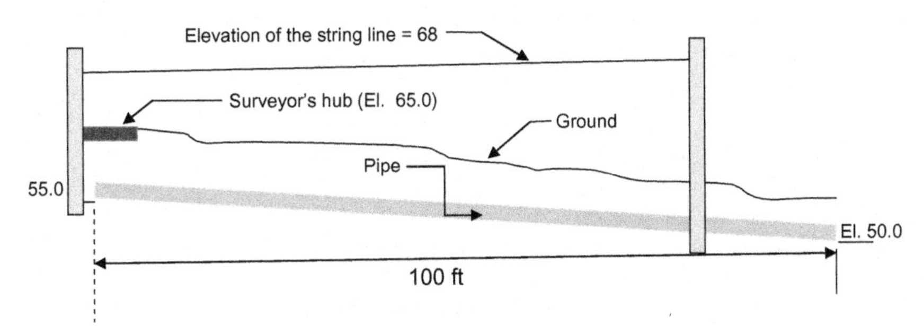

Fig. 9.2 Ground, string line, and trench.

The string line is kept horizontal using a hand level.

Elevation at bottom of pipe at 50 ft is 52.5.

Elevation of the string line is 68. The contractor has to excavate 15.5 ft measured from the string line, at 50 ft distance.

Construction Engineering Design Calculations and Rules of Thumb
http://dx.doi.org/10.1016/B978-0-12-809244-6.00009-3

68 − 52.5 = 15.5 ft. Depth to bottom of trench measured from ground level would vary depending upon the ground elevation.

9.2 TRENCH EXCAVATIONS AND NEARBY BUILDINGS

It is important to note that excavations can cause settlement in nearby buildings. This could happen due to two reasons:

- soil relaxation due to excavation
- groundwater lowering due to excavation

Let us first discuss soil relaxation due to excavations.

Let us assume an excavation was done close to a foundation as shown in Fig. 9.3.

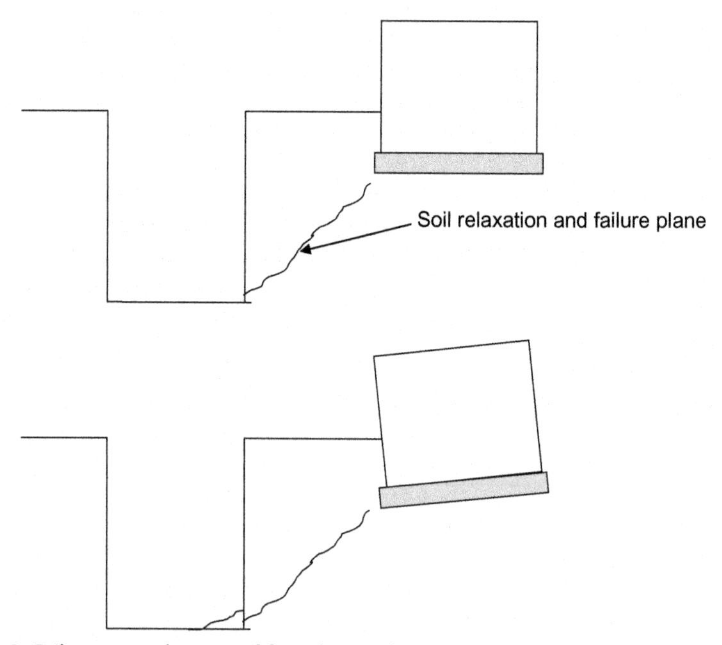

Fig. 9.3 Failure or settlement of foundation due to nearby excavations.

Excavations near foundations should be supported (Fig. 9.4).

In addition, excavations can be done far away from footing. Two horizontals to one vertical (2H:1V) is used as a rule of thumb. Draw a line with 2H:1V starting from the bottom of the footing. If the excavation is within that line, then soil relaxation can be expected (Figs. 9.5 and 9.6).

2H:1V should be increased to 3H:1V for very loose sandy soils (Fig. 9.7).

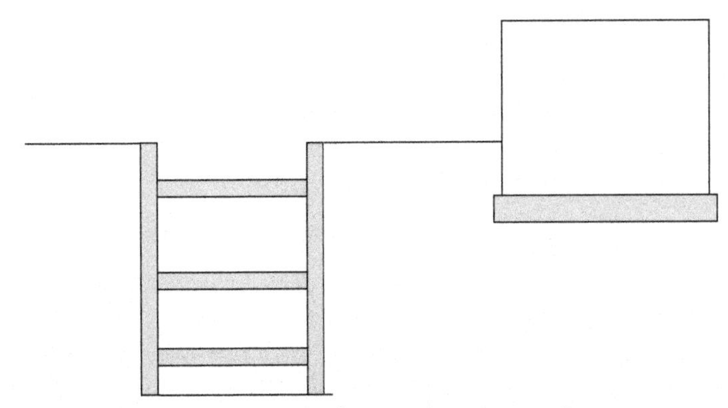

Fig. 9.4 Support the excavations to make sure there is no soil movement.

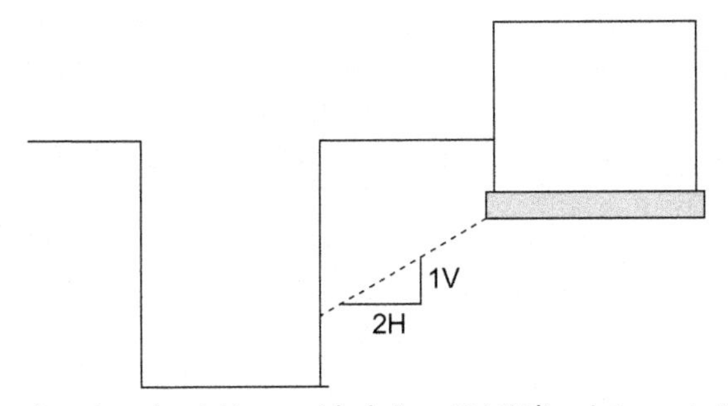

Fig. 9.5 Above situation should be avoided since 2H:1V line intersects the excavation.

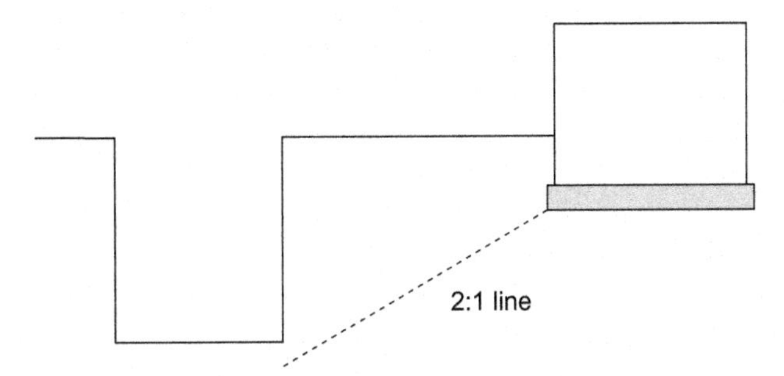

Fig. 9.6 Above situation is acceptable since 2H:1V line is spanning below the excavation.

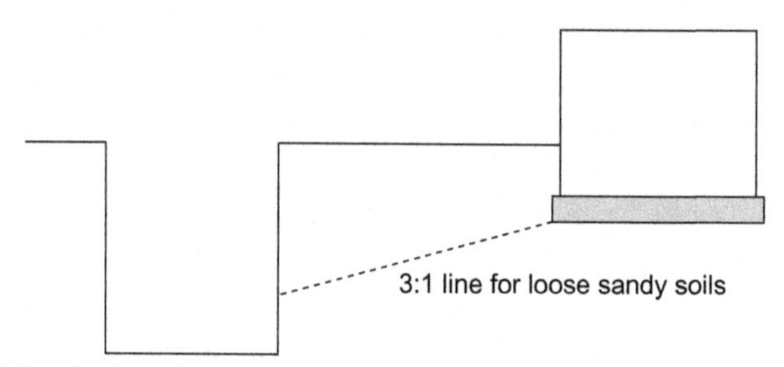

Fig. 9.7 3:1 line for loose soils.

How groundwater can affect footings: Groundwater seeps into excavations. Hence, groundwater level in the vicinity goes down. Lowering of groundwater increases the effective stress and can cause foundation settlement (Fig. 9.8).

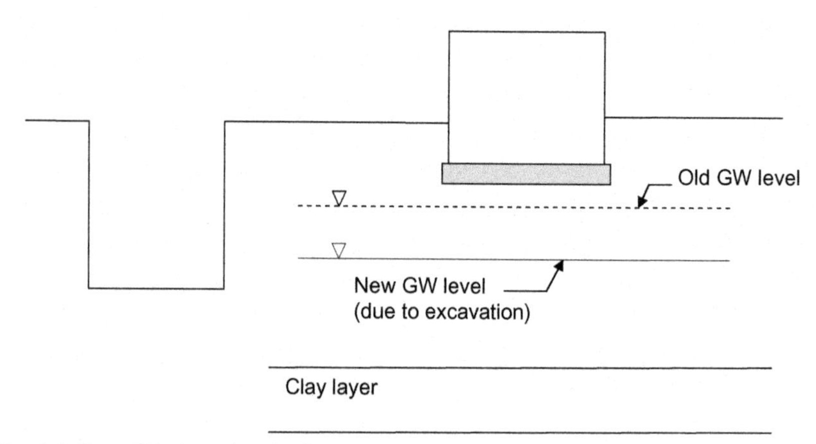

Fig. 9.8 Consolidation of a clay layer due to lowering of groundwater.

Lowering of groundwater can increase the effective stress in clay layers. Larger effective stress can cause foundation settlements.

CHAPTER 10

Construction Stakes and Markings

We all have seen stakes planted in the ground for various construction work. In this chapter, we will look the information these stakes provide and how to read them (Fig. 10.1).

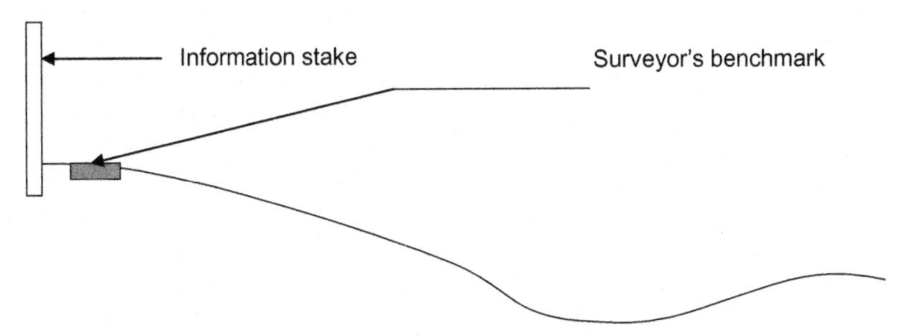

Fig. 10.1 Information stake and benchmark.

Surveyor's benchmark (surveyor's hub): All measurements are taken from the surveyor's benchmark, also known as surveyor's hub. It is a solid plate strongly embedded in the ground.

Information stake: Next to the surveyor's benchmark, there is an information stake. This stake gives the action that needs to be taken at a given point. You should remember that all measurements are taken from the surveyor's benchmark or the hub.

Construction Engineering Design Calculations and Rules of Thumb
http://dx.doi.org/10.1016/B978-0-12-809244-6.00010-X

Practice Problem 10.1

Draw the cut section as for the marking in the following stake.

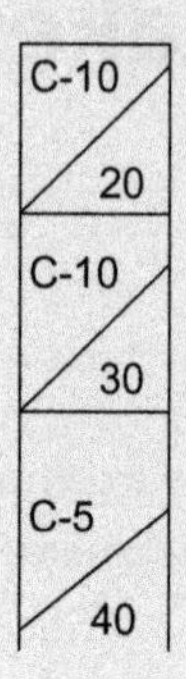

Solution

STEP 1: The top item in the above stake shows C-10/20 This means a cut of 10 ft at a distance of 20 ft (C-10/20). This is shown below.

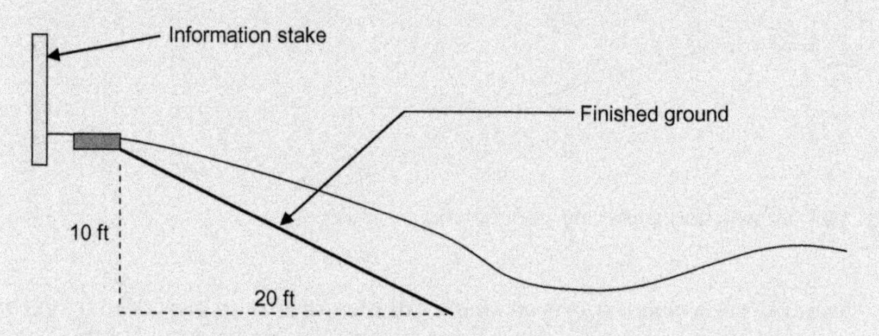

The first point is 20 ft away and 10 ft below the surveyor's hub.

STEP 2: The next item is C-10/30. This means, the next point is 30 ft away (horizontal distance) and 10 ft below the surveyor's hub.

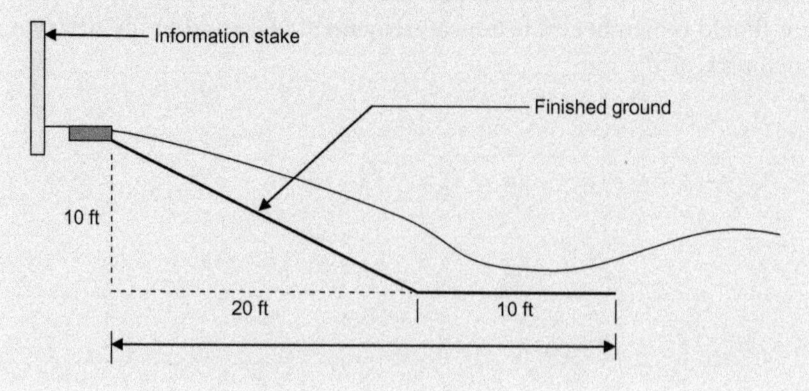

At 30 ft, the finished ground is 10 ft below the surveyor's hub.
STEP3: The last item is C-5/40. This means the next point is 40 ft away and 5 ft below the surveyor's hub.

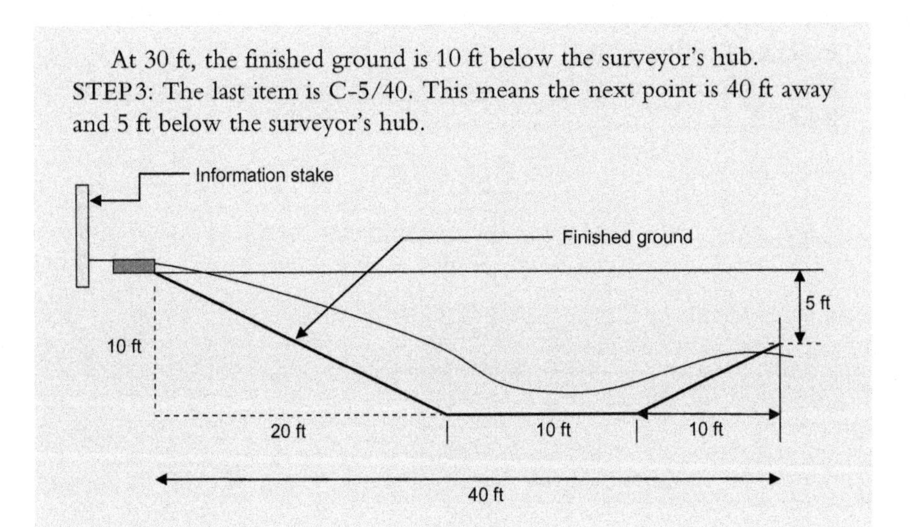

Establishing a reference stake: If you look at the above profile, there is a small problem. Cutting the ground from the surveyor's hub is not a good idea since that could move the surveyor's hub. If the surveyor's hub is moved or lost, then the surveyor needs to come back to the site again. Due to this reason, surveyors provide a reference stake.

Typically, a surveyor would make a note as shown below.

RS
2.0/10.0

This indicates to establish a stake 10 ft away and 2 ft below the surveyor's hub. Let us assume information stake indicates the following:

RS
2.0/10.0

The figure below shows the reference stake. The measurement for the reference stake is taken from the surveyor's hub.

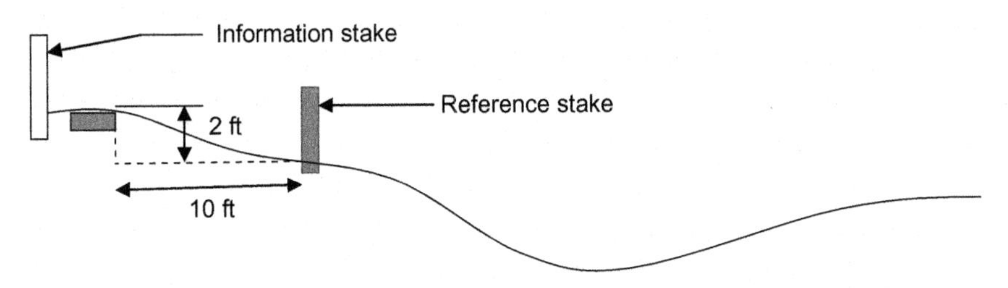

Practice Problem 10.2

Draw the cut section as for the marking in the information stake given below.

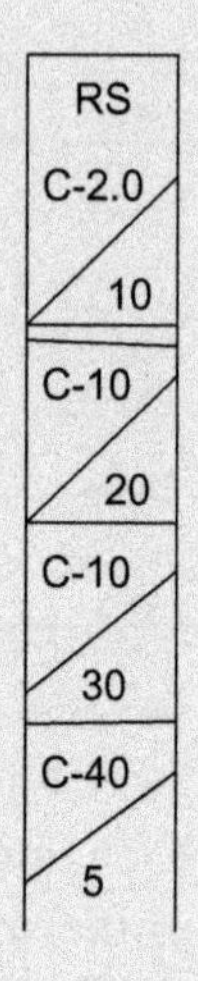

Solution

STEP 1: Establish the reference stake.

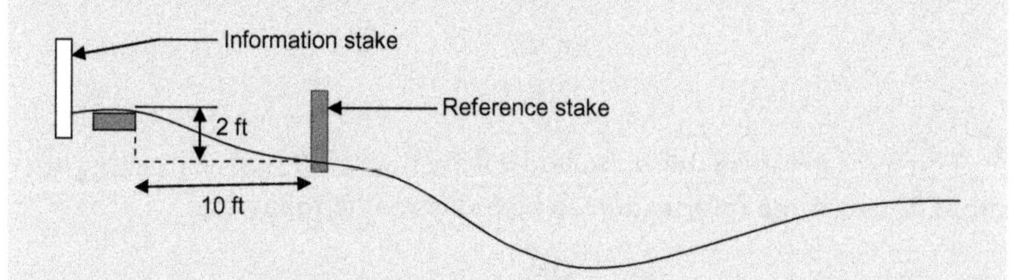

STEP 2: Look at the information stake. The double line after the RS means that all the measurements are to be taken from the reference stake. It is also known as reference point or RP.

$$C-10/20$$

After establishing the reference stake, a cut of 10 ft need to be done at a point 20 ft away from the *reference stake*. Make sure that all measurements are taken from the reference stake.

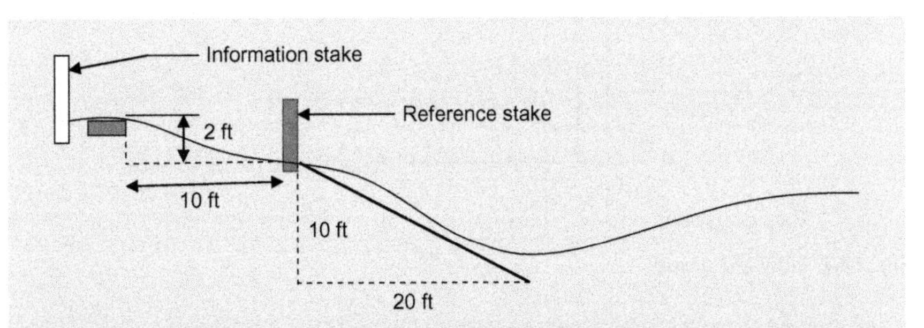

STEP 3: The next item indicates a cut of 10 ft, 30 ft away from the reference stake (C-10/30).

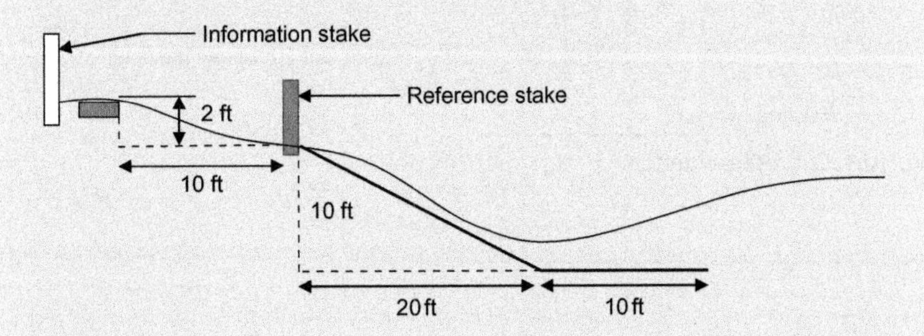

STEP 4: The next item indicates a cut of 5 ft, 40 ft away from the reference stake (C-5/40).

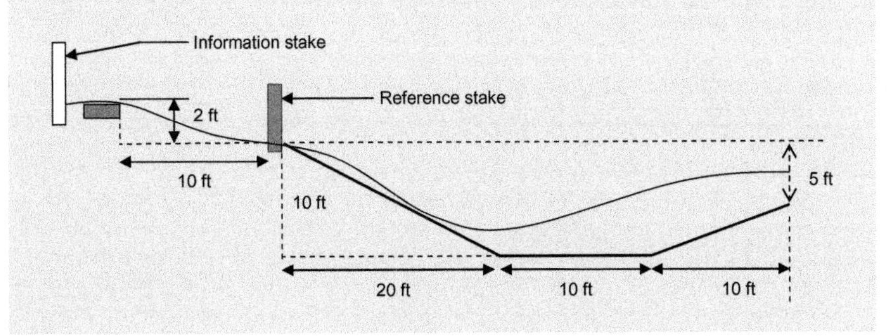

Grading: Grading is the leveling the ground to the desired elevation. A surveyor would provide stakes for the dozer operators (Fig. 10.2).

A circle and an arrow indicate that finished grade should be at the end of the arrow. In this case, soil has to be filled (Fig. 10.3).

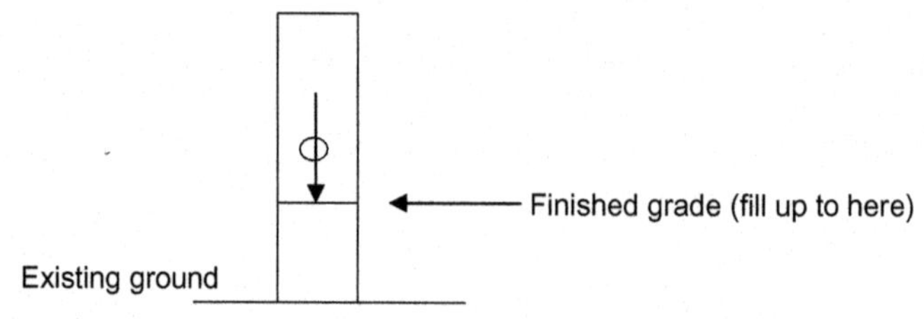

Fig. 10.2 Finished grade.

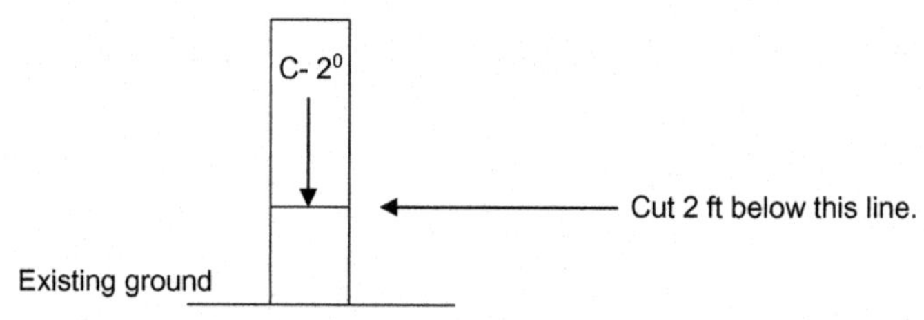

Fig. 10.3 Cut representation.

CHAPTER 11

Earthwork Mass Diagrams

11.1 INTRODUCTION

Earthwork mass diagrams are used to compute the cut and fill quantity required for road construction projects (Fig. 11.1).

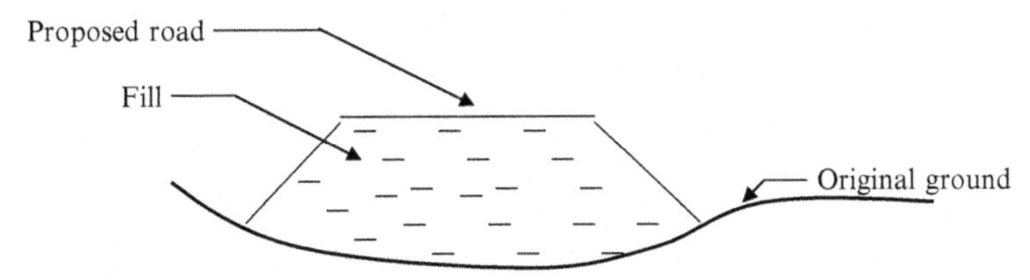

Fig. 11.1 Fill quantity at a station.

During road projects, the road is divided into stations. Typically, stations are given as 0+00, 1+00, 2+00, etc. 1+00 means 100 ft from the reference point and 2+00 means 200 ft from the reference point. Similarly, 2+30 means 230 ft from the reference point. The fill quantity required at each station varies since the ground surface tends to vary.

The fill quantity required at station 1+00 is different from the quantity that is required at station 2+00.

The volume of fill required is obtained by multiplying the average area of two stations by the distance between stations.

$$\boxed{\begin{array}{c}\text{Volume of fill required} = (A1 + A2)/2 \times d \\ d = \text{distance between stations}\end{array}}$$

How does one obtain the area of $A1$ and $A2$?

In the real world, computer programs and planimeters are used to obtain the area of $A1$ and $A2$ since they cannot be computed due to their irregular shapes. In the exam, these areas will be provided.

Fig. 11.2 shows a highway that is being constructed through a mountain range. Such projects involve that a large quantity of soil is cut, transported, and filled.

Construction Engineering Design Calculations and Rules of Thumb
http://dx.doi.org/10.1016/B978-0-12-809244-6.00011-1

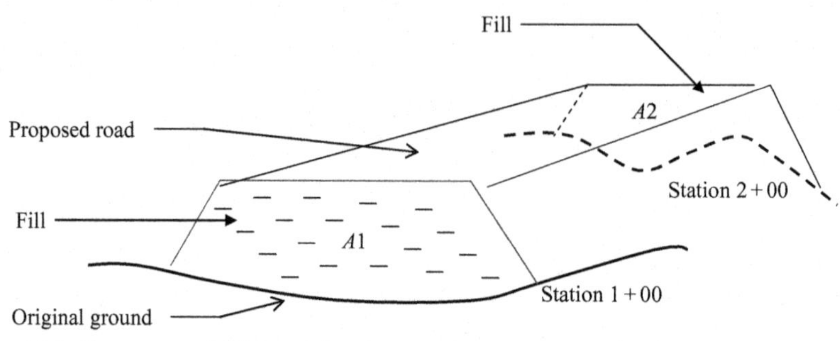

Fig. 11.2 Placement of fill in road construction.

Practice Problem 11.1

Find the volume of fill required from station 1+00 to 1+50. The fill area at station 1+00 is 60 ft^2 and station 1+50 is 80 ft^2.

Solution

$$\begin{aligned}\text{Volume of fill required} &= (A1 + A2)/2 \times \text{distance between stations}\\ &= (60 + 80)/2 \times 50\,\text{ft}^3\\ &= 3500\,\text{ft}^3 = 129.6\,\text{yd}^3\end{aligned}$$

Cut: During road construction, some locations may have to be cut. This happens when the proposed road is at a lower elevation than the existing ground (Fig. 11.3).

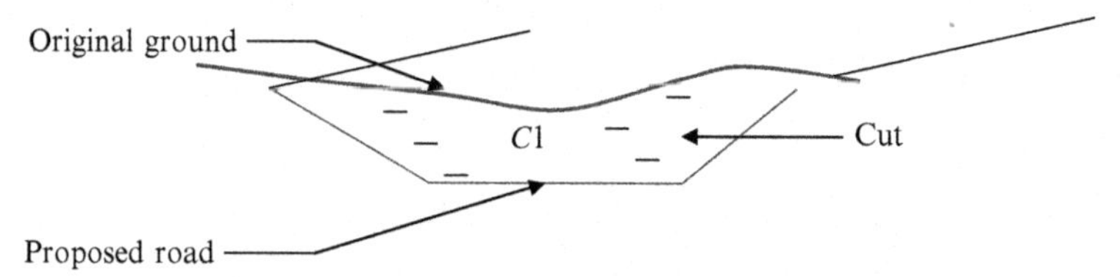

Fig. 11.3 Cut in a station.

In some stations both cut and fill can occur (Fig. 11.4).

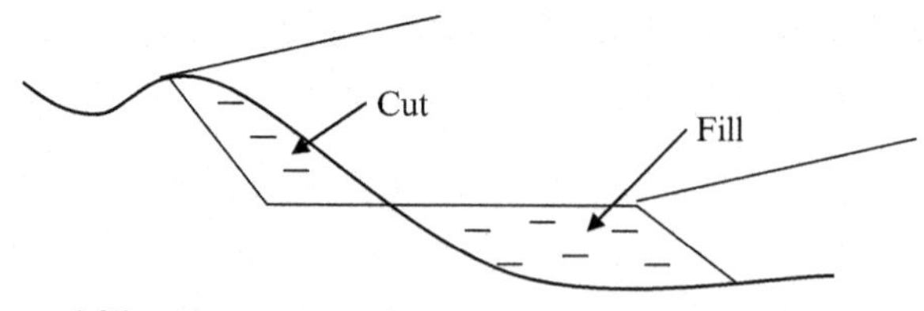

Fig. 11.4 Cut and fill at the same station.

When both cut and fill occur at the same station, a net value is obtained. The process is easily explained using an example.

Practice Problem 11.2

Find the net cut or fill between station 1+50 and 2+00 (Fig. 11.5).

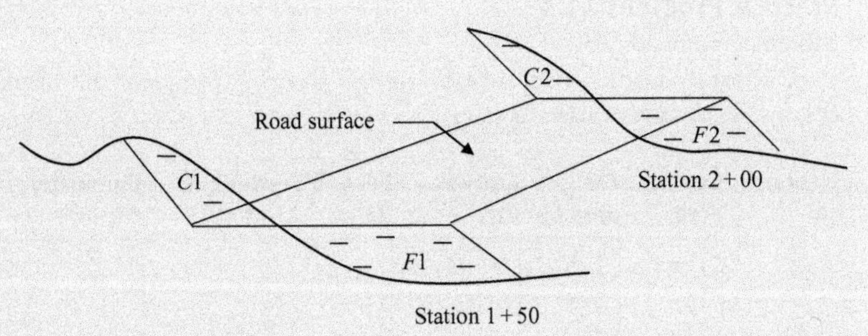

Fig. 11.5 Cut and fill diagram.

Following information provided.

$$\text{Station } 1+50 \quad C1 = 120\,\text{ft}^2 \quad F1 = 170\,\text{ft}^2$$

$$\text{Station } 2+00 \quad C2 = 200\,\text{ft}^2 \quad F2 = 60\,\text{ft}^2$$

Usually this information is tabulated.

Station no.	Cut area	Fill area	Cut vol.	Fill vol.	Net cut (yd³)
1+50	120	170			
2+00	200	60			

Solution

$$\text{Total cut between two stations} = (C1 + C2)/2 \times 50$$
$$= (120 + 200)/2 \times 50 = 8000\,\text{ft}^3 = 296.3\,\text{yd}^3$$

$$\text{Total fill between two stations} = (F1 + F2)/2 \times 50$$
$$= (170 + 60)/2 \times 50 = 5750\,\text{ft}^3 = 213\,\text{yd}^3$$

$$\text{Net amount} = 296.3 - 213 = 83.3\,\text{yd}^3\ (\text{cut})$$

Cut is represented with positive numbers while fill is represented with negative numbers.

Station no.	Cut area	Fill area	Cut vol. (ft³)	Fill vol. (ft³)	Net cut (ft³)
1+50	120	170			
2+00	200	60	296.3	213	83.3

11.2 MASS DIAGRAMS

A mass diagram is drawn using the net cut values. Mass diagrams are easily explained using an example.

Practice Problem 11.3

Mass diagram example:

Cut areas and fill areas are tabulated as shown. Complete the blank columns and draw a mass diagram.

Station no.	Cut area	Fill area	Cut vol. (ft^3)	Fill vol. (ft^3)	Net cut (ft^3)	Cumulative cut (ft^3)
0+00	175	125				
0+50	117	123				
1+00	238	250				
1+50	211	240				
2+00	198	180				
2+50	140	141				
3+00	258	200				

Solution

STEP 1: Complete the "Cut vol." column.

$$\text{Cut volume (1st entry)} = (175 + 117)/2 \times 50 = 7300$$

$$\text{Cut volume (2nd entry)} = (117 + 238)/2 \times 50 = 8875$$

$$\text{Cut volume (3rd entry)} = (238 + 211)/2 \times 50 = 11{,}225$$

$$\text{Cut volume (4th entry)} = (211 + 198)/2 \times 50 = 10{,}225$$

$$\text{Cut volume (5th entry)} = (198 + 140)/2 \times 50 = 8450$$

$$\text{Cut volume (6th entry)} = (140 + 258)/2 \times 50 = 9950$$

Station no.	Cut area	Fill area	Cut vol. (ft^3)	Fill vol. (ft^3)	Net cut (ft^3)	Cumulative cut (ft^3)
0+00	175	125	0	0	0	
0+50	117	123	7300			
1+00	238	250	8875			
1+50	211	240	11,225			
2+00	198	180	10,225			
2+50	140	141	8450			
3+00	258	200	9950			

STEP 2: Complete the "Fill vol." column.

$$\text{Fill volume (1st entry)} = (125 + 123)/2 \times 50 = 6200$$

$$\text{Fill volume (2nd entry)} = (123 + 250)/2 \times 50 = 9325$$

$$\text{Fill volume (3rd entry)} = (250 + 240)/2 \times 50 = 12,250$$

$$\text{Fill volume (4th entry)} = (240 + 180)/2 \times 50 = 10,500$$

$$\text{Fill volume (5th entry)} = (180 + 141)/2 \times 50 = 8025$$

$$\text{Fill volume (6th entry)} = (141 + 200)/2 \times 50 = 8525$$

Station no.	Cut area	Fill area	Cut vol.	Fill vol.	Net cut (ft^3)	Cumulative cut
0+00	175	125	0	0	0	
0+50	117	123	7300	6200		
1+00	238	250	8875	9325		
1+50	211	240	11,225	12,250		
2+00	198	180	10,225	10,500		
2+50	140	141	8450	8025		
3+00	258	200	9950	8525		

STEP 3: Complete the "Net cut" column.

$$\text{Net cut (1st entry)} = 7300 - 6200 = 1100$$

$$\text{Net cut (2nd entry)} = 8875 - 9325 = -450$$

$$\text{Net cut (3rd entry)} = 11,225 - 6200 = -1025$$

$$\text{Net cut (4th entry)} = 7300 - 6200 = -275$$

$$\text{Net cut (5th entry)} = 7300 - 6200 = 425$$

$$\text{Net cut (6th entry)} = 7300 - 6200 = 1425$$

Station no.	Cut area	Fill area	Cut vol.	Fill vol.	Net cut (ft^3)	Cumulative cut
0+00	175	125	0	0	0	0
0+50	117	123	7300	6200	1100	
1+00	238	250	8875	9325	−450	
1+50	211	240	11,225	12,250	−1025	
2+00	198	180	10,225	10,500	−275	
2+50	140	141	8450	8025	425	
3+00	258	200	9950	8525	1425	

Continued

STEP 4: Complete the "Cumulative cut" column.

Cumulative cut is obtained by adding the present net cut to the previous net total.

$$\text{Cumulative cut (1st entry)} = 1100 + 0 = 1100$$

$$\text{Cumulative cut (2nd entry)} = 1100 + (-450) = 650$$

$$\text{Cumulative cut (3rd entry)} = 650 + (-1025) = -375$$

$$\text{Cumulative cut (4th entry)} = -375 + (-275) = -650$$

$$\text{Cumulative cut (5th entry)} = -650 + 425 = -225$$

$$\text{Cumulative cut (6th entry)} = -225 + 1425 = 1200$$

Station no.	Cut area	Fill area	Cut vol.	Fill vol.	Net cut (ft^3)	Cumulative cut
0+00	175	125	0	0	0	0
0+50	117	123	7300	6200	1100	1100
1+00	238	250	8875	9325	−450	650
1+50	211	240	11,225	12,250	−1025	−375
2+00	198	180	10,225	10,500	−275	−650
2+50	140	141	8450	8025	425	−225
3+00	258	200	9950	8525	1425	1200

STEP 5: Draw the mass diagram (Fig. 11.6).

Mass diagram is drawn between the cumulative cut and the stations.

Point A to B: This section is all cut. Hence, the mass diagram keeps going up.

Point B to C: This section is fill. The mass diagram changes direction at point B and start going down.

Point C to D: Point C to D is all cut. The mass diagram changes direction and start going up.

Beyond point D: Beyond point D is fill. The mass diagram changes direction and start pointing down.

Mass diagram implications: Consider point "A" shown with a vertical line. At point "A" the curve crosses the "*X*" axis. Point "A" lies between station 1+00 and 1+50. Let us assume point "A" to be at station 1+30.

At station 1+30, the curve crosses the "*X*" axis.

What does this mean?

It means that between station 0+00 and station 1+30, the total required fill is zero. In other words, all the cut material has been used for fill purposes.

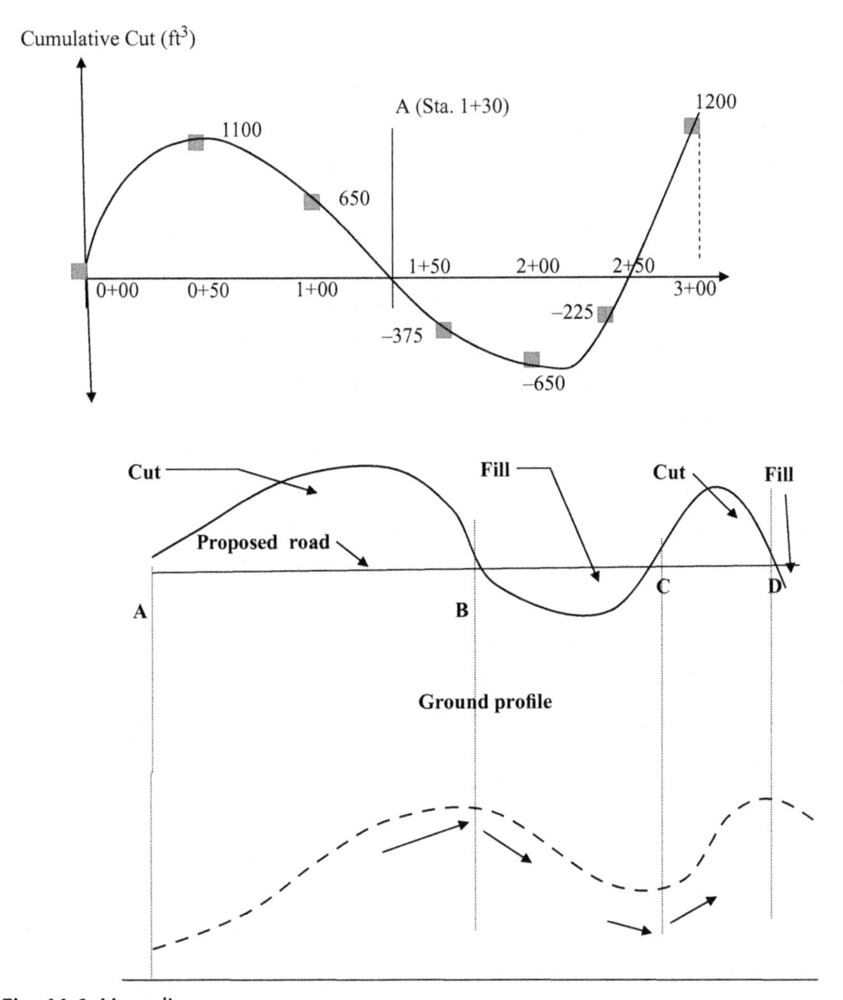

Fig. 11.6 Mass diagram.

Balanced points: The net cut between balanced points is zero. Station 0+00 to 1+30, net cut is zero. Hence, station 0+00 and 1+30 are balanced points. In other words, all the cut material in this region is utilized for fill. Similarly, station 1+30 to 2+60 also are two balanced points.

Volume reduction due to compaction: When soil is compacted, the volume reduces. Let us assume that a contractor has found out that a portion of the road needs 200 yd^3 of fill after compaction. It is common sense for him to bring more than 200 yd^3 of fill since the volume decreases after compaction.

$$\text{Volume reduction factor} = (\text{Additional volume of soil required}) / \text{Volume of soil after compaction}$$

Practice Problem 11.4

The contractor has to fill and compact a portion of the road. The volume of the fill needed (after compaction) is calculated to be 200 yd^3. Calculate the volume of soil need to be brought to the site if the volume reduction factor is 10% (Fig. 11.7).

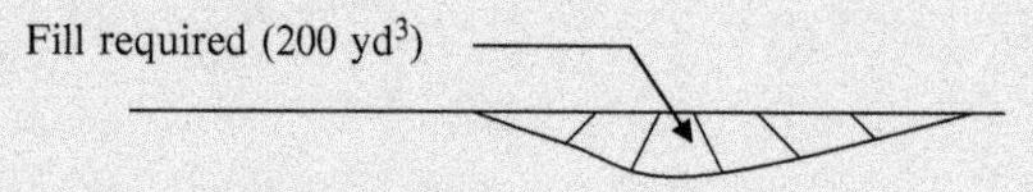

Fig. 11.7 Fill required after compaction.

Solution

$$\begin{aligned}
\text{Volume reduction factor} &= (\text{Additional volume of soil required}) \\
&\quad /\text{Volume of soil after compaction} \\
\text{Volume reduction factor} &= 10\% \\
\text{Volume of soil after compaction} &= 200\,\text{yd}^3 \\
\text{Volume reduction factor} = 0.10 &= (\text{Additional volume of soil required}) \\
&\quad /200 \\
\text{Additional volume of soil required} &= 0.10 \times 200 \\
&= 20\,\text{yd}^3 \\
\text{Total volume of soil required} = 200 + 20 &= 220\,\text{yd}^3
\end{aligned}$$

Expansion: Expansion occurs when soil is cut and removed from ground. Expansion is defined as follows.

$$\text{Expansion} = \text{Additional volume after cut and removal from ground} / \text{Volume in ground}$$

Soil volume increases after cut and removed from ground.

Practice Problem 11.5

During a road project, 150 yd^3 has to be cut and removed from site. Expansion factor of soil is 7%. Each truckload can carry 10 yd^3. Find how many truckloads are required to remove the soil from site.

Solution

$$\text{Expansion} = \text{Additional volume after cut and removal from ground/Volume in ground}$$

$$\text{Expansion} = 7\%$$

$$\text{Volume of soil in ground} = 150\,\text{yd}^3$$

$$0.07 = \text{Additional volume after cut and removal from ground}/150$$

$$\text{Additional volume after cut and removal from ground} = 0.07 \times 150 = 10.5\,\text{yd}^3$$

$$\text{Total volume of soil after cut and removed from ground} = 150 + 10.5 = 160.5$$

Sixteen truckloads will be needed to carry 160 yd^3, since each truck can carry 10 yd^3. One final truck is required to carry last remaining 0.5 yd^3. Hence, total of 17 truckloads are required.

Practice Problem 11.6

Mass diagram example (considering expansion and volume reduction):

Cut areas and fill areas are tabulated as shown. Complete the blank columns and draw a mass diagram. The volume reduction factor is given to be 9% and the expansion factor is given to be 8%.

Solution

A	B	C	D	E	F	G	H	I
							H = E − G	
Station no.	Cut area	Fill area	Cut vol. (in ground)	Cut vol. after expansion	Fill vol. required	Fill volume required considering volume reduction	Net Cut (ft^3)	Cumulative cut (ft^3)
0+00	175	125						
0+50	117	123						
1+00	238	250						
1+50	211	240						
2+00	198	180						
2+50	140	141						
3+00	258	200						

Continued

STEP 1: Complete the "Cut vol. (in ground)" column.

$$\text{Cut volume (1st entry)} = (175 + 117)/2 \times 50 = 7300$$

$$\text{Cut volume (2nd entry)} = (117 + 238)/2 \times 50 = 8875$$

$$\text{Cut volume (3rd entry)} = (238 + 211)/2 \times 50 = 11{,}225$$

$$\text{Cut volume (4th entry)} = (211 + 198)/2 \times 50 = 10{,}225$$

$$\text{Cut volume (5th entry)} = (198 + 140)/2 \times 50 = 8450$$

$$\text{Cut volume (6th entry)} = (140 + 258)/2 \times 50 = 9950$$

A	B	C	D	E	F	G	H	I
							H = E − G	
Station no.	Cut area	Fill area	Cut vol. (in ground)	Cut vol. after expansion	Fill vol. required	Fill volume required considering volume reduction	Net cut (ft³)	Cumulative cut (ft³)
0+00	175	125	0					
0+50	117	123	7300					
1+00	238	250	8875					
1+50	211	240	11,225					
2+00	198	180	10,225					
2+50	140	141	8450					
3+00	258	200	9950					

STEP 2: Complete the "Cut vol. after expansion" column.

As we discussed earlier, when the soil is cut and removed from ground it expands. The expansion factor is given to be 8%.

$$\text{Cut volume in ground} = 7300$$

$$\text{Expansion} = 7300 \times 0.08 = 584$$

$$\text{Volume of soil after expansion} = 7300 + 584 = 7884\,\text{ft}^3$$

$$\text{Simply this can be done by multiplying } 7300 \times (1 + 0.08) = 7884\,\text{ft}^3$$

$$\text{Similarly } 8875 \times (1.08) = 9585$$

$$11{,}225 \times 1.08 = 12{,}123\,\text{ft}^3$$

$$10{,}225 \times 1.08 = 11{,}043\,\text{ft}^3$$

$$8450 \times 1.08 = 9126\,\text{ft}^3$$

$$9950 \times 1.08 = 10{,}746\,\text{ft}^3$$

Input these values in column "E."

A	B	C	D	E	F	G	H	I
							H=E−G	
Station no.	Cut area	Fill area	Cut vol. (in ground)	Cut vol. after expansion	Fill vol. required	Fill volume required considering volume reduction	Net cut (ft^3)	Cumulative cut (ft^3)
0+00	175	125	0	0				
0+50	117	123	7300	7884				
1+00	238	250	8875	9585				
1+50	211	240	11,225	12,123				
2+00	198	180	10,225	11,043				
2+50	140	141	8450	9126				
3+00	258	200	9950	10,746				

STEP 3: Complete the "Fill vol." column.

$$\text{Fill volume (1st entry)} = (125 + 123)/2 \times 50 = 6200$$

$$\text{Fill volume (2nd entry)} = (123 + 250)/2 \times 50 = 9325$$

$$\text{Fill volume (3rd entry)} = (250 + 240)/2 \times 50 = 12,250$$

$$\text{Fill volume (4th entry)} = (240 + 180)/2 \times 50 = 10,500$$

$$\text{Fill volume (5th entry)} = (180 + 141)/2 \times 50 = 8025$$

$$\text{Fill volume (6th entry)} = (141 + 200)/2 \times 50 = 8525$$

A	B	C	D	E	F	G	H	I
							H=E−G	
Station no.	Cut area	Fill area	Cut vol. (in ground)	Cut vol. after expansion	Fill vol. required	Fill volume required considering volume reduction	Net cut (ft^3)	Cumulative cut (ft^3)
0+00	175	125	0	0	0			
0+50	117	123	7300	7884	6200			
1+00	238	250	8875	9585	9325			
1+50	211	240	11,225	12,123	12,250			
2+00	198	180	10,225	11,043	10,500			
2+50	140	141	8450	9126	8025			
3+00	258	200	9950	10,746	8525			

Continued

STEP 4: When fill material is compacted, the volume is reduced. Complete the "Fill volume required considering volume reduction" column.

Volume reduction factor is given to be 9%.

If fill volume required is 6200 ft^3, the contractor has to bring $6200 \times 1.09\ \text{ft}^3 = 6758\ \text{ft}^3$.

Similarly

$$9325 \times 1.09 = 10,164\text{ft}^3$$

$$12,250 \times 1.09 = 13,352\text{ft}^3$$

$$10,500 \times 1.09 = 11,445\text{ft}^3$$

$$8025 \times 1.09 = 8747\text{ft}^3$$

$$8525 \times 1.09 = 9292\text{ft}^3$$

Input these values in column "G."

A	B	C	D	E	F	G	H	I
							H=E−G	
Station no.	**Cut area**	**Fill area**	**Cut vol. (in ground)**	**Cut vol. after expansion**	**Fill vol. required**	**Fill volume required considering volume reduction**	**Net cut (ft^3)**	**Cumulative cut (ft^3)**
0+00	175	125	0	0	0			
0+50	117	123	7300	7884	6200	6758		
1+00	238	250	8875	9585	9325	10,164		
1+50	211	240	11,225	12,123	12,250	13,352		
2+00	198	180	10,225	11,043	10,500	11,445		
2+50	140	141	8450	9126	8025	8747		
3+00	258	200	9950	10,746	8525	9292		

STEP 5: Complete column "H."

$$\text{Net cut (1st entry)} = 7884 - 6758 = 1126\text{ft}^3$$

$$\text{Net cut (2nd entry)} = 9585 - 10,164 = -579\text{ft}^3$$

$$\text{Net cut (3rd entry)} = 12,123 - 13,352 = -1229\text{ft}^3$$

$$\text{Net cut (4th entry)} = 11,043 - 11,445 = -402\text{ft}^3$$

$$\text{Net cut (5th entry)} = 9126 - 8747 = 379\text{ft}^3$$

$$\text{Net cut (6th entry)} = 10,746 - 9292 = 1454\text{ft}^3$$

Input these values in column "H."

A	B	C	D	E	F	G	H	I
							H = E − G	
Station no.	Cut area	Fill area	Cut vol. (in ground)	Cut vol. after expansion	Fill vol. required	Fill volume required considering volume reduction	Net cut (ft^3)	Cumulative cut (ft^3)
0+00	175	125	0	0	0	0	0	
0+50	117	123	7300	7884	6200	6758	1126	
1+00	238	250	8875	9585	9325	10,164	−579	
1+50	211	240	11,225	12,123	12,250	13,352	−1229	
2+00	198	180	10,225	11,043	10,500	11,445	−402	
2+50	140	141	8450	9126	8025	8747	379	
3+00	258	200	9950	10,746	8525	9292	1454	

STEP 6: Complete the "Cumulative cut" column.

The cumulative cut is obtained by adding the present net cut to the previous net total.

$$\text{Cumulative cut (1st entry)} = 1126 + 0 = 1126$$
$$\text{Cumulative cut (2nd entry)} = 1126 + (-579) = 547$$
$$\text{Cumulative cut (3rd entry)} = 547 + (-1229) = -682$$
$$\text{Cumulative cut (4th entry)} = -682 + (-402) = -1084$$
$$\text{Cumulative cut (5th entry)} = -1084 + 379 = -705$$
$$\text{Cumulative cut (6th entry)} = -705 + 1454 = 749$$

Input these values in column "I."

A	B	C	D	E	F	G	H	I
							H = E − G	
Station no.	Cut area	Fill area	Cut vol. (in ground)	Cut vol. after expansion	Fill vol. required	Fill volume required considering volume reduction	Net cut (ft^3)	Cumulative cut (ft^3)
0+00	175	125	0	0	0	0	0	0
0+50	117	123	7300	7884	6200	6758	1126	1126
1+00	238	250	8875	9585	9325	10,164	−579	547
1+50	211	240	11,225	12,123	12,250	13,352	−1229	−682
2+00	198	180	10,225	11,043	10,500	11,445	−402	−1084
2+50	140	141	8450	9126	8025	8747	379	−705
3+00	258	200	9950	10,746	8525	9292	1454	749

Continued

STEP 7: Draw the mass diagram.

The mass diagram is drawn between the "cumulative cut" and the "stations."

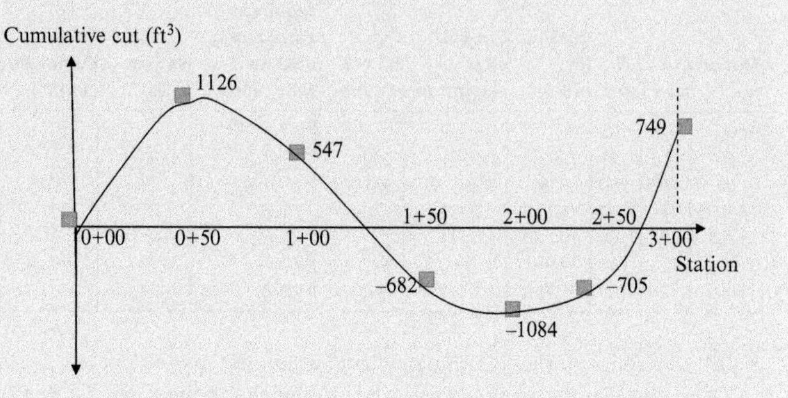

Practice Problem 11.7

Draw the mass diagram for the road construction project shown. The volume of soil cut or filled shown in the drawing.

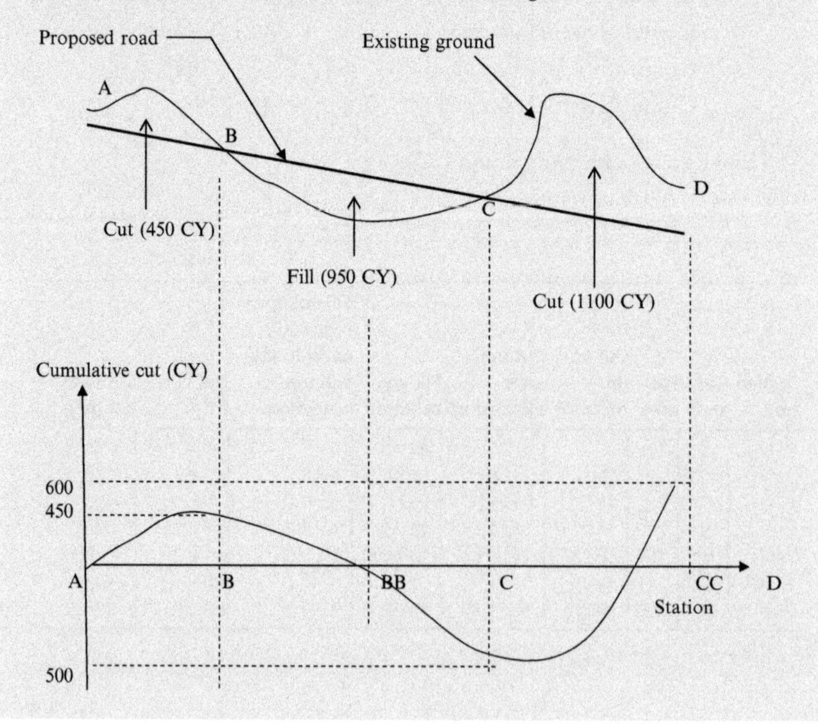

Solution

Point A to B—The ground has to be cut. 450 CY of soil will be cut from point A to B.

Point B to C—The ground has to be filled. 950 CY of soil needed.

Point C to D—The ground has to be cut. 1100 CY of soil will be cut from point C to D.

$$\text{Total cut} = 450 + 1100 = 1550\,\text{CY}$$

$$\text{Total fill} = 950\,\text{CY}$$

$$\text{Waste} = 1550 - 950 = 600\,\text{CY}$$

The cumulative cut volume increases from point A to B. The cumulative cut volume starts to decrease after point B, since filling has started. At point BB, the cut volume will be equal to the fill volume. There is not enough information given in this example to find point BB.

From point BB to C, the fill volume increases. Cutting starts again at point C. Hence, the cumulative cut volume starts to increase.

At point CC, the cut and fill become equal again. There is not enough information given in this example to obtain point CC.

From point CC to D, the cut volume increases. There is excess of 600 CY of soil that would be left at the end of the project.

Free haul distance (FHD): Free haul distance is known as the "distance the soil is moved" without additional compensation. This distance is agreed upon prior to construction.

Free haul distance can be explained using an example.

Practice Problem 11.8

Free haul distance for a cut and fill project is given to be 100 yd. Find the free haul areas of the mass diagram shown.

$$\text{Point A to B} = \text{Cut volume} = 600\,\text{CY}$$

$$\text{Point B to C} = \text{Fill volume} = 600\,\text{CY}$$

$$\text{Point C to D} = \text{Fill volume} = 350\,\text{CY}$$

$$\text{Point D to E} = \text{Cut volume} = 500\,\text{CY}$$

Continued

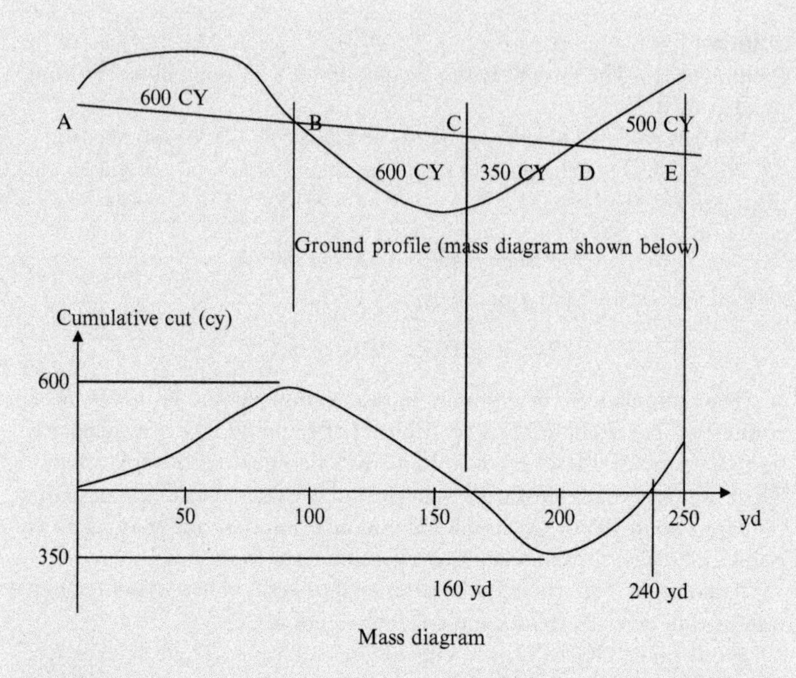

Mass diagram

Solution

The free haul distance is given to be 100 yd. From point A to C all cut material can be used for the fill. The distance between points A to C is 160 yd. This is greater than the free haul distance.

Draw a horizontal line (GH) to a distance of 100 yd as shown below.

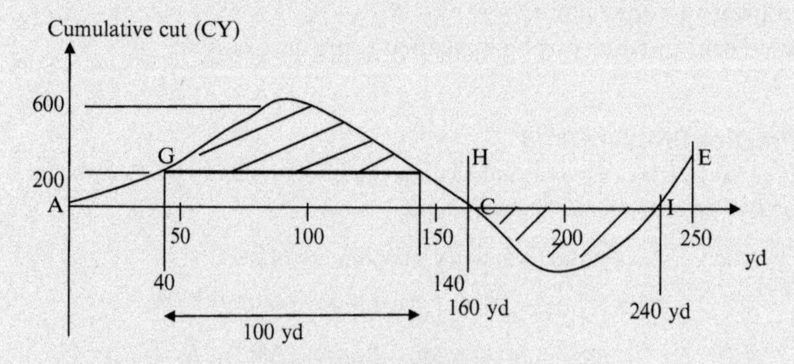

Find the coordinates of point G and H.

Point G = Cumulative cut = 200 CY, station = 40

Point H = Cumulative cut = 200 CY, station = 140

Between points G and H, all the cut material has been used for fill purposes. Hence, the contractor cannot charge additional funds for hauling of material in this section of the road (point G to H is the free haul distance).

Point A to G = Not within the free haul distance. Additional cost will incur for hauling material.

Point H to C = Not within the free haul distance. Additional cost will incur for hauling material.

Point C to I = Distance is 80 yd (240 − 160).

This is less than the free haul distance. Hence, no additional cost will incur for points C to I.

Cut material from I to E is waste.

Overhaul: When soil is moved beyond the free haul distance, it is known as overhaul. The contractor is required to be compensated for overhaul.

Limit of profitable haul (LPH): Soil can be profitably moved from point A to point B. If the distance between two points is larger than the LPH, then it is more profitable to obtain soil from an outside source.

Borrow: Volume of soil obtained from an outside source.

Waste: Volume of soil that has to be discarded.

11.3 HAULING

Soil that is cut has to be hauled to fill areas. Small distances can be hauled using dozers. Typically, distances less than one mile are hauled using scrapers. Scrapers may not be economical for anything beyond one mile. In such situations, trucks are used. It is the judgment of the site supervisor to decide what machines will be used for a given project (Fig. 11.8).

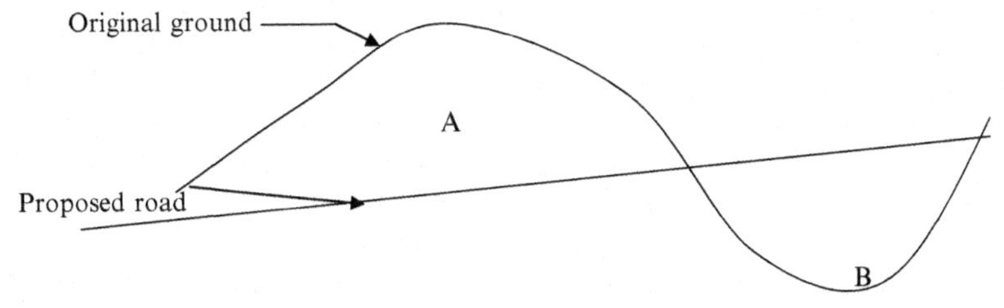

Fig. 11.8 Topographic profile.

Volume A has to be cut and volume B has to be filled. Small haul distances can use dozers.

STEP 1: Cut the volume of soil A1 and fill volume B1. Hauling can be done using dozers since the distance is short (Fig. 11.9).

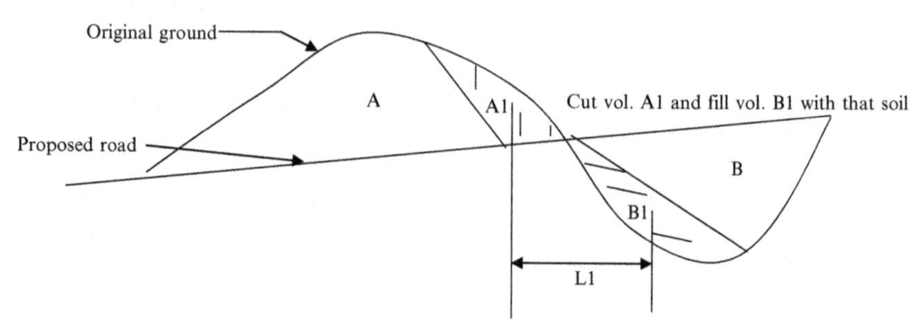

Fig. 11.9 Topographic profile (cut vol. and fill vol.).

L1 is the average transportation distance.

Cut vol. A1 and fill vol. B1 with that soil

L1 = Average transportation distance = Distance from center of gravity of A1 to center of gravity of B1.

At the beginning of a cut and fill operation soil can be moved with dozers. The distance from the cut site to the fill site is small.

STEP 2: Cut a volume of soil A2 and fill volume B2 with the same soil (soil at A2 goes to B2).

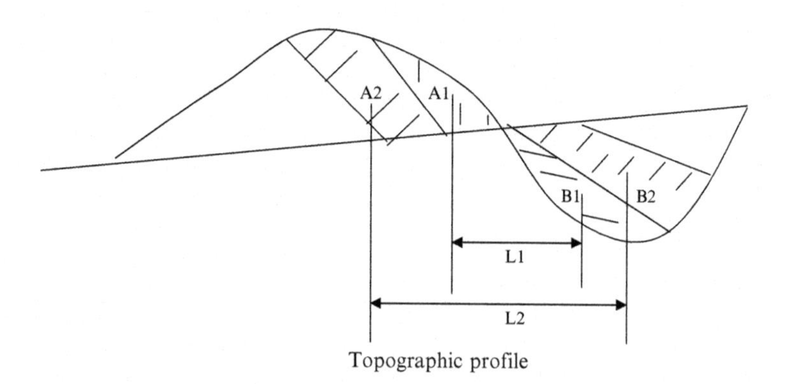

Topographic profile

Cut volume A2 and fill B2 with that soil.

L2 is the average transportation distance from A2 to B2.

One can see soil has to be transported longer distances. L2 > L1

Depending upon the site conditions it may not be efficient to use dozers to move soil. Hence, scrapers can be utilized.

STEP 3: Cut a volume of soil at A3 and fill volume B3.

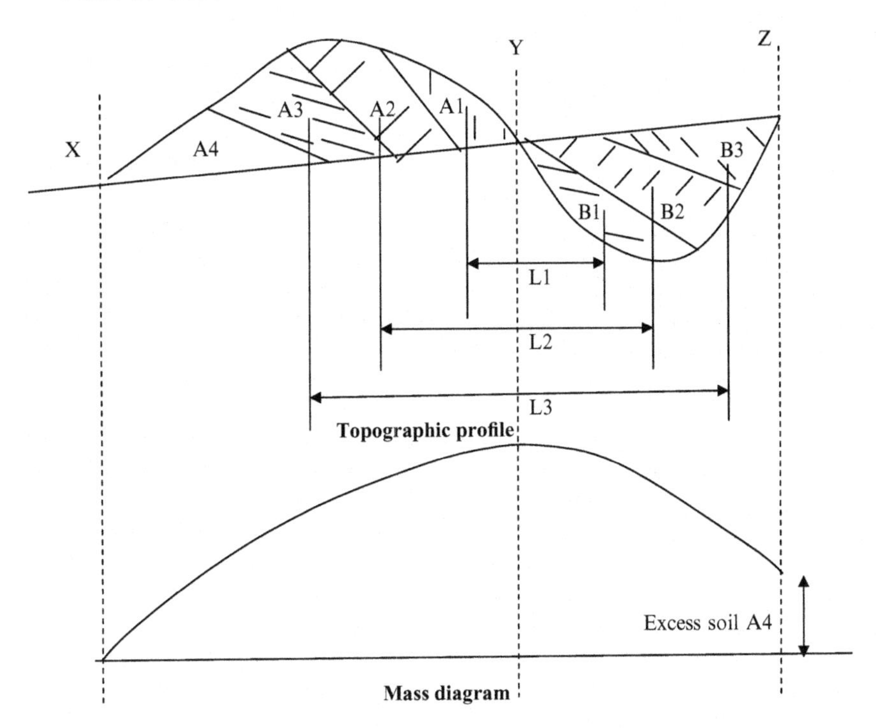

Cutting occurs from X to Y. Hence, the cumulative cut volume increases until Y. After Y, fill begins.

The cumulative cut volume increases from point X to Y. Fill starts at point Y. Therefore the mass diagram will start to bend down. Since there is more cut in this section, there is excess soil at the end that needs to be trucked away.

Moving soil volume A3 to B3 involves transportation of soil for much longer distances. In such situations, trucks may be the most efficient method.

Volume of soil A4 is the surplus. This soil may be cut and transported out of the site.

11.4 MOST EFFICIENT METHOD TO CONDUCT A CUT AND FILL OPERATION

Cutting of soil is mostly done using dozers. Occasionally backhoes may be needed to cut the soil due to steep slopes or loose soil situations where dozers are unable to access the location. Moving soil for shorter distances can be done using dozers efficiently. Dozers become increasingly inefficient when the distance increase. In such situations, scrapers are used to scrape and move the soil. One has to understand that the rental cost of a scraper may be twice as much as a dozer. Trucks are needed for much longer distances. In many situations, trucks may have to use local roads that could have significant traffic. Thus it is advisable to move the soil early in the morning before the rush hour and then after the rush hour. The supervisor of the site has to plan his work in a most efficient manner.

Contour lines: Contour lines represent elevations.

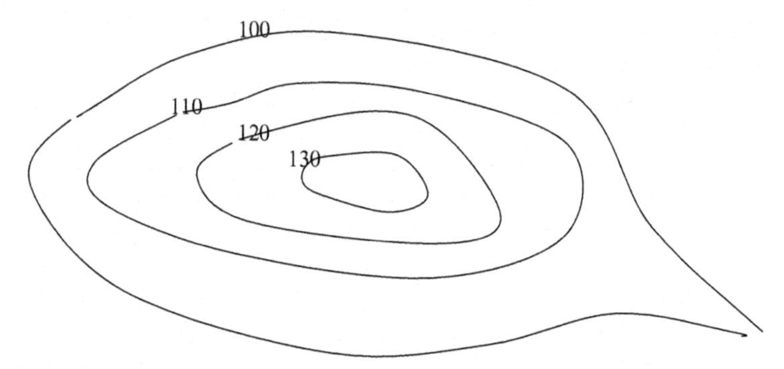

Fig. 11.10 Contour lines.

If you look at Fig. 11.10, contour elevation increases from 100 to 130. This represents a hill. On the other hand, Fig. 11.11 represents a lake or a crater.

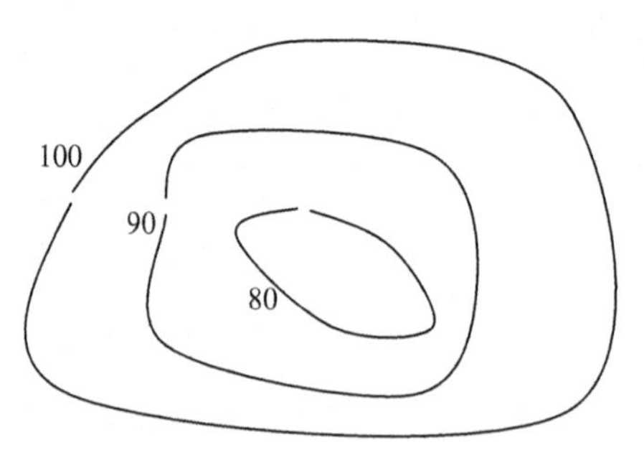

Fig. 11.11 Contour lines (a crater).

11.5 CUT AND FILL COMPUTATIONS USING CONTOUR LINES

Cut and fill can be computed using contour lines. Example below shows how to compute cut and fill volumes using contour lines.

Practice Problem 11.9

The width of the road is 30 ft and BD = 500 ft and DF = 500 ft. Elevation of the proposed road is 98 ft (Fig. 11.12).

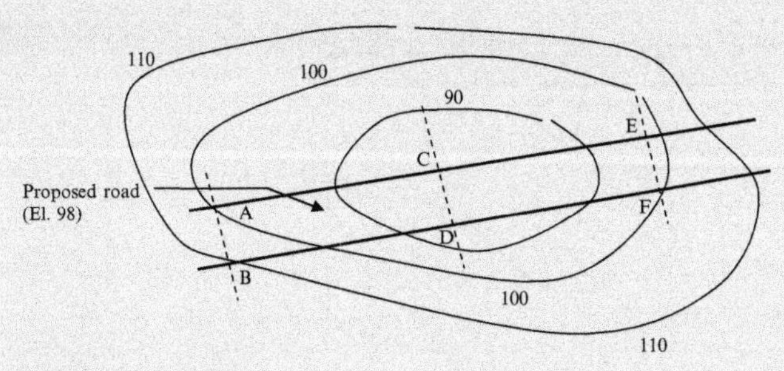

Fig. 11.12 Proposed road.

Find the following;

(A) Cross sectional areas at sections AB, CD, and EF.

(B) Find the cut and fill volumes from B to F.

Solution

STEP 1: Find the cross sectional area at section AB.

$$\text{Elevation at point A} = 100 \text{ (see the contour 100)}$$
$$\text{Elevation at point B} = 110$$
$$\text{Proposed road elevation} = 98 \text{ ft}$$

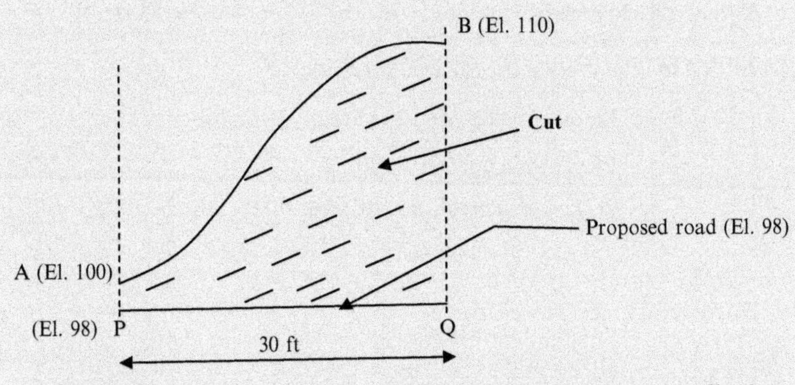

Points P and Q are on the proposed road. Hence, the elevation of both P and Q is 98.0

Continued

$$AP = 2; \quad BQ = 12$$

$$\text{Area of the trapezoid (APQB)} = (2 + 12)/2 \times 30 = 210\,\text{ft}^2$$

STEP 2: Find the cross sectional area at section CD.

Elevation at point C = 85 (elevation at C can be approximated to 85 ft. Both points C and D are inside the contour 90).

Elevation at point D = 87 (point D is closer to 90 ft contour. Hence, it should be around 87).

Proposed road elevation = 98 ft.

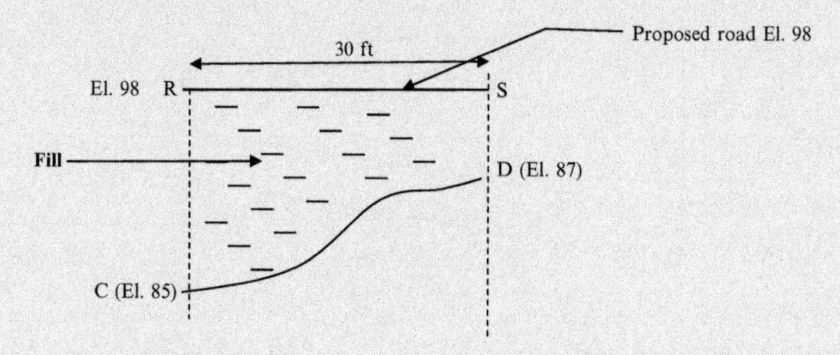

Assume points R and S to be on a proposed road.

$$RC = 13\,\text{ft}, \quad SD = 11\,\text{ft}$$

$$\text{Area of trapezoid (RCDS)} = (13 + 11)/2 \times 30 = 360\,\text{ft}^2$$

$$\text{Cut at station AB} = 210\,\text{ft}^2 \text{ (found in step 1)}$$

$$\text{Fill at station AB} = 0\,\text{ft}^2$$

$$\text{Cut at station CD} = 0\,\text{ft}^2$$

$$\text{Fill at station CD} = 360\,\text{ft}^2$$

$$\text{Average cut between two stations} = (210 + 0)/2 \times 500\,\text{ft}^3 = 1944\,\text{CY}$$

$$\text{Average fill between two stations} = (0 + 360)/2 \times 500\,\text{ft}^3 = 3333\,\text{CY}$$

STEP 3: Find the cross sectional area at section EF.

$$\text{Elevation at point E} = 98 \text{ (approximately)}$$

$$\text{Elevation at point F} = 100$$

$$\text{Proposed road elevation} = 98\,\text{ft}$$

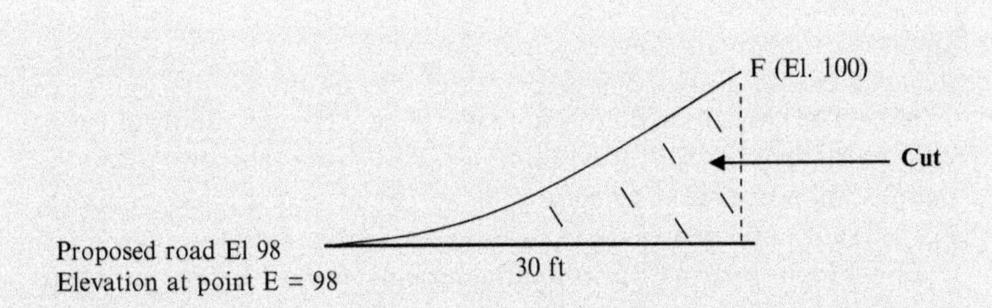

There is no fill at this station.

Cut area can be calculated by considering an approximate triangle.

$$\text{Cut area} = 2 \times 30/2 = 30\,\text{ft}^2$$

Fill area = 0

Average fill volume between stations CD and EF

$= (360 + 0)/2 \times 500\,\text{ft}^3 = 3333\,\text{CY}$

Average cut volume between stations CD and EF $= (0 + 30)/2 \times 500\,\text{ft}^3$
$= 278\,\text{CY}$

Practice Problems 11.10

Three proposed roads are shown. What is the best route to build a road at elevation 90?

(a) Considering cut and fill costs only

(b) Considering drainage infrastructure build up costs only

(c) Considering retaining wall construction costs only

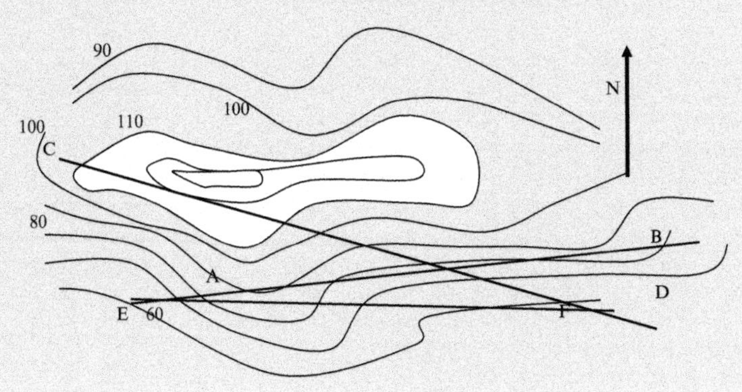

Continued

Solution

Cut and fill costs:

AB—The proposed road elevation is at 90 ft, AB passes mostly through contours 80 and 70. When the proposed road passes through 80 and 70 contours, fill is needed to bring it to 90 ft. The proposed route requires a large amount of fill that needs to be brought in from outside.

EF—Proposed road EF goes mostly through 60 and 70. EF route needs more fill than AB route.

CD—The CD route has plenty of cut (the northwestern end) while the southeast requires fill. It can be argued that the cut from the northwestern end can be utilized for the fill in the southeast end of the road. Since on site cut material could be used, the cost could be cheaper.

Drainage costs:

As per contours given, the hills to the north would drain to the proposed routes AB and EF. Drainage to route EF would be higher than route AB, therefore drainage costs for the route EF would be higher than AB. On the other hand, CD is on high ground. Hence, less drainage can be expected. The drainage infrastructure build up costs would be less for CD.

Retaining wall costs:

The northwestern end of route CD goes through contour 110. Contours 120 and 130 are locate just north of the road CD. A large retaining wall is needed in this section. Contours for the most part are tightly packed along CD. This indicates steep slopes and costly and high retaining walls. Retaining wall heights for proposed routes AB and EF would be moderate compared to CD.

CHAPTER 12

Quantity Takeoff and Cost Estimating

12.1 INTRODUCTION

A cost estimate is needed to assess the cost of a construction project. Owners would like to know the cost before starting a project. The following is a list of items that determine the cost of a project:

Material cost: The cost of concrete, steel, bricks, windows, etc.

Transportation cost: The cost of transportation of materials.

Cost of labor: The cost of labor includes wage for concrete workers, carpenters, steel workers, supervisors, and many other construction personnel.

Cost of engineering: The fee charged by an engineering firm for their design

Quantity takeoff is the process of determining the material cost. To obtain the material cost, the quantity of materials needs to be obtained.

12.2 QUANTITY TAKEOFF

The first step in cost estimating is to conduct a quantity takeoff using design drawings. Quantity takeoff can be explained easily using examples. You are required to obtain quantity of concrete, bricks, steel weight, etc., using design drawings.

Practice Problem 12.1

Find the following quantities using the design drawing given.

(a) Volume of soil to be excavated

(b) Concrete volume for footings

(c) Area of formwork required for footings

Continued

Construction Engineering Design Calculations and Rules of Thumb
http://dx.doi.org/10.1016/B978-0-12-809244-6.00012-3

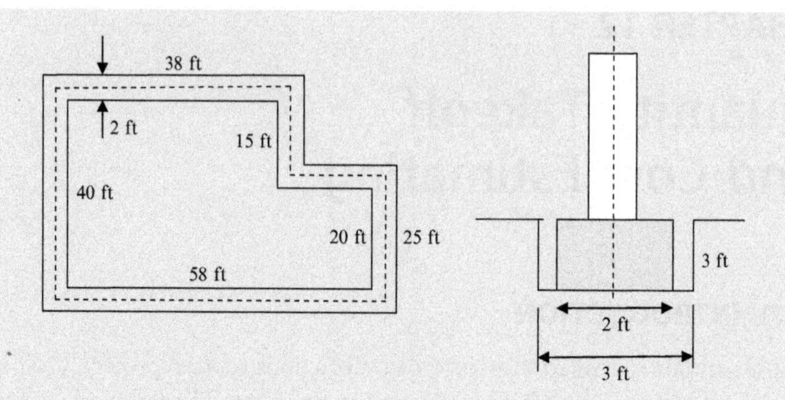

Footing plan (lengths are given along the centerline of the footing)
The thickness of the footing is 2 ft.

Solution

Note: When you are measuring distances, you need to measure along the centerline. Other method is the "In to In and Out to Out" method. In this method, one side is measured "In to In" while the perpendicular side is measured "Out to Out."

STEP 1:

$$\text{Volume of soil to be excavated} = (3 \times 3) \times \text{length of the footing}$$

$$\text{Length of the footing} = 38 + 15 + 20 + 25 + 58 + 40 = 196\,\text{ft}$$

Note: If distances are not given along the centerline of the footing, you have to compute the lengths. In some cases, lengths are given along the outer perimeter.

$$\text{Volume of soil to be excavated} = (3 \times 3) \times 196 = 1764\,\text{ft}^3$$

$$= 65.3\,\text{yd}^3 \left(1\,\text{yd}^3 = 27\,\text{ft}^3\right)$$

STEP 2: Concrete volume of footings $= 2 \times 3 \times 196 = 1176\,\text{ft}^3 = 43.6\,\text{yd}^3$

STEP 3: Area of formwork required for footings:

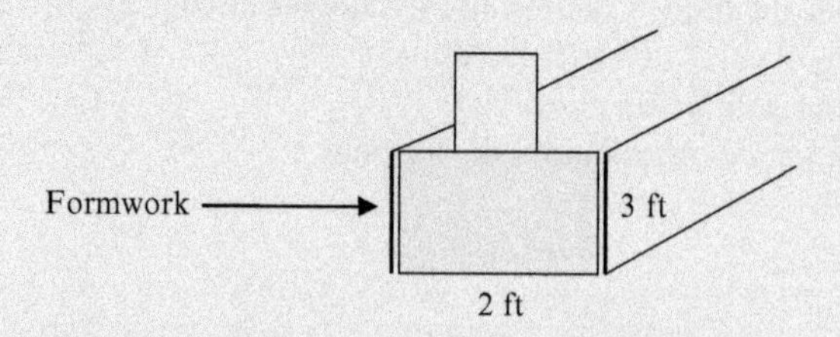

Formwork has to be erected on either side of the footing.

Formwork area for one side of the footing $= (3 \times 196)\,\text{ft}^2 = 588\,\text{ft}^2$

Formwork area for both sides of the footing $= 2 \times 588\,\text{ft}^2 = 1176\,\text{ft}^2$.

Practice Problem 12.2

Find the area of formwork required for the footing shown.

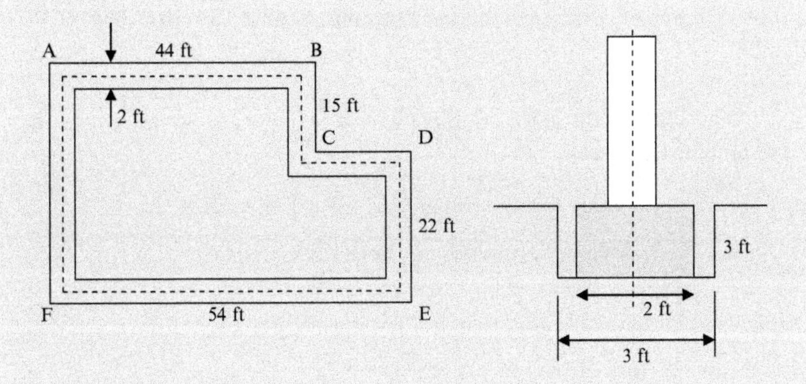

Footing plan (lengths are given along the outer perimeter)

Solution

Find the length of footing along the centerline:

In to In and Out to Out method: Horizontal sides are measured In to In. Vertical distances are measured Out to Out.

$$\text{Distance along AB} = 40\,\text{ft}\ \ (44-4)\ \ (\text{In to In})$$

(measured wall to wall from inside the house)

$$\text{Distance along BC} = 15\ \ (\text{Out to out})$$

(measured wall to wall from outside the house)

$$\text{Distance along CD} = 10\,\text{ft}\ \ (\text{In to In})$$

(measured wall to wall from inside the house)

$$\text{Distance along DE} = 22\,\text{ft}\ \ (\text{Out to out})$$

(measured wall to wall from outside the house)

$$\text{Distance along EF} = 50\,\text{ft}\ \ (\text{In to In})$$

(measured wall to wall from inside the house)

$$\text{Distance along AF} = 37\,\text{ft}\ \ (\text{Out to out})$$

(measured wall to wall from outside the house)

Continued

Total = 174 ft

This can be double-checked using centerline method.

$$\text{Total lengths of footings along the centerline} = 42 + 15 + 10 + 20 + 52 + 35$$
$$= 174\,\text{ft}$$

Note: The lengths in this example have no connection to the lengths in the previous example.

$$\text{Area of formwork per side} = 3 \times 174 = 522\,\text{ft}^2$$

$$\text{Area of formwork for both sides} = 1044\,\text{ft}^2$$

Practice Problem 12.3

Find the quantities for following items for the 200 ft long retaining wall shown.

(1) Excavation volume
(2) Concrete volume
(3) Gravel volume
(4) Backfill volume

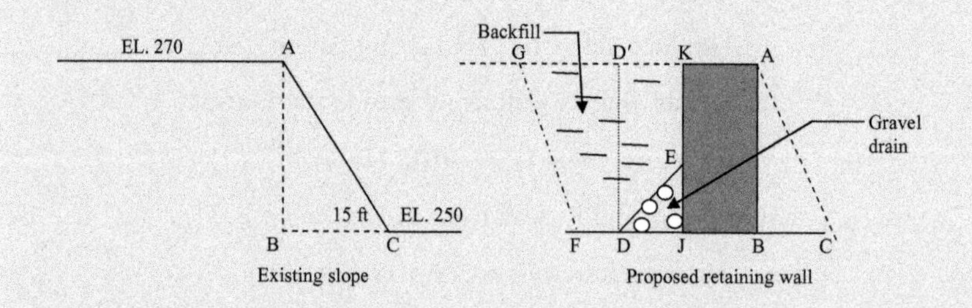

Following distances are given:
BC = 15 ft, JB = 10 ft, DJ = 3 ft, EJ = 2.5 ft, FD = 3 ft, GA = 20 ft
Construction procedure:
STEP 1: Excavate and remove soil volume GFCA
STEP 2: Construct the retaining wall KAJB
STEP 3: Construct the gravel drain EDJ
STEP 4: Backfill GKEDF

Quantity takeoff:
STEP 1:

$$\text{Excavation volume} = \text{Area GFCA} \times 200 \quad (200\,\text{ft is the length})$$

$$\text{Area of GFCA} = (\text{GA} + \text{FC})/2 \times \text{Height KJ}$$

GA is given to be 20 ft.

$$\text{FC} = \text{BC} + \text{FB} = 15 + (10 + 3 + 3) = 31\,\text{ft}$$

$$\text{Area of GFCA} = (\text{GA} + \text{FC})/2 \times \text{Height}$$

$$\text{KJ} = (20 + 31)/2 \times (\text{EL } 270 - \text{El } 250) = (20 + 31)/2 \times 20 = 510$$

$$\text{Excavation volume} = 510 \times 200\,\text{ft}^3 = 102,000\,\text{ft}^3 = 3778\,\text{yd}^3$$

STEP 2: Concrete volume for the retaining wall = Area KAJB × 200

$$= (20 \times 10) \times 200 = 40,000\,\text{ft}^3 = 1481\,\text{yd}^3$$

STEP 3: Volume of gravel required

$$= \text{Area EDJ} \times 200 = (3 \times 2.5)/2 \times 200 = 750\,\text{ft}^3 = 28\,\text{yd}^3$$

STEP 4:

$$\text{Backfill volume} = \text{Area GKEDF} \times 200$$

Area GKEDF can be broken down to GD′DF and D′KED.

$$\text{Area GD}'\text{DF} = (\text{FD} + \text{GD}')/2 \times \text{D}'\text{D} = (7 + 3)/2 \times 20 = 100$$

$$\text{Area D}'\text{KED} = (\text{D}'\text{D} + \text{EK})/2 \times \text{D}'\text{K} = (20 + 17.5)/2 \times 3 = 56.3$$

$$\text{Area GKEDF} = 100 + 56.3 = 156.3$$

$$\text{Backfill volume} = 156.3 \times 200 = 31,260\,\text{ft}^3 = 1158\,\text{yd}^3$$

12.3 MASONRY QUANTITY TAKEOFF

Quantity takeoff for masonry work is considered.

Practice Problem 12.4

Find the mortar volume required per brick in the wall shown. Bricks are 15 in. × 4 in. × 4 in. Mortar thickness is 1 in.

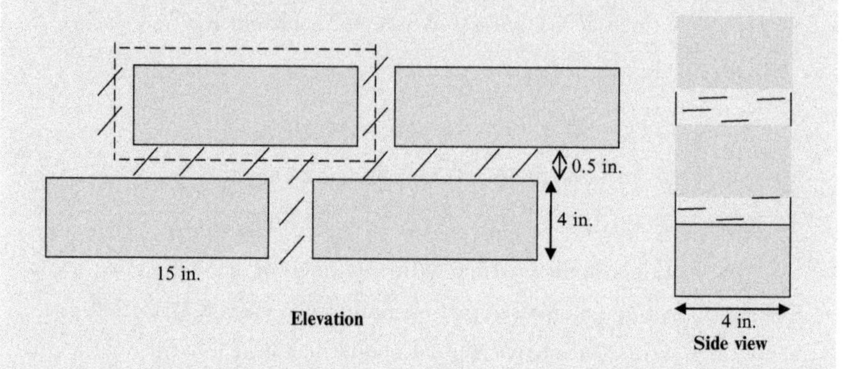

Solution

$$\text{Length of a brick with the mortar} = 15 + 1/2 + 1/2 = 16\text{in.}$$

$$\text{Height of a brick with the mortar} = 4 + 1/2 + 1/2 = 5\text{in.}$$

$$\text{Total volume of a brick with mortar} = 16 \times 5 \times 4 = 320\text{in.}^3$$

$$\text{Volume of the brick without mortar} = 15 \times 4 \times 4 = 240\text{in.}^3$$

$$\text{Volume of mortar per brick} = 80\text{in.}^3$$

Practice Problem 12.5

A wall measuring 100 ft long, 4 in. wide and 11 ft high is constructed using the brick shown above (15 in. × 4 in. × 4 in.) and 0.5 in mortar. Find the following:

(a) Number of bricks required
(b) Volume of mortar required

Solution

$$\text{Total volume of the wall} = (100 \times 12) \times (11 \times 12) \times 4\text{in.} = 633{,}600\ \text{in.}^3$$

$$\text{Total volume of a brick with mortar} = 16 \times 5 \times 4 = 320\ \text{in.}^3$$

(see the previous example)

$$\text{Number of bricks required} = 633{,}600/320 = 1980\text{bricks}$$

Note: To obtain the number of bricks required for a wall, use the volume of a brick with the mortar. If you use bare volume of the brick (in this case 240 in.^{3}), you will get the wrong answer.

$$\text{Volume of brick alone} = 15 \times 4 \times 4 = 240\,\text{in.}^3$$

$$\text{Mortar volume in each brick} = 320 - 240 = 80\,\text{in.}^3$$

$$\text{Mortar volume} = \text{Number of bricks} \times \text{Mortar}$$

$$\text{vol. per brick} = 1980 \times 80 = 158,400\,\text{in.}^3$$

Your answers can be checked by using the following equations:

$$\text{Total volume of the wall} = \text{Volume of bricks} + \text{volume of mortar}$$

$$\text{Volume of bricks only} = 1980 \times 240 = 475,200\,\text{in.}^3$$

$$\text{Volume of bricks} + \text{Volume of mortar} = 475,200 + 158,400 = 633,600\,\text{in.}^3$$

This is equal to the volume of the wall.

Standard bricks: Bricks come in various sizes. Bricks known as standard bricks are 2 ¼ × 3 ¾ × 8 in.

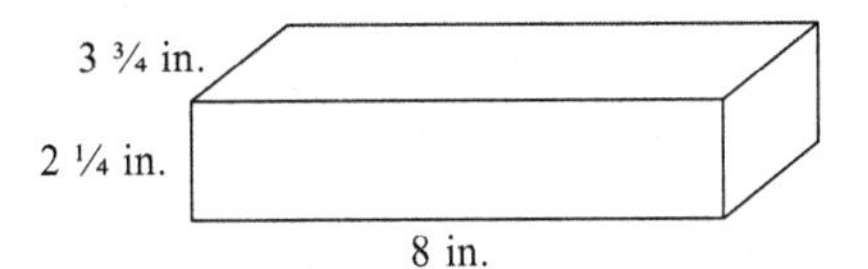

Practice Problem 12.6

A wall is built using standard bricks. The wall is one brick thick. How many bricks are required to build 100 ft^2 of the wall. Assume mortar is one-quarter in thick.

Solution

$$\text{Length of one brick with the mortar} = 8 + 1/8 + 1/8 = 8\tfrac{1}{4}\,\text{in.}$$

$$\text{Height of one brick with mortar} = 2\tfrac{1}{4} + 1/8 + 1/8 = 2\tfrac{1}{2}\,\text{in.}$$

$$\text{Area of one brick with mortar} = (8\tfrac{1}{4}) \times (2\tfrac{1}{2}) = 8.25 \times 2.5 = 20.625\,\text{in.}^2$$

$$\text{Bricks in } 100\,\text{ft}^2 \text{ of wall} = (100 \times 144)/20.625 = 698$$

Practice Problem 12.7

A rectangular building has a length of 120 ft and width 80 ft The building walls are 9 ft high. There are total of 180 ft^2 of openings for windows and doors. The walls are built of standard bricks and are one brick thick. Find the number of bricks and volume of mortar. Standard bricks are 8 in. × 2¼ in. × 3¾ in. Assume mortar thickness to be 0.5 in.

Solution

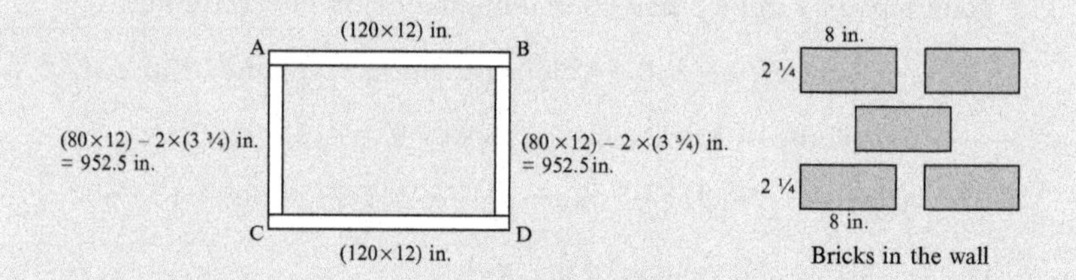

Note: The brick at the edge needs to be deducted to avoid double counting.

STEP 1: *Find the area of brick walls:*

$$\text{Area of side AB} = (120 \times 12) \times \text{height} = (120 \times 12) \times (9 \times 12) = 155{,}520\,\text{in.}^2$$

$$\text{Area of side AC} = \text{width} \times \text{height} = 952.5 \times (9 \times 12) = 102{,}870\,\text{in.}^2$$

$$\text{Total area of four sides} = 2 \times (155{,}520 + 102{,}870) = 516{,}780\,\text{in.}^2$$

$$\text{Area of doors and windows} = 180\,\text{ft}^2 = 180 \times 144 = 25{,}920\,\text{in.}^2$$

$$\text{Area of walls after reducing for openings} = 516{,}780 - 25{,}920 = 490{,}860\,\text{in.}^2$$

$$\text{Effective brick area including the mortar thickness} = (8 + 0.5) \times (2.25 + 0.5) = 23.375\,\text{in.}^2$$

STEP 2: Number of bricks $= 490{,}860/23.375 = 20{,}999$ bricks

STEP 3: Find the mortar volume:

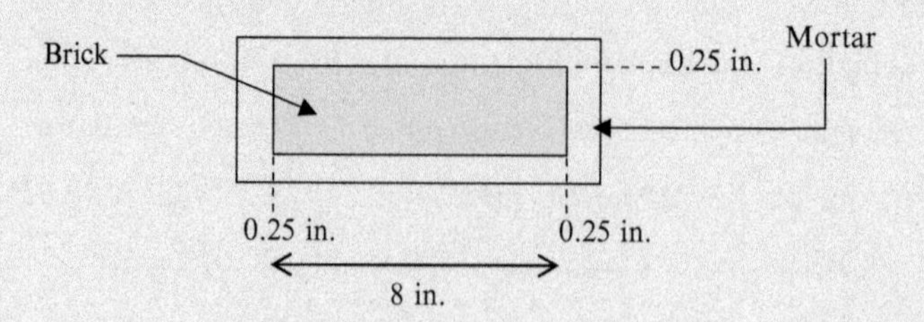

$$\text{Area of mortar per brick} = \text{Total area with mortar} - \text{brick area}$$
$$= (8.5 \times 2.75) - (8 \times 2.25) = 5.375\,\text{in.}^2$$

Wall is one brick thick. The thickness of the brick is 3.75 in.

$$\text{Volume of mortar per brick} = 5.375 \times 3.75 = 20.156\,\text{in.}^3$$
$$\text{Number of bricks} = 20{,}999 \quad \text{(see step 2)}$$
$$\text{Volume of mortar required} = 20{,}999 \times 20.156\,\text{in.}^3$$
$$= 423{,}255\,\text{in.}^3 = 245\,\text{ft}^3 = 9.07\,\text{yd}^3$$

12.4 QUANTITY TAKEOFF (STEEL)

Steel structures contain W sections, S sections, L shapes, Channels, Angles, hollow tubular sections, and hollow rectangular sections.

W shapes (wide flange):

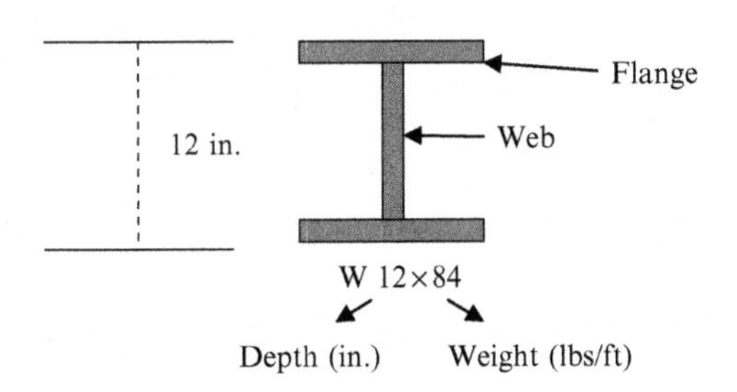

Above 12 indicates the approximate depth. Above 84 indicates weight of the section in lbs per foot.

Channels (American Standard Channel):

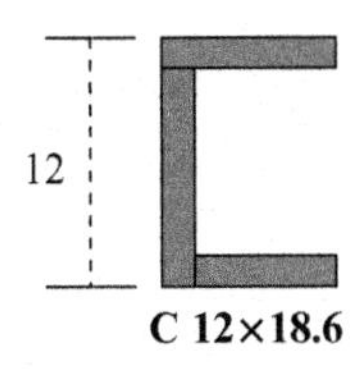

Above 12 indicates the approximate depth. Above 18.6 indicates weight of the channel section in lbs per foot.

S sections: S sections are similar to W sections. They are known as American Standard Steel. W sections have a wider flange than S sections.

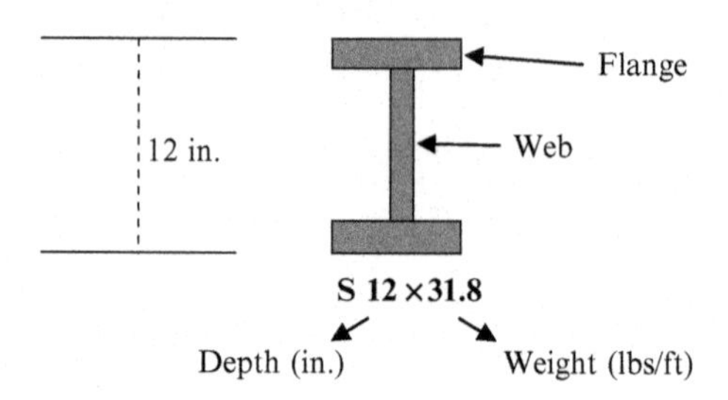

M sections: M sections are known as "Miscellaneous Beam." These beams are manufactured by various steel manufacturers with varying thicknesses and depths.

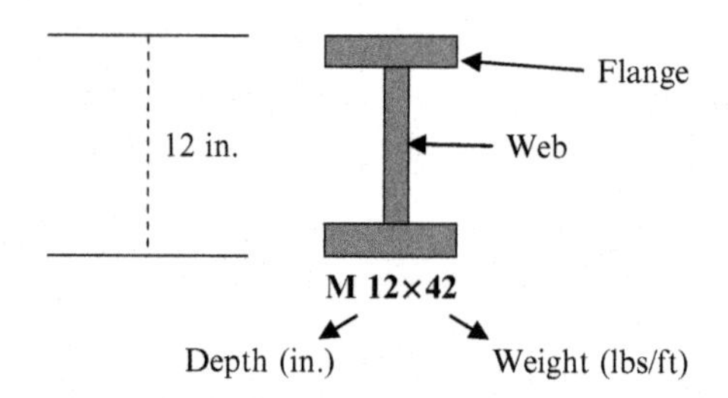

MC sections: MC stands for "Miscellaneous Channel." These channels are produced by various manufacturers with varying thicknesses and depths.

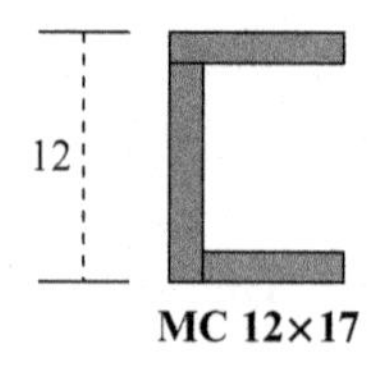

Above 12 indicates the approximate depth. Above 17 indicates weight of the channel section in lbs per foot.

L angles: Angles are represented with "L." Angles may have equal or unequal legs.

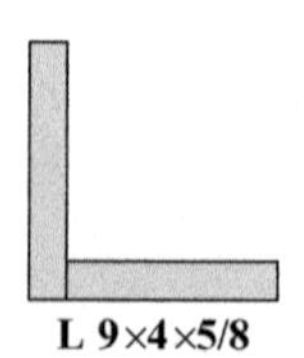

Above 9 indicates the length of one leg in inches and 4 indicates the length of the other leg in inches and 5/8 represents the thickness of the section. Weight of the L section has to be obtained from steel tables. Steel angles are shown below.

HP-piles (or simply H-piles): HP sections are used for piles.

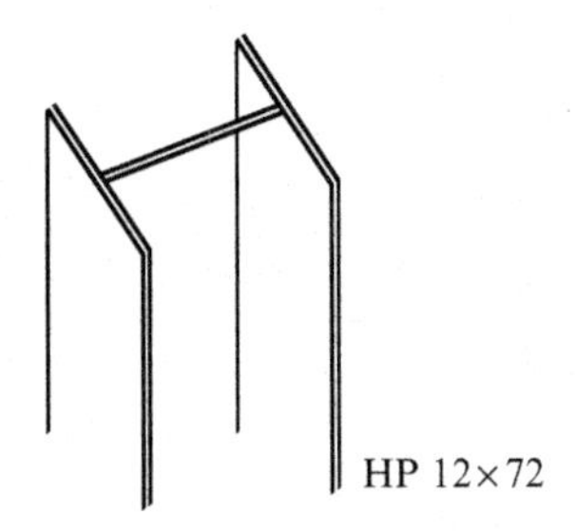

12 is the depth and 72 is the weight of the pile per linear foot given in lbs.

WT sections: MT sections are cut from W (wide flange) sections. Typically, a W section is obtained and one flange is cut off to obtain a T shape.

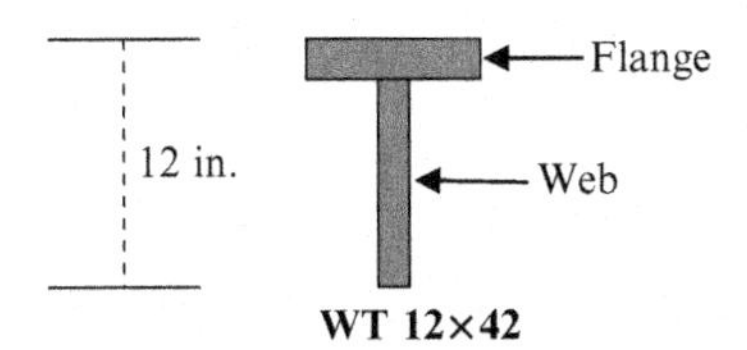

Above 12 is the depth and 42 is the weight per foot in lbs.

MT sections: M section is obtained and one flange is removed to obtain a MT section. Similar to WT sections.

ST sections: S section is obtained and one flange is removed to obtain a ST section. Similar to WT sections.

Practice Problem 12.8

Find the weight of steel in the building shown.

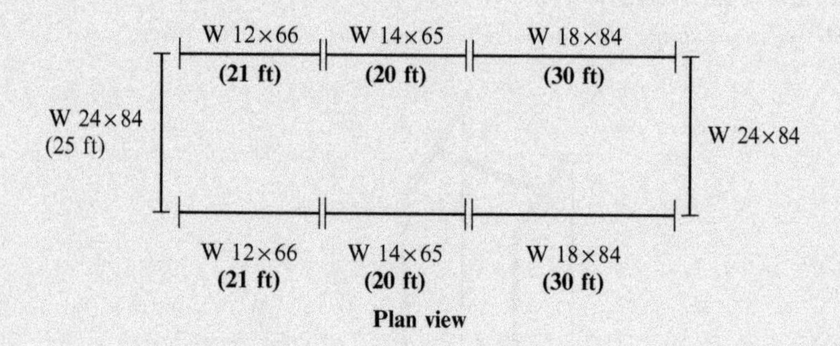

STEP 1: Find the weight of each beam:

$$\text{W } 12 \times 66\,(21\,\text{ft}):\ \text{Weight} = 21 \times 66\,\text{lbs} = 1386\,\text{lbs}$$

$$\text{W } 14 \times 65\,(20\,\text{ft}):\ \text{Weight} = 20 \times 65\,\text{lbs} = 1300\,\text{lbs}$$

$$\text{W } 18 \times 84\,(30\,\text{ft}):\ \text{Weight} = 30 \times 84\,\text{lbs} = 2520\,\text{lbs}$$

$$\text{Subtotal} = 5206\,\text{lbs}$$

Multiply by 2 to get the weight of other side. (This is possible since the structure is symmetrical.)

$$\text{Subtotal} = 10{,}412\,\text{lbs}$$

Two side beams:

$$\text{W } 24 \times 84\,(25\,\text{ft}):\ \text{Weight} = 25 \times 84\,\text{lbs} = 2100\,\text{lbs}$$

$$\text{Subtotal} = 4200\,\text{lbs}$$

$$\text{Total weight of beams} = 10{,}412 + 4200 = 14{,}612\,\text{lbs} = 7.306\,\text{tons}$$

12.5 QUANTITY TAKEOFF—REINFORCEMENT BARS (REBARS)

Concrete itself does not have tensile strength. Tensile strength for concrete is provided by reinforcing bars or simply known as rebars. The table below provides rebar sizes and weight per foot.

Bar no.	Nominal diameter (in.)	Nominal diameter (mm)	Nominal weight (lb/ft)
3	0.375 (3/8)	9.5	0.376
4	0.500 (4/8)	12.7	0.668
5	0.625 (5/8)	15.9	1.043
6	0.750 (6/8)	19.1	1.502
7	0.875 (7/8)	22.2	2.044
8	1.000	25.4	2.670
9	1.128	28.7	3.400
10	1.270	32.3	4.303
11	1.410	35.8	5.313
14	1.693	43	7.650

Practice Problem 12.9

A construction site needs 2000 ft of #3 rebars and 150 ft of #7 rebars. Find the total weight of rebars.

Solution

$$\#3\,\text{rebars}\ (0.376\text{lbs/ft}):\ 2000 \times 0.376 = 752\text{lbs}$$

$$\#7\,\text{rebars}\ (2.044\text{lbs/ft}):\ 150 \times 2.044 = 306.6$$

$$\text{Total weight} = 1058.6\text{lbs}$$

Practice Problem 12.10

A wall measuring 60 ft long 7.5 ft high needs rebars as shown. Find the total rebars needed in tons.

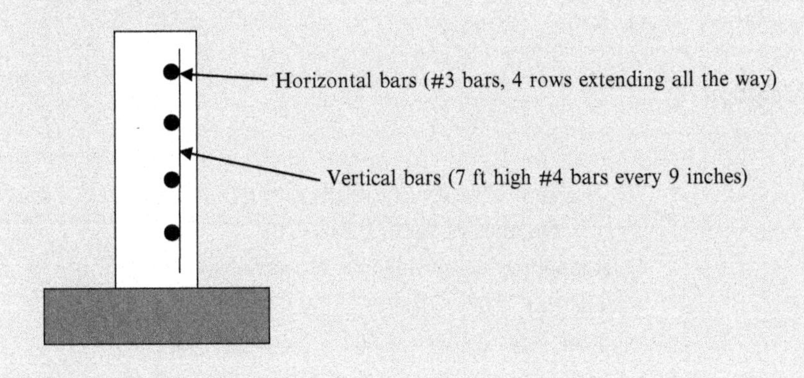

Continued

Solution

Horizontal bars: The wall is 60 ft long. Four rows of horizontal bars needed as shown in the drawing. Weight of #3 bars is 0.376 lbs/ft.

$$4 \times 60 \times 0.376 = 90.24\text{lbs}$$

Vertical bars: The wall is 60 ft long. That is 720 in. Vertical bars are placed every 9 in.

$$\text{Number of vertical bars} = 720/9 + 1 = 81$$

Each vertical bar is 7 ft long.

$$\text{Total length of vertical bars} = 7 \times 81 = 567\text{ft}$$

$$\text{Weight of 567ft of \#4 bars} = 567 \times 0.668 = 378.8\text{lbs}$$

$$\text{Total weight of rebars} = 90.24 + 378.8 = 469\text{lbs} = 0.234\text{tons}$$

Practice Problem 12.11

Find the weight of the steel rebars in the drilled shaft shown. The drilled shaft is 35 ft long and ties are attached every 5 ft. There are 8 ties. The diameter of the shaft is 2 ft

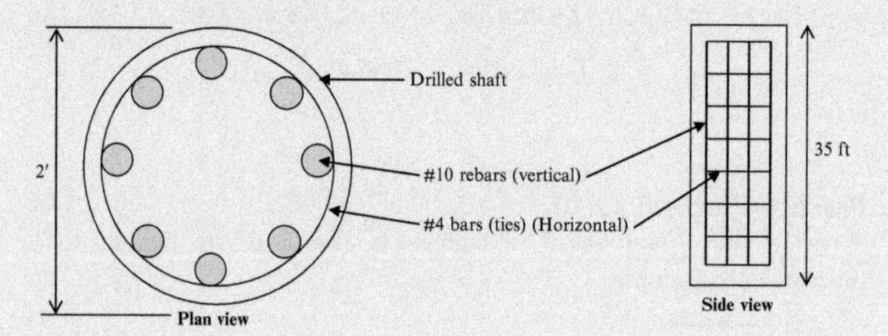

Solution

STEP 1:

$$\text{Length of \#10 vertical bars} = 35\text{ft}$$

$$\text{Number of vertical bars} = 8$$

$$\text{Total length of vertical bars} = 35 \times 8 = 280\text{ft}$$

$$\text{Weight of \#10 vertical bars} = 4.303\text{lbs/ft}$$

$$\text{Weight of vertical bars} = 4.303 \times 280 = 1205\text{lbs}$$

STEP 2:

$$\text{Length of ties} = \text{Circumference of the drilled shaft (approximately)}$$
$$\text{Length of ties} = \pi \times \text{Diameter} = \pi \times 2 = 6.283\,\text{ft}$$
$$\text{Number of ties} = 8$$
$$\text{Total length of ties} = 8 \times 6.283 = 50.3\,\text{ft}$$
$$\text{Weight of ties (\#4 bars)} = 0.668\,\text{lbs/ft}$$
$$\text{Total weight of ties} = 0.668 \times 50.3 = 33.6\,\text{lbs}$$
$$\text{Total weight of rebars} = 1205 + 33.6 = 1238.6\,\text{lbs}$$

12.6 COST ESTIMATING

Cost estimating is a very important function in any project.

The basic steps involved in cost estimating are:

- quantity takeoff (covered in the previous chapter)
- assess the cost of labor, equipment, and material
- assess overhead and profits

The estimator should take into account the location, time of the year, and any other factors that could impact the cost. The cost of a project during the winter could be higher than in the summer. A construction project on top of a hill or in the middle of a major city could also cost more due to efficiency, traffic, and various other factors.

The estimator needs to pay attention to the complexity of a project. For example, the construction of a car dealership may be straightforward compared to a waste water plant. Each company has their own specialty. A company that specializes in building construction may not be able to construct a highway in a cost effective manner. In such situations, it may be profitable to have a subcontractor on board.

The average construction project always starts with site clearing. Then the contractor may decide to build temporary roads for truck traffic and construction vehicles. Almost all construction projects will have some concreting. In today's world concrete is the most common construction material followed by steel. Here are the major subject areas that can be seen in an average building project:

(1) site clearing (includes cutting down trees, removal of debris)
(2) earthwork (grading)
(3) excavation for footings
(4) pile driving (in some building projects, piles may not be necessary)

(5) construction of footings
(6) construction of walls and columns
(7) construction of slabs
(8) construction of upper floors
(9) plumbing (sewer, water, drainage, fire standpipes)
(10) mechanical and electrical work

Labor: The crew hour rate is defined as cost per crew hour. Let's assume the formwork crew consists of 3 carpenters plus 2 laborers. Let's say carpenters earn $50 per hour while laborers earn $30 per hour.

The crew rate would be $(3 \times 50 + 3 \times 30) = \$240$ per hour. In other words if the crew cannot work for some delay, the contractor will lose $240 per hour.

The crew production rate is the output per hour by the crew. The output can be increased by increasing the crew. If the contractor adds more carpenters, he may be able to get more formwork done. On the other hand, the cost of the crew will also increase.

Practice Problem 12.12

Which crew is better?

$$\text{Crew A}: 3\,\text{carpenters} + 3\,\text{laborers}\ \left(\text{Production } 100\,\text{ft}^2 \text{ of formwork per hour}\right)$$
$$\text{Crew B}: 5\,\text{carpenters} + 3\,\text{laborers}\ \left(\text{Production } 125\,\text{ft}^2 \text{ of formwork per hour}\right)$$
$$\text{Carpenter} = \$60\,\text{per hour},\quad \text{Laborer} = \$30\,\text{per hour}$$

Solution

$$\text{Crew A} - \text{Cost per crew hour} = (3 \times 60) + (3 \times 30) = \$270$$
$$\text{Crew A} - \text{Production} = 100\,\text{ft}^2 \text{ per crew hour}$$
$$\text{Cost per ft}^2 = 270/100 = \$2.7\,\text{per ft}^2 \text{ of formwork}$$
$$\text{Crew B} - \text{Cost per crew hour} = (5 \times 60) + (3 \times 30) = \$390$$
$$\text{Crew B} - \text{Production} = 125\,\text{ft}^2 \text{ per crew hour}$$
$$\text{Cost per ft}^2 = 390/125 = \$3.12\,\text{per ft}^2 \text{ of formwork}$$

The contractor has to pay $3.12 for each square feet of formwork if he uses crew B. Obviously, crew A is cheaper than crew B. On the other hand, crew B produces 125 ft^2 per hour compared to crew A, who produce only 100 ft^2 per hour. If the project has to be completed sooner, then crew B should be used even though that crew costs more.

Labor hour (LH): Examples given by NCEES guidebook uses labor hours (LH). Hence, it is important that you should be familiar with LH. Labor hour is obtained by dividing the crew hour by number of workers.

Practice Problem 12.13

A concreting crew consists of 3 masons and 2 laborers.

$$\text{Mason} = \$70\,\text{per Laborer hour} = \$30\,\text{per hour}$$
$$\text{Crew production} = 10\,\text{yd}^3\ \text{per hour}$$

Find crew hour and labor hour (LH).

Solution

$$\text{Crew hour} = (3 \times 70) + (2 \times 30) = \$270$$

Labor hour is obtained by dividing the cost of the crew hour by the total number of workers in the crew. In this case, there are total of 5 workers.

$$\text{Hence cost of labor hour} = \$270/5 = \$54$$

Crew production is given to be 10 yd^3 per hour.

The production per labor hour is obtained by dividing the production per crew hour by number of workers in that crew.

$$\text{Production per labor hour} = 10/5 = 2\,\text{yd}^3\ \text{per labor hour (LH)}$$

Labor rates: Workers need to be paid various insurances beyond the base wages. It is important to understand benefits and fringes involved in payment to workers.

Base pay: Base pay is the starting pay rate for a worker. Base wages are negotiated or agreed upon with the union.

Social security tax: All employers and employees have to pay social security tax by national law. Social security tax goes to a collective fund maintained by the federal government.

Unemployment insurance: When workers are unemployed, they are eligible for unemployment benefits. Unemployment insurance will be maintained by the state.

Workers compensation insurance: Workers are entitled for payment in the case of an injury while at work. Workers compensation insurance covers injury to workers.

General liability insurance: General liability insurance covers any harm or injury to a third party due to the action of the worker. This insurance may cover property damage to public or any other entity.

Fringe benefits: Fringe benefits covers workers health, vacation, and pension plans.

Overtime pay: The normal workday in US is 8 h per day. Typically any hours worked beyond 8 h is paid at 50% extra known as time and a half. Usually workers get paid double time on Sundays and holidays. Some workers get double time on Saturdays while others get time and a half. These rates are dependent upon union agreements in the locality.

Practice Problem 12.14

A mason works 11 h per day from Monday to Friday. He works 7 h on Saturday and 7 h on Sunday. As per union agreement, a worker is entitled to time and a half on any hours beyond 8 h during weekdays. A worker is entitled to time and half on Saturdays and double time on Sundays.

(a) A worker is entitled to how many hours of pay?

Solution

$$\text{Monday to Friday (Hours per day)} = (8\,\text{Standard time} + 3\,\text{Overtime})$$
$$= 8 + (3 \times 1.5) = 12.5\text{h}$$
$$\text{Total hours for Monday to Friday} = 5 \times 12.5 = 62.5\text{h}$$
$$\text{Saturday} = 7 \times 1.5 = 10.5\text{h}$$
$$\text{Sunday} = 7 \times 2 = 14\text{h}$$
$$\text{Total hours} = 62.5 + 10.5 + 14 = 87\text{h}$$

Practice Problem 12.15

A carpenter's base wage is \$45 per hour. Unemployment insurance is 3% of his actual wages. Social security tax is 6% of actual wages. Worker's compensation insurance is 7% of base wages. General liability insurance is 4.5% of base wages. Fringe benefits are \$5.30 per hour. The carpenter works 40 h a week (all standard hours).

(a) What is the weekly cost of the carpenter?

(b) What is the hourly rate of the carpenter?

Solution

STEP 1: Find the base wages per week:

$$\text{Base wages per week} = \$45 \times 40 = \$1800$$

STEP 2: Add insurance and benefits:

$$\text{Unemployment insurance 3\% of actual wages} = 3/100 \times 1800 = \$54$$
$$\text{Social security tax 6\% of actual wages} = 6/100 \times 1800 = \$108$$
$$\text{Worker's compensation insurance 7\% of base wages} = 7/100 \times 1800 = \$126$$
$$\text{General liability insurance 4.5\% of base wages} = 4.5/100 \times 1800 = \$81$$

Fringe benefits are \$5.30 per hour. Carpenter works 40 h a week. Hence fringe benefits are $40 \times 5.30 = \$212$

(a) Total cost per week to hire a carpenter $= 1800 + 54 + 108 + 126 + 81 + 212 + \$2381$

(b) Hourly cost of the carpenter $= \$2381/40 = \$59.525$

Practice Problem 12.16

To meet schedule obligations, a contractor is planning to have the carpenter in Practice Problem 12.15 work an additional 8 h on Saturday. As per union agreement, the contractor is required to pay time and a half for Saturdays.

Data given: The carpenter's base wage is \$45 per hour. Unemployment insurance is 3% of his actual wages. Social security tax is 6% of actual wages. Worker's compensation insurance is 7% of base wages. General liability insurance is 4.5% of base wages. Fringe benefits are \$5.30 per hour. Carpenter works 40 h a week from Monday to Friday and 8 h on Saturday.

(a) What is the new weekly cost of the carpenter?

(b) What is the new hourly rate of the carpenter?

Solution

STEP 1: Find the base wages per week:

The carpenter works total 48 h including the 8 h on Saturday.

$$\text{Base wages per week} = \$45 \times 48 = \$2160$$

STEP 2: Find the actual wages per week without insurance and benefits:

The carpenter works 40 h during the week and 8 h on Saturday. Saturday 8 h has to be paid at time and half.

$$\text{Hours carpenter should get paid} = 40 + 1.5 \times (8) = 52\text{h}$$
$$\text{Actual wages without insurance and benefits} = \$45 \times 52 = \$2340$$

STEP 3: Add insurance and benefits:

Continued

$$\text{Unemployment insurance 3\% of actual wages} = 3/100 \times 2340 = \$70.2$$
$$\text{Social security tax 6\% of actual wages} = 6/100 \times 2340 = \$140.4$$
$$\text{Worker's compensation insurance 7\% of base wages} = 7/100 \times 2160 = \$151.2$$
$$\text{General liability insurance 4.5\% of base wages} = 4.5/100 \times 2160 = \$97.2$$

Fringe benefits are \$5.30 per hour. Carpenter works 48h a week.

Hence fringe benefits are $48 \times 5.30 = \$254.4$

(a) Total cost per week to hire

$$\text{a carpenter} = 2340 + 70.2 + 140.4 + 151.2 + 97.2 + 254.4 = \$3053.4$$

(b) New hourly rate of the carpenter $= \$3053.4/48 = \$63.6$

Note: The carpenter is paid for 52 h but he works only 48 h.

Alternative Solution

This problem can be solved using base rate and actual rate. Following is the solution using rates.

STEP 1:

$$\text{Base rate} = \$45/\text{h}$$

STEP 2: Find the actual rate:

Carpenter works 48 h but is paid for 52 h, since he get paid time and half for Saturday.

$$\text{Actual rate} = \text{Base rate} \times \text{hours paid}/\text{hours worked}$$
$$\text{Actual rate} = 45 \times 52/48 = \$48.75$$

STEP 3: Add insurance and benefits to the actual rate:

$$\text{Hourly rate (actual)} = \$48.75$$
$$\text{Unemployment insurance 3\% of actual rate} = 3/100 \times 48.75 = \$\ 1.4625$$
$$\text{Social security tax 6\% of actual rate} = 6/100 \times 48.75 = \$2.925$$
$$\text{Worker's compensation insurance 7\% of base rate} = 7/100 \times 45 = \$3.15$$
$$\text{General liability insurance 4.5\% of base rate} = 4.5/100 \times 45 = \$2.025$$
$$\text{Fringe benefits} = \$5.30 \text{ per hour}$$
$$\text{Total hourly rate} = 48.75 + 1.4625 + 2.925 + 3.15 + 2.025 + 5.30 = 63.61$$
$$\text{Total weekly wages} = 48 \times 63.61 = 3053$$

12.6.1 Soil Excavation

Soil is a very important construction material. Soil is used for roads, buildings, and retaining walls. Some of the terminology in soil excavation needs to be understood.

12.6.1.1 Bank Volume, Loose Volume, and Compacted Volume

Bank volume: The volume of naturally existing soil (Fig. 12.1)

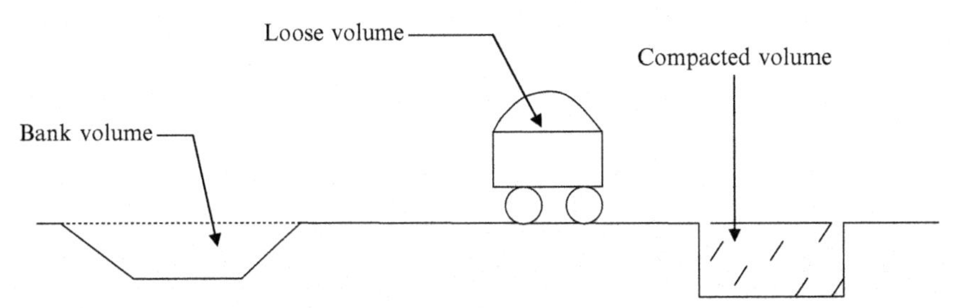

Fig. 12.1 Bank volume, loose volume, and compacted volume.

Loose volume: When natural soil is excavated and loaded into trucks, new volume is known as loose volume.

It is easy to see that loose volume is greater than bank volume.

$$\boxed{\text{Loose volume} = (1 + S_w) \times \text{Bank volume}}$$

S_w = swell factor.

When the soil is compacted, volume is reduced.

$$\boxed{\text{Compact volume} = (1 - S_H) \times \text{Bank volume}}$$

S_H = Shrinkage factor

$$\boxed{\text{Compact volume} = \text{Loose volume} \times (1 - S_H)/(1 + S_w)}$$

Practice Problem 12.17

(a) Derive an equation for compact volume using loose volume.

(b) Ten trucks, each carrying 10 yd^3 of soil, was compacted. Find the bank volume and compact volume.

Swell factor is 20% and shrinkage factor is 15%.

Solution

(a)

$$\text{Loose volume} = (1 + S_w) \times \text{Bank volume} \quad (12.1)$$

$$\text{Compact volume} = (1 - S_H) \times \text{Bank volume} \quad (12.2)$$

From (12.1)

$$\text{Bank volume} = \text{Loose volume}/(1 + S_w)$$

Replace Bank volume in Eq. (12.2).

Continued

$$\text{Compact volume} = (1 - S_H) \times \text{Loose volume}/(1 + S_w)$$
$$= \text{Loose volume} \times (1 - S_H)/(1 + S_w)$$

(b)

$$\text{Loose volume} = 10 \times 10 = 100\,\text{yd}^3$$

$$\text{Loose volume} = (1 + S_w) \times \text{Bank volume}$$

$$100 = (1 + 0.2) \times \text{Bank volume}$$

$$\text{Bank volume} = 83.33\,\text{yd}^3$$

$$\text{Compact volume} = (1 - S_H) \times \text{Bank volume} = (1 - 0.15) \times 83.33 = 71\,\text{yd}^3$$

Installation of underground pipes: The installation of underground sewer pipes, water pipes, electrical duct banks, drain pipes, and cables are a very common construction activity.

Activities involved in installation of an underground pipe:

- excavation for the trench
- stockpiling of excavated material (usually usable soil and un-usable soil are separated).
- shoring of trench sides
- installation of the pipe
- backfill and compaction

Practice Problem 12.18

A contractor is excavating a 5 ft deep, 3 ft wide, and 100 ft long trench to install a 6 in. pipe. The contractor is planning to reuse the excavated soil for backfill. If the swell factor is 15% and shrinkage factor is 12% will he be able to backfill the trench with excavated soil? If not, how many truckloads of soil does he need? Assume each truck carries 3 yd^3 of soil.

Solution

STEP 1:

$$\text{Find the volume of the soil in the trench (Bank volume)} = 5 \times 3 \times 100$$
$$= 1500\,\text{ft}^3$$

STEP 2: What is the volume of this soil after compacted?

$$\text{Compact volume} = (1 - S_H) \times \text{Bank volume} = (1 - 0.12) \times 1500$$
$$= 1320\,\text{ft}^3$$

STEP 3: Find the volume of soil needed (after compacted) (Fig. 12.2).

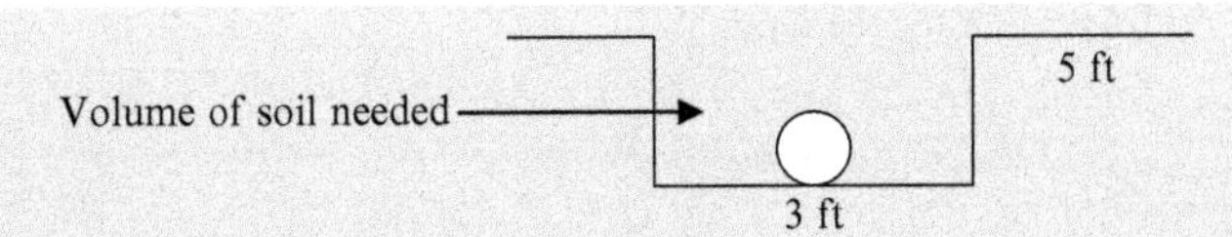

Fig. 12.2 Excavation.

$$\begin{aligned}\text{Volume of compacted soil needed} &= \text{Volume of the trench} \\ &\quad - \text{Volume of the pipe} \\ &= 5 \times 3 \times 100 - \left(\pi \times d^2/4\right) \times 100 \\ &= 1500 - \pi \times 0.5^2 \times 100 = 1480\,\text{ft}^3\end{aligned}$$

From step 2, we know we could get only 1320 ft^3 from the excavated soil.

$$\begin{aligned}\text{Volume of soil contractor needs} &= 1480 - 1320 \\ &= 160\,\text{ft}^3 \quad \text{(after compacted).}\end{aligned}$$

Convert this compacted volume to bank volume.

$$\begin{aligned}\text{Compact volume} &= (1 - S_H) \times \text{Bank volume} \\ 160 &= (1 - 0.12) \times \text{Bank volume} \\ \text{Bank volume} &= 182\,\text{ft}^3\end{aligned}$$

Convert bank volume to lose volume;

$$\begin{aligned}\text{Loose volume} &= (1 + S_w) \times \text{Bank volume} = (1 + 0.15) \times 182 = 209.3\,\text{ft}^3 \\ &= 7.75\,\text{yd}^3\end{aligned}$$

Since each truck carries 3 yd^3, the contractor needs 2.58 trucks.

12.6.2 Estimating Concrete Work

Concrete is the most used building construction material in the world today. Prior to concreting, formwork is erected and rebars are placed. Next concreting is conducted. After the concrete is set, the formwork is removed. Reinforcing bars are placed by ironworkers. Formwork is erected by carpenters and concreting is conducted by concrete masons. The stripping of the formwork is typically done by laborers. In most cases, formwork is reused. During reuse, a certain percentage of formwork could be damaged and has to be discarded. It is fair to assume 10% of formwork will be damaged during each use.

Concreting steps for a footing: In the case of footings, the sides have to be backfilled (Fig. 12.3).

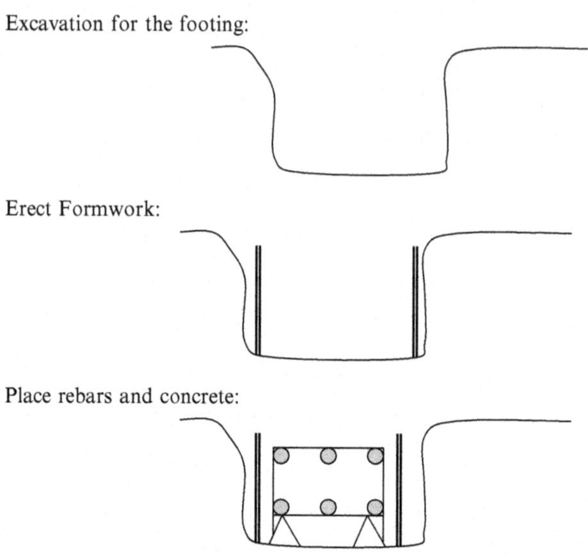

Fig. 12.3 Concreting of a footing.

Practice Problem 12.19

Cost estimating for concreting.

Find the cost involved in concreting the walls and footings of the building shown.

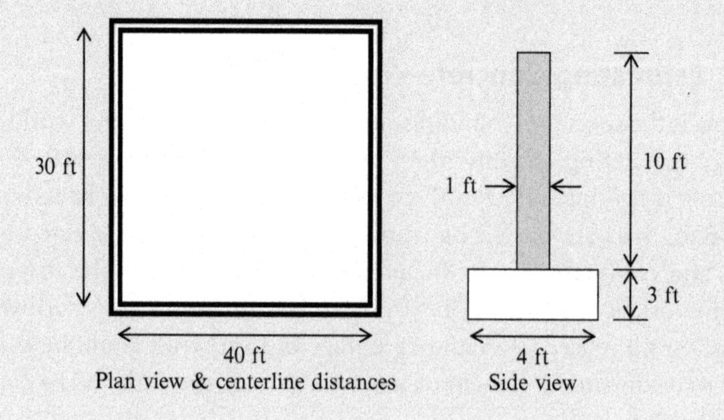

Following information is given:

$$\text{Concreting crew}: \quad 1-\text{Foreman}, \quad 3-\text{Laborers}$$
$$\text{Foreman wages} = \$60/\text{h}, \quad \text{Laborer} = \$30/\text{h}$$

The crew can concrete 30 yd^3 per day (8 h shift)
Find the total cost involved in concreting.

Solution

STEP 1: Find the total quantity of concrete involved.

Quantity of concrete in footings = Perimeter length × footing cross sectional area

$$\text{Length along center lines} = 2 \times (40 + 30) = 140\,\text{ft}$$

$$\text{Footing cross sectional area} = 3 \times 4 = 12\,\text{ft}^2$$

$$\text{Concrete volume in footings} = 140 \times 12 = 1680\,\text{ft}^3 = 1680/27\,\text{yd}^3 = 62\,\text{yd}^3$$

$$\text{Quantity of concrete in walls} = \text{Perimeter length} \times \text{wall cross sectional area}$$

$$\text{Concrete volume in walls} = 2 \times (40 + 30) \times 10 \times 1 = 1400\,\text{ft}^3 = 1400/27\,\text{yd}^3$$
$$= 52\,\text{yd}^3$$

$$\text{Total concrete volume} = 52 + 62 = 114\,\text{yd}^3$$

STEP 2: Find the crew rate

$$\text{The cost for crew hour} = (1 \times 60) + (3 \times 30) = \$150$$

The crew consists of 1 supervisor and 3 laborers.

STEP 3:

$$\text{Number of days required} = \text{Concrete volume/Concrete production}$$
$$\text{per day} = 114/30 = 3.8\,\text{days}$$

Concrete production is given to be 30 yd^3 per day.

$$\text{Number of hours} = 3.8 \times 8 = 30.4\,\text{h}$$

Each hour costs \$150 for the crew (see step 2).

$$\text{Total cost} = 30.4 \times 150 = \$4560$$

Please note that in this example the cost, cost of forming, overhead, and equipment costs were neglected to keep the example simple.

Practice Problem 12.20

Cost estimating for formwork:

Find the cost for formwork required for walls and footings.

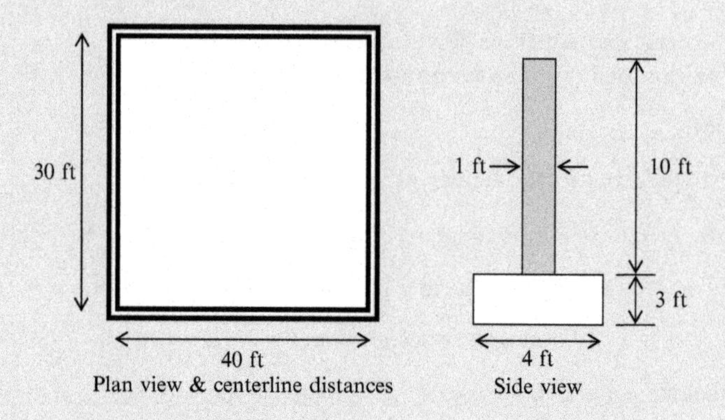

Following information is provided.

$$\text{Crew required for formwork}: \quad 2\,\text{Carpenters}, \quad 3\,\text{laborers}$$

$$\text{Carpenter wages} = \$50/\text{h}, \quad \text{Laborer} = \$30/\text{h}$$

The crew can form 60 ft^2 per hour.

Find the total cost involved in formwork.

Solution

STEP 1: Find the total quantity of formwork involved.

Quantity of formwork in footings = Perimeter length × footing side area

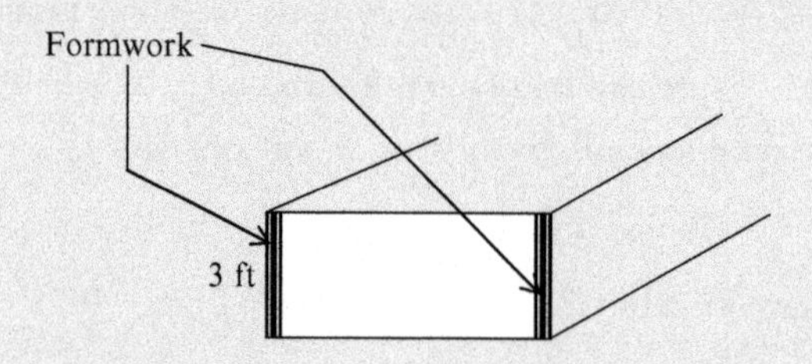

$$\text{Centerline length} = 2 \times (40 + 30) = 140\,\text{ft}$$

$$\text{Depth} = 3\,\text{ft}$$

$$\text{Formwork required (one side)} = 140 \times 3 = 420\,\text{ft}^2$$

$$\text{Formwork required (both sides)} = 840\,\text{ft}^2$$

Quantity of formwork in walls = Perimeter length × wall side area

$$\text{Quantity of formwork in walls (one side)} = 2 \times (40 + 30) \times 10 = 1400\,\text{ft}^2$$

$$\text{Quantity of formwork in walls (both sides)} = 2 \times 1400 = 2800\,\text{ft}^2$$

$$\text{Total formwork for walls and footings} = 840 + 2800 = 3640\,\text{ft}^2$$

STEP 2: Find the crew rate

$$\text{The cost for the crew per hour (2 carpenters + 3 laborers)} = (2 \times 50) + (3 \times 30) = \$190$$

Production rate of the crew is given to be 60 ft^2 per hour.

$$\text{Number of hours required} = \text{Formwork area/Formwork production per hour} = 3640/60 = 61\,\text{h}$$

$$\text{Total cost} = 61 \times 190 = \$11,590$$

Practice Problem 12.21

Concreting:

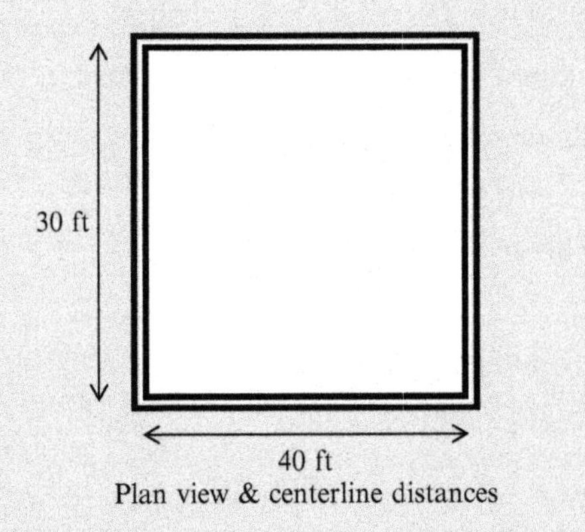

Plan view & centerline distances

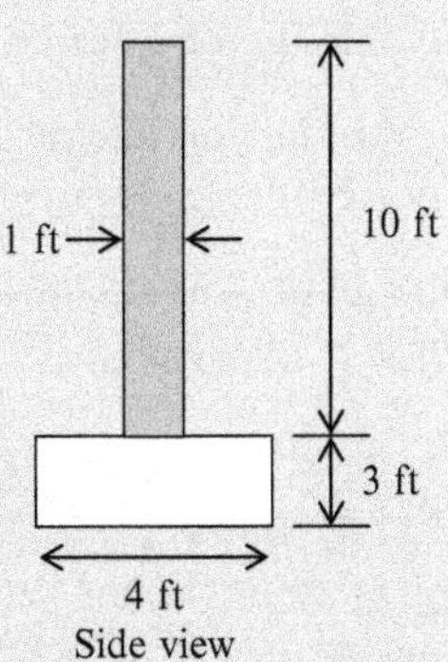

Side view

Following information is provided.

$$\text{Crew required for concreting:}\quad 2\,\text{Masons},\ 3\,\text{laborers}$$

$$\text{Mason wages} = \$55/\text{h},\quad \text{Laborer} = \$30/\text{h}$$

The crew can concrete 9 yd^3 per 8 h shift

$$\text{Concrete material cost} = \$110\,\text{per yd}^3\ (\text{assume } 10\%\ \text{waste})$$

Find the total cost involved in concreting.

Solution

Volume of concrete for footings:

Continued

$$= \text{Centerline length of the footing} \times \text{footing cross section}$$
$$= 2 \times (40 + 30) \times (4 \times 3) = 1680\,\text{ft}^3$$
$$= 1680/27\,\text{yd}^3 = 62.2\,\text{yd}^3$$

Volume of concrete for walls:

$$\text{Perimeter length of the wall} \times \text{Wall cross section}$$
$$= 2 \times (40 + 30) \times (1 \times 10) = 1400\,\text{ft}^3$$
$$= 1400/27\,\text{yd}^3 = 51.9\,\text{yd}^3$$

$$\text{Total concrete volume} = 62.2 + 51.9 = 114.1\,\text{yd}^3$$

STEP 2: Find the crew rate

$$\text{The cost for the crew per hour (2 masons + 3 laborers)}$$
$$= (2 \times 55) + (3 \times 30) = \$200$$
$$\text{Crew production rate per 8 h} = 9\,\text{yd}^3$$
$$\text{Crew production per hour} = 9/8 = 1.125\,\text{yd}^3 \text{ per hour}$$
$$\text{Number of hours required} = \text{Concrete volume}/\text{Concrete}$$
$$\text{production rate per hour} = 114.1/1.125 = 101.4\,\text{h}$$
$$\text{Total cost} = 101.4 \times 200 = \$20,280$$

\$200 is the crew rate per hour.
STEP 3: Material cost:

$$\text{Concrete volume required} = 114.1$$
$$\text{Add 10\% for waste} = 114.1 + (10\% \times 114.1) = 125.5\,\text{yd}^3$$
$$\text{Material cost} = 125.5 \times 110 = \$13,805$$
$$\text{Total cost} = \text{Labor} + \text{Material} = 20,280 + 13,805 = \$34,085$$

Labor hour method: The previous examples were done using crew hour method. Some estimators perform calculations using labor hours. It is prudent to know both methods since it is possible that exam questions could be in labor hours.

Labor hour rate = Crew hour rate/Number of workers in the crew.

Practice Problem 12.22

Steel erection crew consists of one supervisor, four steel workers, and one laborer.

Find the crew hour rate and labor hour rate.

Following hourly rates are available.

Supervisor—$60, Steel worker—$50, Laborer—$35

Solution

$$\text{Crew hour rate} = (1 \times 60) + (4 \times 50) + (1 \times 35) = \$295$$

$$\text{Labor hour rate} = \text{Crew hour rate}/\text{Number of workers in the crew}$$

$$\text{Labor hour rate} = 295/6 = \$49.2$$

Practice Problem 12.23

Formwork erection is done by one supervisor and three carpenters.

Supervisor—$60, Carpenter—$45

Productivity of formwork erection is 7.2 ft^2/LH (LH = labor hour)

Find the cost of erection of formwork if the project consists of 1000 ft^2 of formwork.

Solution

$$\text{Crew hour rate} = (1 \times 60) + (3 \times 45) = \$195$$

$$\text{Labor hour rate} = \text{Crew hour rate}/\text{Number of workers in the crew}$$

$$\text{Labor hour rate} = 195/4 = \$48.75$$

$$\text{Total sq. ft of the project} = 1000\,\text{ft}^2$$

$$\text{Productivity of formwork erection is } 7.2\,\text{ft}^2/\text{LH}$$

$$\text{Labor hours required} = 1000/7.2 = 138.9$$

$$\text{Total cost} = 138.9 \times 48.75 = \$6770$$

Formwork reuse: In most cases, the contractor will not buy formwork for the whole project. He will buy only a certain percentage of formwork and reuse them. If the project requires 10,000 ft^2 of formwork, the contractor may decide to purchase 3000 ft^2 of formwork and reuse them. Typically, 10% of the formwork may not be reusable due to damage.

Practice Problem 12.24

A project requires to complete 5000 ft^2 of formwork. The contractor purchases 2500 ft^2 of formwork at a price of $3 per ft^2 and plans to reuse them. The contractor believes he may have to replace 10% of formwork due to damage. Find the cost of the formwork for the project.

Solution

$$\text{Cost for purchasing } 2500\,\text{ft}^2 \text{ of formwork} = 3 \times 2500 = \$7500$$

$$\text{Loss of formwork due to damage} = 10\% \times 2500 = 250\,\text{ft}^2$$

$$\text{Cost of replacement of damaged formwork} = 3 \times 250 = \$750$$

$$\text{Total cost of formwork} = 7500 + 750 = \$8250$$

Practice Problem 12.25

A wall measuring 12 ft high, 200 ft long will be concreted in three pours. The contractor plans to reuse the formwork. The initial cost of formwork is $5.10 per ft^2. Subsequently, the contractor will purchase formwork only to replace the damaged formwork. The contractor believes 10% of the formwork will be damaged during every use.

Estimate the cost of the formwork.

Solution

STEP 1: Cost of formwork for pour 1:

Since the 12 ft high wall is concreted in three pours, each pour is 4 ft high.

Formwork for the first pour:

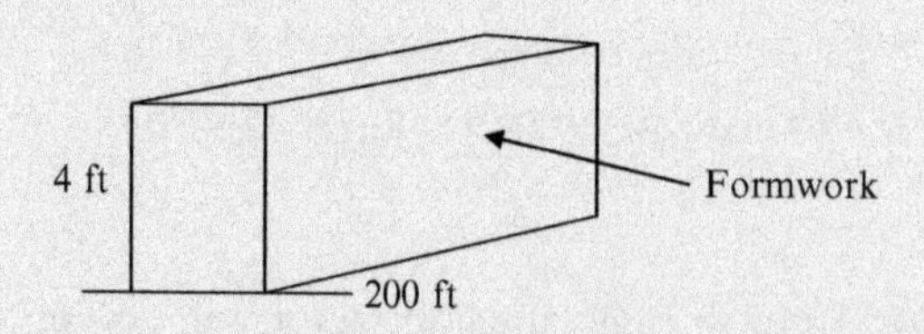

$$\text{Formwork for the first pour (per side)} = 4 \times 200 = 800\,\text{ft}^2$$

$$\text{Formwork for the first pour (both sides)} = 1600\,\text{ft}^2$$

$$\text{Cost of formwork for the first pour} = 5.10 \times 1600 = \$8160$$

Formwork for the second pour:

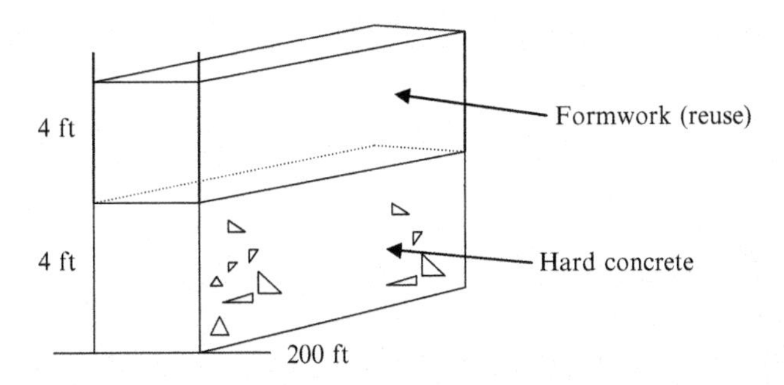

$$\text{Formwork for the second pour (both sides)} = 2 \times 4 \times 200 = 1600\,\text{ft}^2$$

Formwork from the first pour will be reused. However, 10% of the formwork will be damaged. Hence, contractor has to buy 160 ft^2 of formwork.

$$\text{Formwork for the second pour (both sides)} = 160\,\text{ft}^2$$

$$\text{Additional cost of formwork for the second pour} = 5.10 \times 160 = \$816$$

Formwork for the third pour:

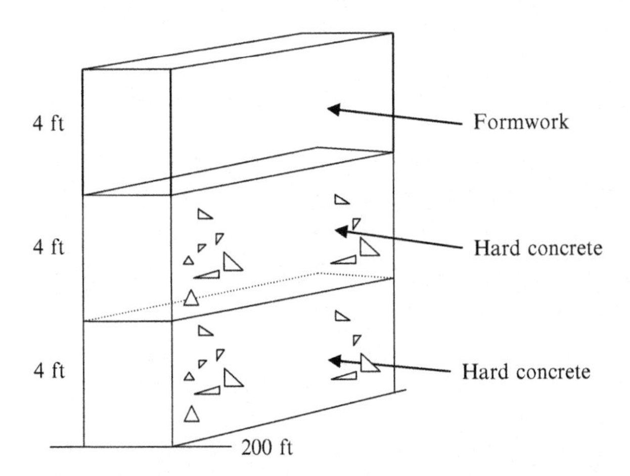

$$\text{Formwork for the third pour (both sides)} = 2 \times 4 \times 200 = 1600\,\text{ft}^2$$

Formwork from the second pour will be reused. However, 10% of the formwork will be damaged. Hence, contractor has to buy another 160 ft^2 of formwork.

$$\text{Formwork for the third pour (both sides)} = 160\,\text{ft}^2$$

$$\text{Additional cost of formwork for the third pour} = 5.10 \times 160 = \$816$$

$$\text{Total cost of formwork} = \text{Cost for the first pour} + \text{Cost for the second pour}$$

$$+\text{Cost for the third pour} = 8160 + 816 + 816 = \$9792$$

Alternative method of formwork reuse cost computation (unit cost method): A contractor may compute the formwork cost using unit costs. The unit cost of formwork for the first pour is the purchasing cost of the formwork. The unit cost for the second pour is much smaller since he is planning to reuse formwork from first pour. Hence, based on experience, he may assign a unit cost for formwork for the second pour and third pour, etc. This can be demonstrated using an example.

Practice Problem 12.26

Contractor is planning to concrete a 12 ft high 200 ft long wall in three pours.

The purchasing cost of the formwork is \$5.10 per ft^2. The contractor is planning to reuse the formwork.

The unit cost of the formwork for the second and third pours is 10% of the initial cost.

Solution

STEP 1: Cost of formwork for the first pour $= 2 \times 4 \times 200 \times 5.10 = \$8160$

Note that the formwork unit cost for the first pour is 5.10 per ft^2 and the unit cost for the second and third pours are 0.51 per ft^2.

STEP 2: Cost of formwork for the second pour $= 2 \times 4 \times 200 \times 0.51 = \$816$

STEP 3: Cost of formwork for the third pour $= 2 \times 4 \times 200 \times 0.51 = \$816$

Total cost of formwork $= 8160 + 816 + 816 = \$9792$

12.6.3 Estimating Formwork Cost

Forms are needed to hold the wet concrete in place until it hardens. The formwork cost in concrete work is significant. Here are some of the different timber sizes used in practice to build forms (Fig. 12.4):

It is important that you understand all the elements involved in formwork and the function of each element.

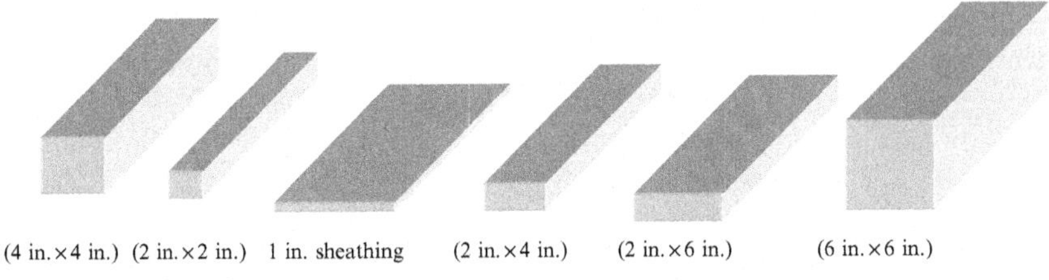

Fig. 12.4 Timber shapes.

Measure of timber (fbm): Timber quantities are measured using fbm (footboard measure). Footboard measure is also known as board feet (BF). Some abbreviate it as fbm and some other authors abbreviate it as BF.

$$\boxed{1\,\text{fbm (or BF)} = 144\,\text{in.}^3}$$

To find the fbm of a timber multiply length, width, and depth using inches and then divide by 144.

Practice Problem 12.27

Find the fbm of a 10 ft long, 2 in. × 4 in. timber.

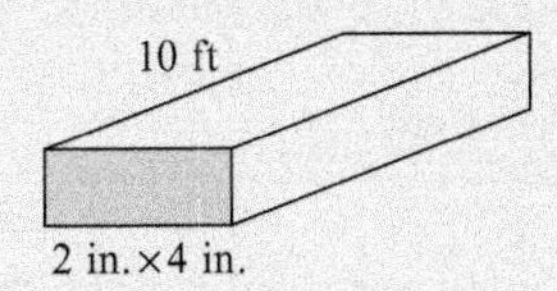

Solution

To obtain fbm multiply length, width, and depth using inches and divide by 144.

$$(10 \times 12) \times 2 \times 4/144 = 6.67\,\text{fbm}$$

Practice Problem 12.28

Find the fbm of 9 pieces of 10 ft long, 2 in. × 4 in. timber and 13 pieces of 20 ft long, 2 in. × 2 in. timber.

Solution

Multiply length, width, and depth using inches and divide by 144.

$$9 \times (10 \times 12) \times (2 \times 4)/144 = 60\,\text{fbm}$$
$$13 \times (20 \times 12) \times (2 \times 2)/144 = 86.7\,\text{fbm}$$
$$\text{Total} = 146.7\,\text{fbm}$$

Formwork for columns: The formwork for the columns are built using 1 in sheathing or ¾ in. plywood. The typical arrangement is shown below.

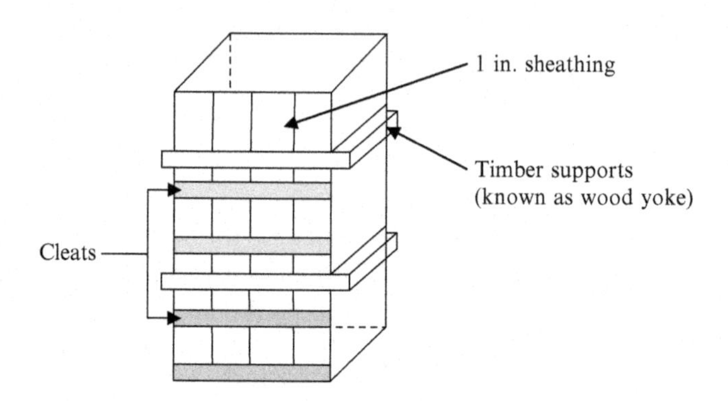

Elements in a column formwork:

1 in sheathing: 1 in. sheathing comes in vertical strips as shown.

Cleats: Cleats are nailed to the sheathing. Typically 1 in. thick and 4 in. wide timber is used for cleats.

Timber supports (yoke): Yoke holds the structure together. Yoke is built using 4 in. × 4 in. timber. Typically, timber supports are held together with bolts as shown below.

Column configuration (plan view):

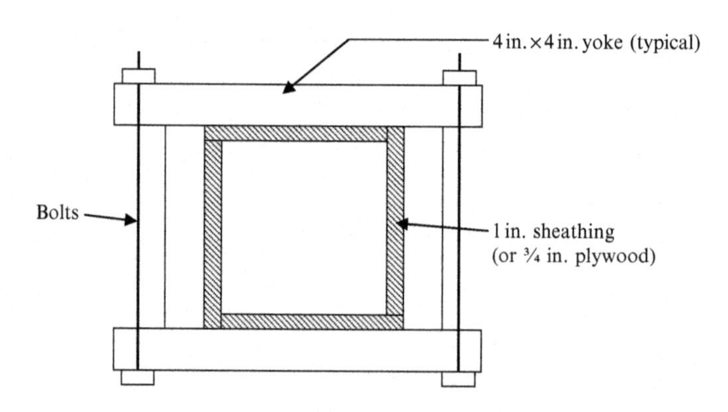

Computation of material for a column formwork: You need to compute the square footage of 1 in. sheathing, and the quantity of timber required for cleats and wood yokes. The labor required for construction has to be computed based on labor wage rates and production rate.

Practice Problem 12.29

Compute the quantity of timber required for formwork for a 2 ft × 2 ft column with a height of 10 ft. The following information is given:

Formwork is prepared using 1 in. sheathing.
1 in. × 6 in. wood cleats are provided every 18 in.
4 in. × 4 in. wood yokes are provided every 2 ft (six)
Wood yokes are held together with ½ in. diameter bolts.

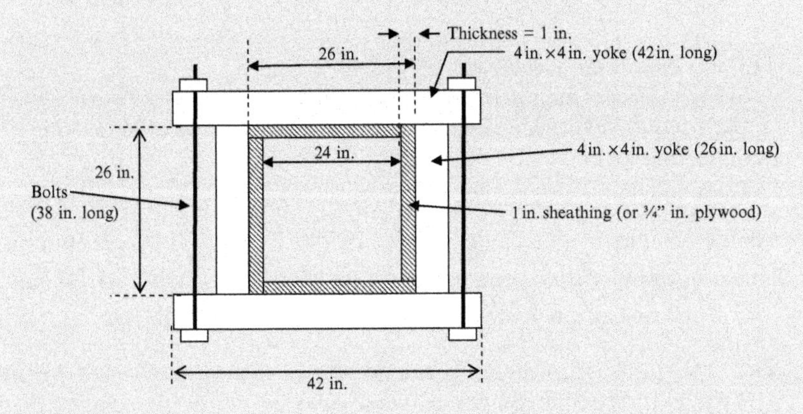

Solution

STEP 1: Compute the 1 in. sheathing required:

1 in. sheathings have to be 25 in. wide (not 24 in.).

$$\text{Required 1 in. sheathing per side} = 25\,\text{in.} \times 10\,\text{ft} = 25 \times 120 \times 1/144\,\text{fbm}$$
$$= 20.83\ \text{fbm}$$

$$\text{Required 1 in. sheathing for four sides} = 4 \times 20.83 = 83.3\,\text{fbm}$$

STEP 2: *Compute the quantity of wood cleats:*

$$\text{Height of the column} = 10\,\text{ft} = 120\,\text{in.}$$

Cleats are provided every 18 in.

$$\text{Number of wood cleats} = 120/18 = 6.67 = 7\,\text{wood cleats per side.}$$

The total column requires 28 wood cleats. Each cleat is 1 in. thick, 6 in. wide, and 26 in. long. (The outside width of the formwork considering the thickness of the 1 in. sheathing.)

Continued

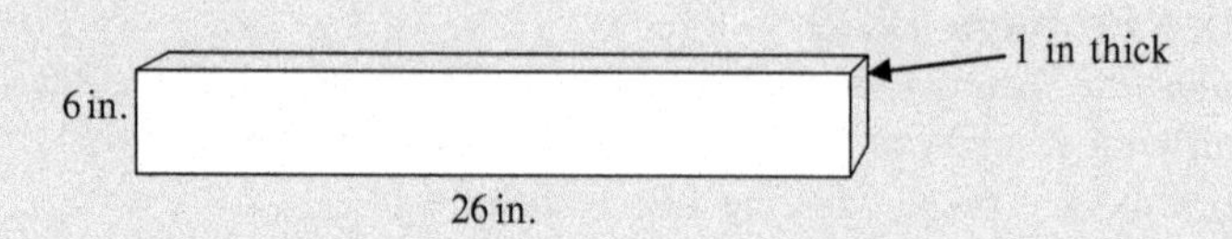

$$\text{Quantity of required wood cleats} = 28 \times (1 \times 6 \times 26)/144\text{fbm}$$
$$= 30.3\text{fbm}$$

STEP 3: *Compute the quantity of wood yokes:*

Six wood yokes are provided for the column.

Each wood yoke needs two 42 in. long pieces and two 26 in. long pieces (see the figure).

Required 42 in. long pieces for six wood yokes = 12

$$\text{Timber quantity of 42in. long pieces} = 12 \times 42 \times (4 \times 4)/144 = 56\text{fbm}$$
$$\text{Timber quantity of 26in. long pieces} = 12 \times 26 \times (4 \times 4)/144 = 34.7\text{fbm}$$
$$\text{Total timber required} = 83.3 + 30.3 + 56 + 34.7 = 204.3\text{fbm}$$

STEP 4: *Compute the quantity of bolts required:*

Two bolts are required per one wood yoke.

Total bolts required are 12, since there are six wood yokes.

Each bolt is 38 in. long.

$$\text{Total length of bolts} = 12 \times 38\text{in.} = 38\text{ft}$$

Practice Problem 12.30

Compute the cost of labor required for the above problem. Following information is given.

$$\text{Carpenter wages: \$70 per hour,} \quad \text{Helper wages: \$40 per hour}$$

$$\text{Saw cutting cleats and wood yokes per } 100\text{ft}^2 \text{ of contact area}$$
$$= (0.5 \text{ carpenter hours} + 2.0 \text{ helper hours})$$
$$\text{Nailing wood cleats per } 100\text{ft}^2 \text{ of contact area}$$
$$= (0.5 \text{ carpenter hours} + 0.5 \text{ helper hours})$$
$$\text{Erecting of wood york per } 100\text{ft}^2 \text{ of contact area}$$
$$= (1.0 \text{ carpenter hours} + 1.0 \text{ helper hours})$$

Solution

STEP 1: Contact area means the contact area between column forms and concrete.

In this case, it is 2 ft × 10 ft per side. There are four sides.

$$\text{Hence contact area} = 20 \times 4 = 80\,\text{ft}^2$$

STEP 2: *Saw cutting of cleats and wood yokes:*

$$\text{Labor cost for } 100\,\text{ft}^2 \text{ of contact area} = (0.5 \times 70 + 2.0 \times 40) = \$115$$
$$(0.5\,\text{carpenter hours} + 2\,\text{helper hours})$$
$$\text{Labor cost for } 80\,\text{ft}^2 \text{ of contact area} = \$115/100 \times 80 = \$92$$

STEP 3: *Nailing wood cleats:*

$$\text{Labor cost for } 100\,\text{ft}^2 \text{ of contact area} = (0.5 \times 70 + 0.5 \times 40) = \$55$$
$$(0.5\,\text{carpenter hours} + 2\,\text{helper hours})$$
$$\text{Labor cost for } 80\,\text{ft}^2 \text{ of contact area} = \$55/100 \times 80 = \$44$$

STEP 4: *Erecting wood yokes:*

$$\text{Labor cost for } 100\,\text{ft}^2 \text{ of contact area} = (1.0 \times 70 + 1.0 \times 40) = \$110$$
$$(1.0\,\text{carpenter hours} + 1.0\,\text{helper hours})$$
$$\text{Labor cost for } 80\,\text{ft}^2 \text{ of contact area} = \$110/100 \times 80 = \$88$$
$$\text{Total cost of labor for the column form} = 92 + 44 + 88 = \$224$$

Elements in a wall formwork:

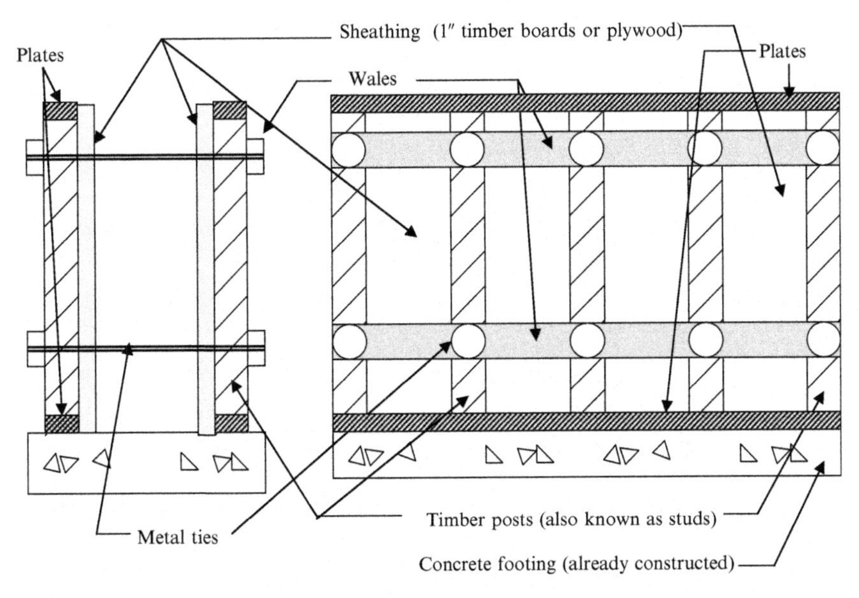

Practice Problem 12.31

Find the quantity of timber required to construct formwork for a 70 ft long and 8 ft high wall. Assume the following. (See the section under "Wall Form General Configuration" for in length description.)

Sheathing: 1 in. thick

Studs: Use 8 ft high studs. Ignore the thickness of timber plates. (2 × 4) timber posts placed every 2 ft

Wales: (2 × 6) timber is used. Two wales per side are used.

Plates: (2 × 4) plates are used top and bottom of studs

Braces: 11 ft long 2 × 6 braces are provide every 10 ft on either side as shown below.

Wastage: 10%

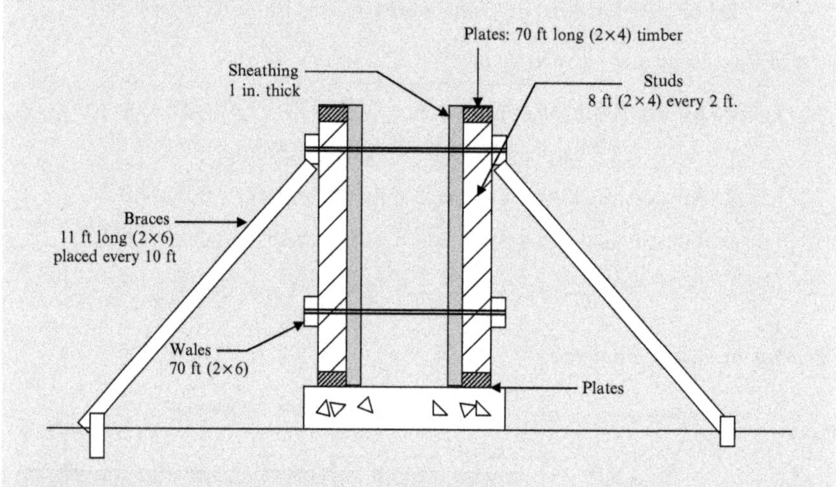

Solution

STEP 1: Sheathing: Length = 70 ft, Height = 8 ft, thickness = 1 in.

$$(70 \times 12) \times (8 \times 12) \times 1/144 = 560\text{fbm for one side}$$

STEP 2: Studs: Studs are used every 2 ft. Hence, (35 + 1) studs are needed for one side.

$$36 \times (8 \times 12) \times (2 \times 4)/144 = 192\text{fbm per one side.}$$

STEP 3: Wales; Length of wales = 70 ft. Two wales are needed for one side.

$$2 \times (70 \times 12) \times (2 \times 6)/144 = 140\text{fbm per one side.}$$

STEP 4: Plates: Length of plates = 70 ft. Two plates are needed per side (top and bottom).

$$2 \times (70 \times 12) \times (2 \times 4)/144 = 93.3\text{fbm per side}$$

STEP 5: Braces: Braces are provide every 10 ft. Hence, (7 + 1) braces are needed per side.

$$8 \times (11 \times 12) \times (2 \times 6)/144 = 88\text{fbm per one side.}$$

$$\text{Total timber quantity per side} = 560 + 192 + 140 + 93.33 + 88 = 1.073.3\text{fbm}$$

$$\text{Total timber quantity needed for both sides} = 2146.6\text{fbm}$$

$$10\% \text{ waste of material anticipated.}$$

$$\text{Hence total timber required} = 1.1 \times 2146.6 = 2361.6\text{fbm}$$

Labor costs for formwork: Formwork is done by carpenters and helpers. Typically, helpers will be moving wood, hoisting and saw cutting pieces. Carpenters will be measuring timber and erecting the formwork.

Practice Problem 12.32

12 columns each with height 11 ft has to be constructed.

Following labor rates are known:

$$\text{Carpenter} = \$70/\text{h; Iron worker} = \$60/\text{h; Concrete mason} = \$70/\text{h}$$
$$\text{Laborer} = \$40/\text{h}$$
$$\text{Formwork crew} = 2\,\text{carpenter} + 3\,\text{laborers}$$
$$\text{Rebar crew} = 2\,\text{iron workers} + 1\,\text{laborer}$$
$$\text{Concreting crew} = 2\,\text{concrete masons} + 5\,\text{laborers}$$

Productivity:

$$\text{Formwork} = 6\text{ft}^2/\text{LH}$$
$$\text{Rebar placement} = 100\text{lbs}/\text{LH}$$
$$\text{Concreting} = 0.5\text{yd}^3/\text{LH}$$

Note: LH = labor hour.

Material cost:

$$\text{Formwork for 4 columns (initial erection)} = \$2.1/\text{ft}^2$$
$$\text{Formwork reuse cost} = \$0.3/\text{ft}^2$$
$$\text{Concrete} = \$120/\text{yd}^3$$
$$\text{Rebar cost} = 0.50/\text{lb}$$

Find the cost of the project.

Continued

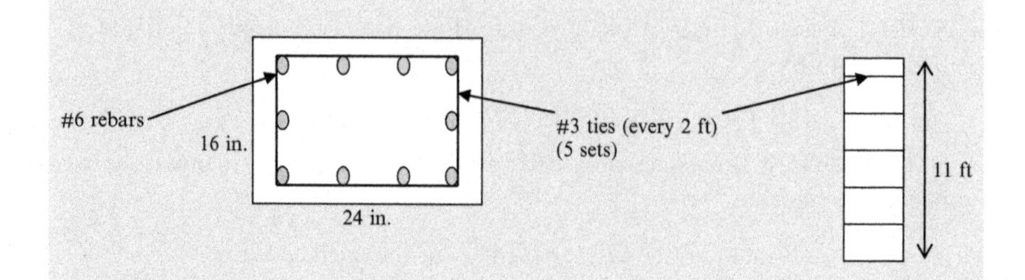

Solution

STEP 1: Find quantities for concrete, rebar, and formwork:

Find the quantity of #6 bars:

$$\text{Number of } \#6\,\text{bars per column} = 10$$

$$\text{Length of each bar} = 11\,\text{ft}$$

$$\text{Total length of } \#6\,\text{bars per column} = 110\,\text{ft}$$

$$\text{Total length of } \#6\,\text{bars for } 12\,\text{columns} = 110 \times 12 = 1320\,\text{ft}$$

$$\text{Total weight of } \#6\,\text{bars} = 1320 \times 1.502\,\text{lbs} = 1983\,\text{lbs}$$

STEP 2: Find the quantity of #3 ties:

$$\text{Length of one } \#3\,\text{tie bar} = 24 + 24 + 16 + 16 = 80\,\text{in. (approximately)}$$

$$\text{Number of tie bars per column} = 5$$

$$\text{Length of } \#3\,\text{bars per column} = 5 \times 80/12 = 33.33\,\text{ft}$$

$$\text{Total length of } \#3\,\text{bars for } 12\,\text{columns} = 33.33 \times 12 = 400\,\text{ft}$$

$$\text{Total weight of } \#3\,\text{bars} = 400 \times 0.376\,\text{lbs} = 150\,\text{lbs}$$

$$\text{Total weight of rebars} = 1983 + 150 = 2133\,\text{lbs}$$

STEP 3: Material cost of rebars $= 2133 \times 0.5 = \$1067$

STEP 4: Formwork (material cost):

$$\text{Formwork per column} = \text{Perimeter} \times \text{Height} = (24 + 24 + 16 + 16)/12 \times 11 = 73.33\,\text{ft}^2$$

$$\text{Formwork quantity for } 12\,\text{columns} = 12 \times 73.33 = 880\,\text{ft}^2$$

$$\text{Formwork for first four columns bought at } \$2.1/\text{ft}^2 = 4 \times 73.33 \times 2.1 = \$616$$

$$\text{Formwork reuse cost} = 0.3/\text{ft}^2$$

$$\text{Area of formwork for remaining } 8\,\text{columns} = 8 \times 73.33 = 586.6\,\text{ft}^2$$

$$\text{Cost of formwork for remaining } 8\,\text{columns} = 0.3 \times 586.6 = \$176$$

$$\text{Total material cost of formwork} = 616 + 176 = \$792$$

STEP 5: Concrete (material cost):

$$\text{Volume of concrete per column} = (16 \times 24)/144 \times 11 = 29.33\,\text{ft}^3 = 1.09\,\text{yd}^3$$

$$\text{Concrete volume for 12 columns} = 12 \times 1.09 = 13.1\,\text{yd}^3$$

$$\text{Cost of concrete for 12 columns} = 12 \times 1.09 \times 120 = \$1570$$

STEP 6: Total material cost $= 1067 + 616 + 176 + 1570 = \$3429$

STEP 7: Labor (formwork):

$$\text{Formwork crew} = 2\ \text{carpenter} + 3\ \text{laborers}$$

$$\text{Cost of crew hour} = (2 \times 70) + (3 \times 40) = 260$$

$$\text{Labor hour} = 260/\text{number of workers} = 260/5 = \$52$$

$$\text{Productivity for Formwork} = 6\,\text{ft}^2/\text{LH}$$

$$\text{Total formwork quantity} = 880\,\text{ft}^2\ (\text{see step 4})$$

$$\text{Required labor hours} = 880/6 = 147$$

$$\text{Cost of labor for formwork} = 147 \times 52 = \$7644$$

STEP 8: Labor (rebars)

$$\text{Rebar crew} = 2\ \text{iron workers} + 1\ \text{laborer}$$

$$\text{Cost of crew hour} = (2 \times 60) + (1 \times 40) = 160$$

$$\text{Labor hour} = 160/\text{number of workers} = 160/3 = \$53.3$$

$$\text{Productivity for rebars} = 100\,\text{lbs}/\text{LH}$$

$$\text{Total rebar quantity} = 2133\,\text{lbs}\ (\text{see step 2})$$

$$\text{Required labor hours} = 2133/100 = 21.3$$

$$\text{Cost of labor for formwork} = 21.3 \times 53.3 = \$1135$$

STEP 9: Labor (concreting)

$$\text{Concreting crew} = 2\ \text{concrete masons} + 5\ \text{laborers}$$

$$\text{Cost of crew hour} = (2 \times 70) + (5 \times 40) = 340$$

$$\text{Labor hour} = 340/\text{number of workers} = 340/7 = \$48.6$$

$$\text{Productivity for concreting} = 0.5\,\text{yd}^3/\text{LH}$$

$$\text{Total concrete quantity } 13.1\,\text{yd}^3\ (\text{see step 5})$$

$$\text{Required labor hours} = 13.1/0.5 = 26.2$$

$$\text{Cost of labor for concreting} = 26.2 \times 48.6 = \$1273$$

$$\text{Total labor cost} = 7644 + 1135 + 1273 = \$10,052$$

$$\text{Total material cost} = \$3429\ (\text{see step 6})$$

$$\text{Total project cost} = \$13,481$$

Concreting of an upper slab:

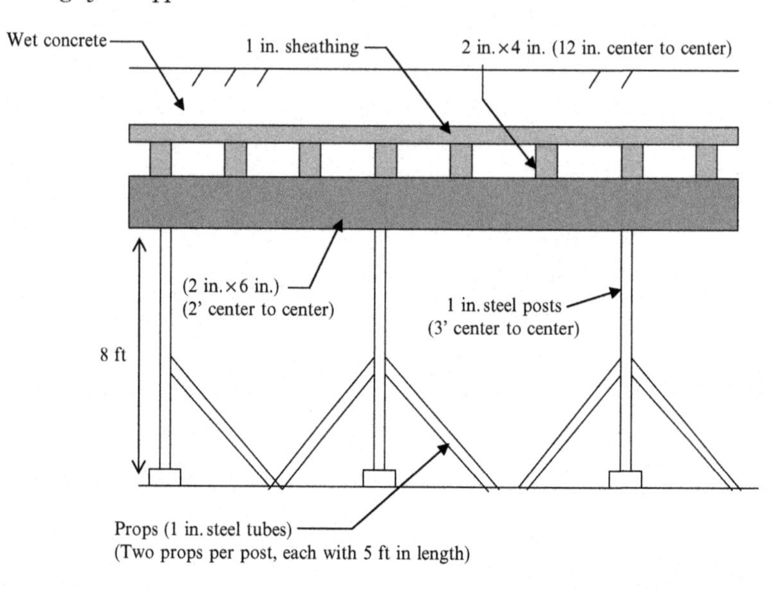

Practice Problem 12.33

50 ft by 50 ft slab has to be concreted. Formwork and shoring is as shown above. (Please see the sketch and the photograph above.)

(a) Compute the total fbm of timber

(b) Find the total length of steel posts

(c) Find the total length of steel props

$$\text{Labor costs: Mason} = \$65/\text{h}, \quad \text{Laborer} = \$40/\text{h}$$

$$\text{Crew: } 2\,\text{masons} + 4\,\text{laborers}$$

Productivity:

$$1\,\text{in. sheathing} = 10\,\text{ft}^2/\text{LH}$$
$$2\,\text{in.} \times 4\,\text{in. beams} = 9\,\text{ft}/\text{LH}$$
$$2\,\text{in.} \times 6\,\text{in. beams} = 8\,\text{ft}/\text{LH}$$
$$\text{Posts and props} = 1\,\text{post and } 2\,\text{props}/\text{LH} \;\; (2\,\text{props per post})$$

Find the total cost.

Solution

STEP 1: *Find the fbm of timber:*

Calculate the fbm of 1 in. sheathing;

$$1\,\text{in. sheathing} = (50 \times 12) \times (50 \times 12) \times 1/144\,\text{fbm} = 2500\,\text{fbm}$$

Explanation: fbm is calculated by calculating the total volume of timber by 144.

Total volume of timber sheathing is area × thickness.

Calculate the fbm of (2 in. × 4 in.) lumber:

Length is 50 ft (Length = 50 × 12 in.)

They are placed 1 ft center to center. Hence, there are 51 pieces.

$$51 \times (50 \times 12) \times (2 \times 4)/144\,\text{fbm} = 1700\,\text{fbm}$$

Calculate the fbm of (2 in. × 6 in.) lumber:

Length is 50 ft (Length = 50 × 12 in.)

They are placed 2 ft center to center. Hence, there are 26 pieces.

$$26 \times (50 \times 12) \times (2 \times 6)/144\,\text{fbm} = 1300\,\text{fbm}$$

$$\text{Total fbm} = 2500 + 1700 + 1300 = 5500\,\text{fbm}$$

STEP 2: *Find the number of steel posts and props:*

Steel posts: Steel posts are placed along 2 × 6 lumber.

There are 26 pieces of (2 × 6) lumber.

Each (2 × 6) lumber is 50 ft long and posts are located every 3 ft

$$\text{Number of posts per one length of } (2 \times 6)\,\text{lumber} = (50/3) = 16.667$$

$$\text{Number of posts per one length of } (2 \times 6)\,\text{lumber cannot be a fraction}$$

$$\text{hence number of posts} = 17$$

$$\text{Total number of posts} = (17) \times 26 = 442$$

$$\text{Number of props} = 2 \times 442 = 884$$

STEP 3: Find the cost of labor hour (LH):

$$\text{Crew} = 2\,\text{masons} + 4\,\text{laborers}$$

$$\text{Cost of crew hour} = (2 \times 65) + (4 \times 40) = \$290$$

$$\text{Cost of labor hour (LH)} = 290/6 = \$48.3$$

Continued

STEP 4: Find the total cost:

$$1\,\text{in. sheathing} = 10\,\text{ft}^2/\text{LH}$$
$$\text{LH required} = 2500/10 = 250$$
$$\text{Labor cost of } 1\,\text{in. sheathing} = 250 \times \$48.3 = \$12,075$$
$$2\,\text{in.} \times 4\,\text{in. beams} = 9\,\text{ft/LH}$$
$$\text{Total length} = 51 \times 50 = 2550$$
$$\text{LH required} = 2550/9 = 283.3$$
$$\text{Cost of } 2 \times 4\,\text{beams} = 283.3 \times \$48.3 = \$13,685$$
$$2\,\text{in.} \times 6\,\text{in. beams} = 8\,\text{ft/LH}$$
$$\text{Total length} = 26 \times 50 = 1300$$
$$\text{LH required} = 1300/8 = 162.5$$
$$\text{Cost of } 2 \times 6\,\text{beams} = 162.5 \times \$48.3 = \$7,848.8$$
$$\text{Posts and props} = 1\,\text{post and } 2\,\text{props/LH}$$
$$\text{Total number of posts} = 442$$
$$\text{Number of props} = 884\ (\text{see step 2})$$

One post has two props. Let's call it a post-prop unit.

One labor hour is required to install one unit (one unit = one post and 2 props).

$$\text{Number of posts and prop units} = 442$$
$$\text{Labor hours required to install } 442\,\text{units} = 442\,\text{LH}$$
$$\text{Cost} = 442 \times 48.3 = \$21,348.6$$
$$\text{Total cost} = \$12,075 + \$13,685 + \$7848 + \$21,348.6 = \$54,956.6$$

Practice Problem 12.34

Compute the cost of labor for the erection of the formwork in the previous example. Following information is given.

Carpenter wages: \$70 per hour, Helper wages: \$40 per hour
Saw cutting timber (for 100 fbm): (1.0 carpenter hours + 0.5 helper hours)
Measurement and erection (for 100 fbm):
(2.0 carpenter hours + 1.0 helper hours)

Solution

STEP 1: *Cost for saw cutting:*

$$\text{Cost for 100fbm of saw cutting} = (1 \times 70) + (0.5 \times 40) = \$90$$
$$\text{Total fbm in the formwork} = 2146.6\text{fbm (not including wastage)}$$
$$\text{Cost for 2361fbm} = 90/100 \times 2146.6 = \$1932$$

STEP 2: *Cost for measurement and erection:*

$$\text{Cost for 100fbm of measurement and erection} = (2 \times 70) + (1.0 \times 40) = \$180$$

Total fbm in the formwork = 2146.6 fbm.

$$\text{Cost for 2361fbm} = 180/100 \times 2146.6 = \$3864$$
$$\text{Total cost} = 1932 + 3864 = \$5796$$

12.6.4 Roofing Estimates

Roofing is a very important aspect of construction. Some of the concepts involved in roofing are discussed here in this chapter. Most roofing frames are built using lumber. Roof material could be shingles or tiles (Fig. 12.5).

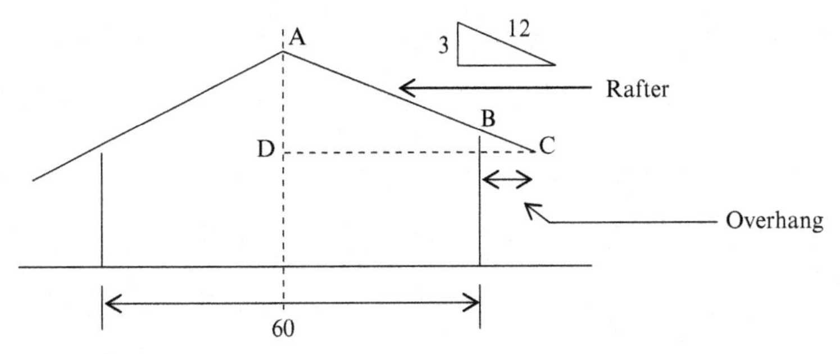

Fig. 12.5 Roof.

If the overhang is 2 ft what is the length of the rafter if the slope is 3–12. This is a simple trigonometric problem (Fig. 12.6).

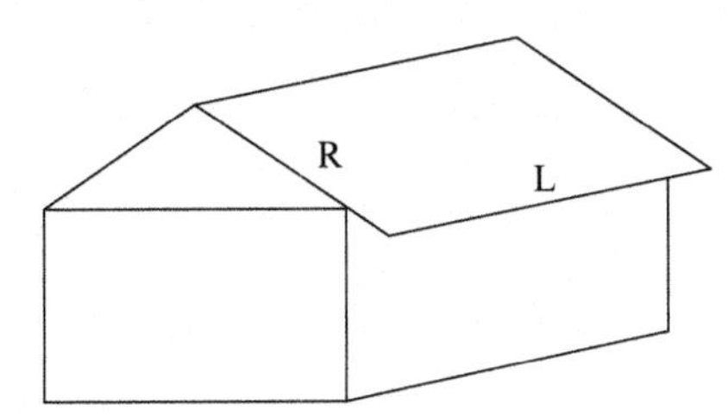

Fig. 12.6 Roof elements.

Length of the rafter is AC.

$$DC = 30 + \text{Overhang} = 32\text{ft}$$

$$\text{Tan (angle ACD)} = 3/12 = 0.25$$

$$\text{Angle ACD} = 14°.$$

$$\text{AC Cos}(14°) = DC = 32$$

$$AC = \text{Length of the rafter} = 32/\text{Cos}(14°) = 32/0.970 = 32.98\text{ft}$$

$$\text{Roofing area} = 2 \times (\text{Length of the rafter} \times \text{Length of the building})$$

Practice Problem 12.35

Find the roofing cost of the building shown below. Overhang of the roof is 2 ft. Roofing material cost \$6.90 per ft^2

Roofing installation crew: 2 roofers + 4 laborers

Wages: Roofer = \$60/h, laborer = \$40/h.

Productivity = 2.5 ft^2/LH (LH = labor hour)

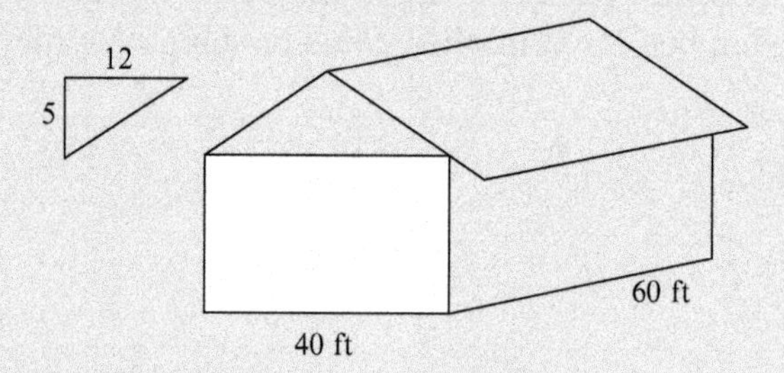

Solution

STEP 1: Find the length of the rafter

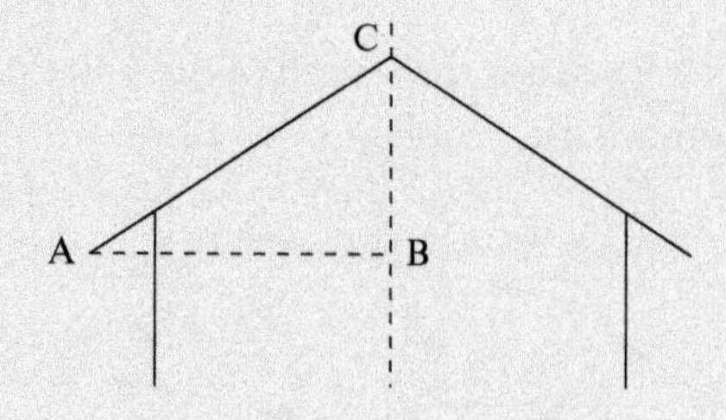

$$AB = 20 + \text{Overhang} = 22$$

$$\text{Tan (Angle CAB)} = 5/12$$

$$\text{Angle CAB} = 22.5°$$

$$AC\ \text{Cos}(22.5°) = AB = 22$$

$$AC = \text{Length of the rafter} = 22/\text{Cos}(22.5°) = 23.8\,\text{ft}$$

STEP 2: Find the area of the roof:

$$\text{Roof area} = 2 \times (\text{Length of the rafter} \times \text{Length of the building})$$
$$= 2 \times 23.8 \times 60 = 2856\,\text{ft}^2$$

STEP 3: Find the cost of roofing material:

$$\text{Cost of roofing material} = 6.90 \times 2856 = \$19,706.4$$

STEP 4: Find the cost of labor hour:

$$\text{Crew} = 2\,\text{roofers} + 4\,\text{laborers}$$
$$\text{Cost of crew hour} = (2 \times 60) + (4 \times 40) = \$280$$
$$\text{Cost of labor hour} = \text{Cost of crew hour/number of workers in the crew}$$
$$= 280/6 = \$46.7$$

STEP 5: Find the labor hours required:

$$\text{Labor hours required} = 2856/2.5 = 1142$$

Note: 2.5 is the productivity. 2.5 ft^2 of roof is installed per labor hour (LH)

STEP 6: Find the total cost of labor:

$$\text{Cost of labor} = \text{Labor hours required} \times \text{Cost of labor hour}$$
$$\text{Cost of labor} = 1142 \times 46.7 = \$53,350$$

STEP 7: Find the total cost of the project:

$$\text{Total cost of the project} = \text{Cost of material} + \text{Cost of labor}$$
$$= 19,706.4 + 53,350 = \$73,056$$

Roof trusses: Roof trusses are needed to support roofs. Most roof trusses are built of lumber or metal (Fig. 12.7).

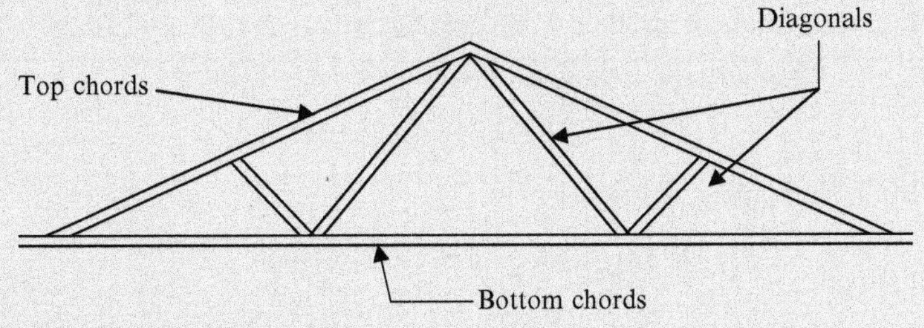

Fig. 12.7 Chords and diagonals.

12.6.5 Estimating Borehole Construction Cost

Following example shows how to estimate borehole or boring construction cost.

Practice Problem 12.36 (Borings)

Boring program is planned by a consulting company. Local building code requires a boring to be conducted every 400 ft^2 of the building footprint. Borings need to 15 ft below the bottom of footings. Building is 80 ft × 100 ft and bottom of footing is 3 ft below the ground. Following information available:

Mobilization = 2 h. Demobilization = 2 h

Material:

Drilling mud mix: 20 bags (cost $10 per bag)

Sampling spoons: 5 (cost $100 per spoon)

Equipment:

$$\text{Drill rig rental} = \$1000 \text{ per day}$$

Crew:

$$\text{Drilling crew} = 1 - \text{Drill rig operator} + 1 \text{ Helper}$$

$$\text{Wages: Drill rig operator} = \$75/\text{h}, \quad \text{Helper} = \$55/\text{h}$$

Productivity: 80 ft/day

Find the cost of the project.

Solution

STEP 1: Total length of drilling required:

$$\text{Building area} = 80 \times 100 = 8000 \text{ft}^2$$

Boring is required for every 400 ft^2.

$$\text{Number of borings required} = 8000/400 = 20$$

$$\text{Depth of borings} = 15\text{ft below the depth of footings} = 3 + 15 = 18$$

$$\text{Total length of borings} = 18 \times 20 = 360\text{ft}$$

STEP 2: Number of days required:

$$\text{Number of days of drilling} = 360/\text{productivity} = 360/80 = 4.5 \text{ days}$$

$$\text{Mobilization and demobilization} = 4\text{h} = 1/2 \text{ day}$$

$$\text{Total project duration} = 5 \text{ days}$$

STEP 3: Equipment rental:

$$\text{Drill rig rental} = 5 \times 1000 = \$5000$$

STEP 4: Material:

$$\text{Drilling mud} = 20 \times 10 = \$200$$

$$\text{Sampling spoons} = 5 \times 100 = \$500$$

$$\text{Total material cost} = \$700$$

STEP 5: Labor:

$$\text{Drilling crew} = 1 - \text{Drill rig operator} + 1\ \text{Helper}$$

$$\text{Crew hourly rate} = 75 + 55 = \$130$$

$$\text{Number of hours} = 5 \times 8 = 40$$

$$\text{Labor} = 130 \times 40 = \$5200$$

$$\text{Total direct cost (without overhead and profit)} = 5000 + 700 + 5200 = \$10,900$$

12.6.6 Estimating Basement Excavation Cost

Practice Problem 12.37 (Excavation)

The excavation for a basement is 150 ft long, 80 ft wide and 40 ft deep. Following activities are identified.

(a) Excavation
(b) Hauling material out of site
(c) Driving of sheet piles
(d) Construction of wales and struts

Overhead is agreed at 15% and profits at 10%. Find the cost of above items.

Following information is provided.

(a) Excavation: \$10 per CY
(b) Hauling material out of site: \$120 per truck (truck capacity = 10 CY)
(c) Driving of sheet piles: Sheet piles are driven 2 ft below the bottom of excavation and 2 ft section above the ground level will be left alone.
Sheet pile material cost = \$15/ft^2
Rental cost of sheet pile driver = \$1200/day
Pile driving crew; 1—Operator, 3—Laborers
Wage rates: operator = \$70/h, Laborer = \$45/h
Productivity of sheet pile installation: 60 ft^2/h
(d) Construction of wales and struts: W26 × 82 sections would be used for wales and struts.
Material cost of W26 × 82 section = \$2 per lb.
Wales and struts are installed by a crew of 7 ironworkers and 5 laborers.
Wage rates: Iron worker = \$60/h, Laborer = \$45/h
Productivity = 1.2 ft/LH (LH = labor hour)
Total length of wales and struts = 800 ft

Continued

Solution

STEP 1: Find the cost of excavation:

$$\text{Excavation volume} = 150 \times 80 \times 40 = 480,000\,\text{ft}^3 = 17,778\,\text{CY}$$
$$\text{Cost of excavation} = 17,778 \times 10 = \$177,780$$

STEP 2: Cost of hauling material:

$$\text{Cost of hauling material} = 17,778/10 \times 120 = \$213,333$$

STEP 3: Material cost of sheetpiles:

$$\text{Length of sheet piles} = 40 + 2 + 2 = 44\,\text{ft}$$
$$\text{Perimeter length of the excavation} = (2 \times 150) + (2 \times 80) = 460\,\text{ft}$$
$$\text{Total square feet of sheetpiles} = 460 \times 44 = 20,240\,\text{ft}^2$$
$$\text{Sheetpile material cost} = 15 \times 20,240 = \$303,600$$

STEP 4: Rental cost of sheetpile driver:

$$\text{Duration of driving sheetpiles} = \text{square footage of sheetpiles/productivity}$$
$$= 20,240/60 = 338\,\text{h} = 42.25\text{ days (8h per day)}$$
$$\text{Rental cost of sheetpile driver} = 42.25 \times \$1200 = \$50,700$$

STEP 5: Labor cost of driving sheet piles:
Sheetpile driving crew: 1—Operator, 3—Laborers

$$\text{Cost of crew hour} = (1 \times 70) + (3 \times 45) = \$205/\text{h}$$
$$\text{Number of hours required} = 338 \text{ (see step 4)}$$
$$\text{Cost of labor} = 338 \times 205 = \$69,290$$

STEP 6: Material cost of wales and struts:

$$\text{Total length of wales and struts} = 800\,\text{ft}$$

W26 × 82 sections weighs 82 lbs per linear foot.

$$\text{Total weight of wales and struts} = 800 \times 82 = 65,600\,\text{lbs}$$
$$\text{Material cost of wales and struts} = \$2 \times 65,600 = \$131,200$$

STEP 7: Labor cost of installation of wales and struts:

$$\text{Cost of crew hour} = (7 \times 60) = (5 \times 45) = \$645/\text{h}$$
$$\text{Cost of labor hour (LH)} = 645/\text{number of workers in the crew}$$
$$= 645/12 = \$53.75$$
$$\text{Productivity} = 1.2\,\text{ft/LH}$$
$$\text{Labor hours required} = 800/1.2 = 667$$
$$\text{Cost of labor for installation of wales and struts} = 667 \times 53.75 = \$35,851$$

STEP 8:

$$\text{Total direct cost} = \$177,780 + \$213,333 + \$303,600 + \$50,700$$
$$+\$69,290 + \$131,200 + \$35,851 = \$981,754 \text{ Overhead} = 0.15 \times \$981,754$$
$$= \$147,263$$
$$\text{Cost including overhead} = \$1,129,017 \text{Profit} = 0.1 \times \$1,129,017$$
$$= \$112,901.7 \text{ Total cost including overhead and profit}$$
$$= \$1,241,919$$

12.6.7 Estimating Wooden Floor Construction Cost

Wooden floors are common in many parts of the world. Wooden floors are built above the ground using timber (Fig. 12.8).

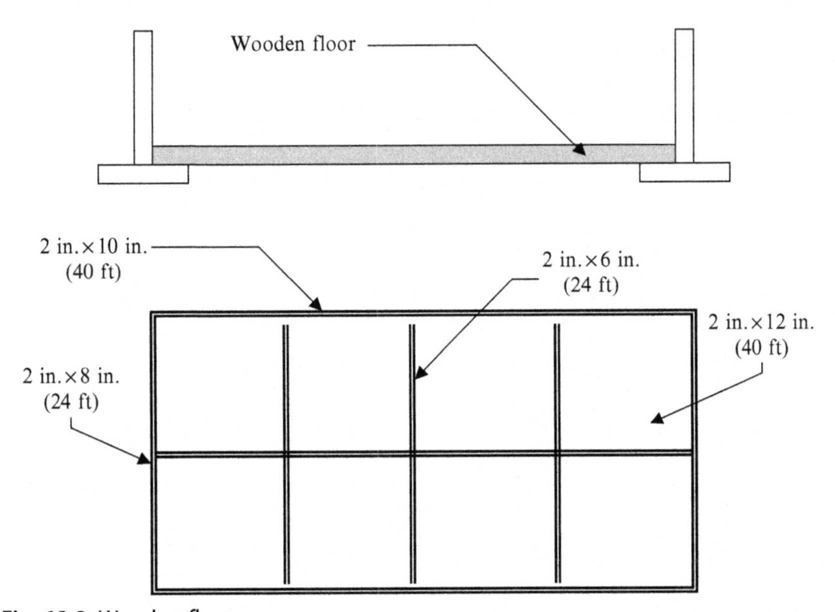

Fig. 12.8 Wooden floor.

Practice Problem 12.38

Find the costs associated with construction of the wooden floor shown above. Following information is available.

Cost of lumber = \$4.2 per fbm
Framing crew consists of 2 carpenters and 3 laborers.
Wages: Carpenter = \$75/h, Laborer = \$55/h
Productivity = 2.1 fbm/LH (LH = labor hour)
Fork lift rental rate = \$200/day
Assume 8 h workday and full crew is working all the time.

Solution

STEP 1: *Find the cost of lumber:*

$$2\,\text{pieces of } 2\text{in.} \times 10\text{in.} \text{ (length 40ft)} = 2 \times (2 \times 10) \times (40 \times 12)/144 = 133.3\,\text{fbm}$$

$$2\,\text{pieces of } 2\text{in.} \times 8\text{in.} \text{ (length 24ft)} = 2 \times (2 \times 8) \times (24 \times 12)/144 = 64\,\text{fbm}$$

$$3\,\text{pieces of } 2\text{in.} \times 6\text{in.} \text{ (length 24ft)} = 3 \times (2 \times 6) \times (24 \times 12)/144 = 72\,\text{fbm}$$

$$1\,\text{piece of } 2\text{in.} \times 12\text{in.} \text{ (length 40ft)} = 1 \times (2 \times 12) \times (40 \times 12)/144 = 80\,\text{fbm}$$

$$\text{Total} = 349\,\text{fbm}$$

$$\text{Cost of lumber} = 4.2 \times 349 = \$1467$$

STEP 2: *Find the cost of labor:*

$$\text{Cost of crew hour} = (2 \times 75) + (3 \times 55) = \$315$$

$$\text{Cost of labor hour (LH)} = \text{Cost of crew hour/number of workers in the crew} = 315/5 = \$63$$

Productivity is given to be 2.1 fbm per labor hour (LH)

$$\text{Number of labor hours needed} = 349/2.1 = 166.2$$

$$\text{Cost of labor} = \text{Number of labor hours} \times \text{cost of labor hour} = 166.2 \times 63 = \$10,470$$

STEP 3: *Find the time required for construction:*

At any given time 2 carpenters and 3 laborers are working. That is a total of 5 workers.

$$\text{Productivity of crew hour} = \text{number of workers in the crew} \times \text{productivity of labor hour}$$

$$\text{Productivity of crew hour} = 5 \times 2.1 = 10.5\,\text{fbm/crew hour}$$

$$\text{Number of crew hours needed} = 349/10.5 = 33.2$$

Since the workers are working 8 h per day

$$\text{Number of days needed} = 33.2/8 = 4.15$$

Fork lift is needed for 5 days.

$$\text{Equipment rental cost} = 5 \times 200 = \$1000$$

$$\text{Total cost} = 1467 + 10,470 + 1000 = \$12,937$$

12.6.8 Estimating Steel Structures

Steel structures are common in the world. Main activities involved in erection of steel structures are:

(1) *Fabrication of steel members:* Contractor has to buy columns, beams, connections, angles, channels, and plates. These steel members come in standard sizes. Contractor has to cut the beams, angles, and channels to the correct length, drill holes for bolts, taper steel members for installation. These activities are known as fabrication.

Fig. 12.9 shows fabrication of a gusset plate using factory manufactured standard plate. Sometimes beams are fabricated using angles or channels.

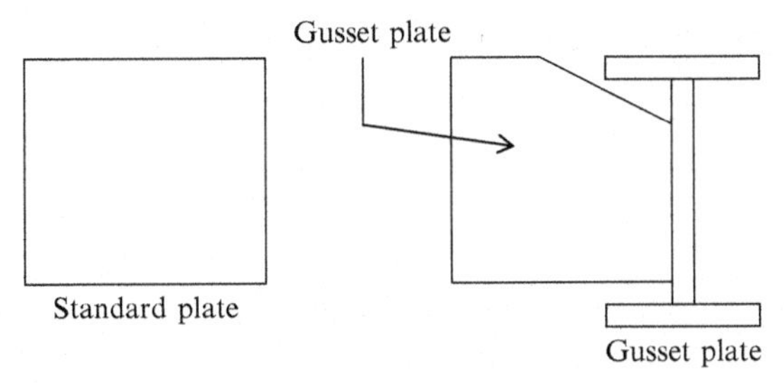

Fig. 12.9 Steel members.

Fabrication of a beam using two channel sections using welding (Fig. 12.10).

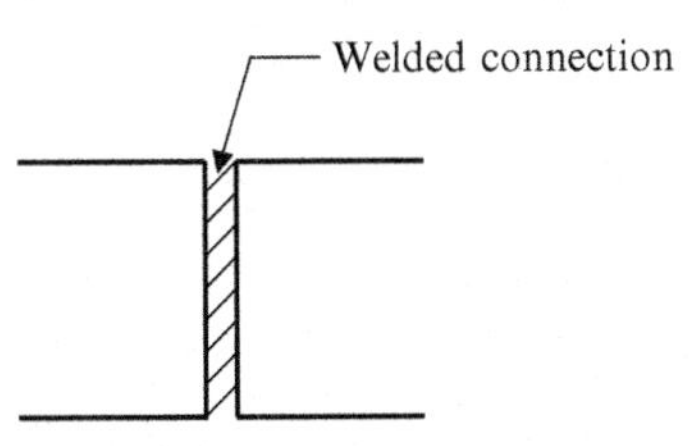

Fig. 12.10 Welded connection.

(2) *Transportation to the site and storage:* After fabrication of the steel members, they are transported and stored in the site.

(3) *Steel erection:* Erection of steel requires lifting to the correct location using a crane and making connections. The erection of steel is a highly specialized profession. Bolts are done by drilling holes. Some holes are factory drilled and some other holes are field drilled.

(4) *Member connections:* Steel members are connected using bolts or welding. In the past, rivets were used for steel connections. Today most riveted connections are converted to bolts.

CHAPTER 13

Equipment Depreciation

Depreciation is the loss of value due to use. Most of us have bought cars. Imagine you bought a car for $20,000. One year later, you would like to sell the car. You advertise in newspapers and the best price you can get is $15,000. You have lost $5000 in 1 year. Depreciation of your car is $5000. If you decide to sell the car in 2 years instead of 1 year, you will find that the value of the car has depreciated further. Let us say 2 years later, the maximum you can get for your car is $12,000. Total depreciation for 2 years is $8000. If you wait for 3 years, let us say value of the car has further dropped to $10,000. Now total depreciation for 3 years is $10,000 (Table 13.1).

Table 13.1 Depreciation

	Value	Total depreciation	Depreciation per year
Purchase price	20,000		
Value after 1 year	15,000	5000	5000
Value after 2 years	12,000	8000	4000
Value after 3 years	10,000	10,000	3333

Notice how the value drops rapidly during initial years.

13.1 STRAIGHT LINE DEPRECIATION

There are many methods to compute depreciation. The easiest method is the straight line depreciation. Straight line depreciation assumes that depreciation will be same every year. This method can be explained using an example.

Practice Problem 13.1 (straight line depreciation)

A contractor has bought a dozer for $100,000. He estimates that the salvage value of the equipment after 4 years is $20,000. Find the depreciation of the equipment after each year.

Continued

Construction Engineering Design Calculations and Rules of Thumb
http://dx.doi.org/10.1016/B978-0-12-809244-6.00013-5

Solution

STEP 1: Find the total depreciation:

$$\text{Purchase price} = 100,000$$

$$\text{Salvage value after 4 years} = 20,000$$

$$\text{Total depreciation after 4 years} = 80,000$$

STEP 2: Find the depreciation per year:

$$\text{Depreciation per year} = 80,000/4 = 20,000$$

STEP 3: Find depreciation end of each year:

$$\text{Purchase price} = 100,000$$

$$\text{Depreciation after 1 year} = 20,000$$

$$\text{Hence, value after 1 year} = 100,000 - 20,000 = 80,000$$

$$\text{Depreciation after 2 years} = 2 \times 20,000 = 40,000$$

$$\text{Value after 2 years} = 100,000 - 40,000 = 60,000$$

$$\text{Depreciation after 3 years} = 3 \times 20,000 = 60,000$$

$$\text{Value after 3 years} = 100,000 - 60,000 = 40,000$$

$$\text{Depreciation after 4 years} = 4 \times 20,000 = 80,000$$

$$\text{Value after 4 years} = 100,000 - 80,000 = 20,000$$

Note

Straight line depreciation is not realistic. In the real world, equipment depreciates faster at the initial years as you saw with the example of the car. To alleviate this problem, other methods, such as the sum of the years digit method and the declining balance method, have been introduced.

13.2 SUM OF THE YEARS DIGITS METHOD

This method is devised to have a higher rate of depreciation during the initial years. As the name indicates, add the sum of each year. If we are planning to depreciate the equipment for 2 years, sum of the digits is $(1+2=3)$.

If we are planning to depreciate for 3 years, sum of the digits: $1+2+3=6$

If we are planning to depreciate for 4 years, sum of the digits: $1+2+3+4=10$

If we are planning to depreciate for 5 years, sum of the digits: $1+2+3+4+5=15$

If we are planning to depreciate for 6 years, sum of the digits: $1+2+3+4+5+6=21$

This method can be better explained using an example:

Practice Problem 13.2 (sum of the years digit method)

Find the depreciation of the equipment in the previous problem using sum of years digit method. A contractor has bought a dozer for $100,000. He estimates that the salvage value of the equipment after 4 years is $20,000. Find the depreciation of the equipment after each year using sum of the years digit method.

Solution

STEP 1: Find the total depreciation:

$$\text{Purchase price} = 100,000$$
$$\text{Salvage value after 4 years} = 20,000$$
$$\text{Total depreciation after 4 years} = 80,000$$

STEP 2: Find sum of the digits:

$$\text{Depreciation is done for 4 years;} \quad 1 + 2 + 3 + 4 = 10$$

STEP 3: Find the depreciation after each year:

$$\begin{aligned}
\text{Depreciation first year} &= 4/10 \times (\text{Total depreciation}) \\
&= 4/10 \times 80,000 = 32,000
\end{aligned}$$

(4 is the last number in the series and 10 is the sum of the series)

$$\begin{aligned}
\text{Value after 1 year} &= \text{Purchase price} - \text{Depreciation} \\
&= 100,000 - 32,000 = 68,000 \\
\text{Depreciation second year} &= 3/10 \times (\text{Total depreciation}) \\
&= 3/10 \times 80,000 = 24,000 \\
\text{Value after 2 years} &= \text{Remaining value} - \text{Depreciation} \\
&= 68,000 - 24,000 = 44,000 \\
\text{Depreciation third year} &= 2/10 \times (\text{Total depreciation}) \\
&= 2/10 \times 80,000 = 16,000 \\
\text{Value after 3 years} &= \text{Remaining value} - \text{Depreciation} \\
&= 44,000 - 16,000 = 28,000 \\
\text{Depreciation fourth year} &= 1/10 \times (\text{Total depreciation}) \\
&= 1/10 \times 80,000 = 8000 \\
\text{Value after 4 years} &= \text{Remaining value} - \text{Depreciation} \\
&= 28,000 - 8000 = 20,000
\end{aligned}$$

Practice Problem 13.3 (sum of the years digit method)

Contractor has bought a backhoe for $200,000. He estimates that the salvage value of the equipment after 6 years is $10,000. Find the depreciation of the equipment after each year using sum of the years digit method.

Continued

Solution

STEP 1: Find the total depreciation:

$$\text{Purchase price} = 200,000$$

$$\text{Salvage value after 6 years} = 10,000$$

$$\text{Total depreciation after 6 years} = 190,000$$

STEP 2: Find sum of the digits:

$$\text{Depreciation is done for 6 years;}\quad 1+2+3+4+5+6=21$$

STEP 3: Find the depreciation after each year:

$$\begin{aligned}
\text{Depreciation first year} &= 6/21 \times (\text{Total depreciation}) \\
&= 6/21 \times 190,000 = 54,286 \\
\text{Value after 1 year} &= \text{Purchase price} - \text{Depreciation} \\
&= 200,000 - 54,286 = 145,714 \\
\text{Depreciation second year} &= 5/21 \times (\text{Total depreciation}) \\
&= 5/21 \times 190,000 = 45,238 \\
\text{Value after 2 years} &= \text{Remaining value} - \text{Depreciation} \\
&= 145,714 - 45,238 = 100,476 \\
\text{Depreciation third year} &= 4/21 \times (\text{Total depreciation}) \\
&= 4/21 \times 190,000 = 36,190 \\
\text{Value after 3 years} &= \text{Remaining value} - \text{Depreciation} \\
&= 100,476 - 36,190 = 64,286 \\
\text{Depreciation fourth year} &= 3/21 \times (\text{Total depreciation}) \\
&= 3/21 \times 190,000 = 27,143 \\
\text{Value after 4 year} &= \text{Remaining value} - \text{Depreciation} \\
&= 64,286 - 27,143 = 37,143 \\
\text{Depreciation fifth year} &= 2/21 \times (\text{Total depreciation}) \\
&= 2/21 \times 190,000 = 18,095 \\
\text{Value after 5 years} &= \text{Remaining value} - \text{Depreciation} \\
&= 37,143 - 18,095 = 19,048 \\
\text{Depreciation sixth year} &= 1/21 \times (\text{Total depreciation}) \\
&= 1/21 \times 190,000 = 9048 \\
\text{Value after 6 years} &= \text{Remaining value} - \text{Depreciation} \\
&= 19,048 - 9048 = 10,000
\end{aligned}$$

13.3 DECLINING BALANCE DEPRECIATION

In this method, the average depreciation rate is obtained per year. The same rate is used each year to compute the depreciation. This method is better explained using an example.

Practice Problem 13.4

A contractor buys a loader for $100,000. He is planning to use the equipment for 5 years. Find the depreciation of the equipment after each year using the declining balance method.

Solution

$$\boxed{\text{Average depreciation rate} = 100/\text{number of years}}$$

STEP 1: Find the average depreciation rate

$$\text{Average depreciation rate} = 100/\text{number of years} = 100/5 = 20\% = 0.2$$

It is assumed that 100% depreciation would occur in 5 years. End of 5 years, there would be a book value for the equipment. In this method, salvage value does not come into equations.

STEP 2: Find the depreciation after each year:

$$\begin{aligned}
\text{Depreciation first year} &= 0.2 \times (\text{Value of the equipment}) \\
&= 0.2 \times 100,000 = 20,000 \\
\text{Value after 1 year} &= \text{Value of the equipment} - \text{Depreciation} \\
&= 100,000 - 20,000 = 80,000 \\
\text{Depreciation second year} &= 0.2 \times (\text{Value of the equipment}) \\
&= 0.2 \times 80,000 = 16,000 \\
\text{Value after second year} &= \text{Value of the equipment} - \text{Depreciation} \\
&= 80,000 - 16,000 = 64,000 \\
\text{Depreciation third year} &= 0.2 \times (\text{Value of the equipment}) \\
&= 0.2 \times 64,000 = 12,800 \\
\text{Value after third year} &= \text{Value of the equipment} - \text{Depreciation} \\
&= 64,000 - 12,800 = 51,200 \\
\text{Depreciation fourth year} &= 0.2 \times (\text{Value of the equipment}) \\
&= 0.2 \times 51,200 = 10,240 \\
\text{Value after fourth year} &= \text{Value of the equipment} - \text{Depreciation} \\
&= 51,200 - 10,240 = 40,960 \\
\text{Depreciation fifth year} &= 0.2 \times (\text{Value of the equipment}) \\
&= 0.2 \times 40,960 = 8192 \\
\text{Value after fifth year} &= \text{Value of the equipment} - \text{Depreciation} \\
&= 40,960 - 8192 = 32,768
\end{aligned}$$

Book value of the equipment after 5 years 32,768

13.4 200% DECLINING BALANCE DEPRECIATION

This method is a variation of the previous method. In this method, the average depreciation is obtained per year and then doubled and applied to the remaining value of the equipment.

$$\boxed{\text{Double average depreciation rate} = 2 \times (100/\text{number of years})}$$

Practice Problem 13.5

A contractor buys a loader for $100,000. He is planning to use the equipment for 5 years. Find the depreciation of the equipment after each year using the declining balance method.

Solution

STEP 1: Find the average depreciation rate:

$$\text{Average depreciation rate} = 100/\text{number of years} = 100/5 = 20\% = 0.2$$

It is assumed that 100% depreciation would occur in 5 years. Then the rate is doubled. Value at the end of 5 years is known as book value.

STEP 2:

$$\text{Double the average depreciation rate} = 2 \times 20\% = 40\% = 0.4$$

STEP 3: Find the depreciation after each year:

$$\text{Depreciation first year} = 0.4 \times (\text{Value of the equipment})$$
$$= 0.4 \times 100,000 = 40,000$$
$$\text{Value after 1 year} = \text{Value of the equipment} - \text{Depreciation}$$
$$= 100,000 - 40,000 = 60,000$$
$$\text{Depreciation second year} = 0.4 \times (\text{Value of the equipment})$$
$$= 0.4 \times 60,000 = 24,000$$
$$\text{Value after second year} = \text{Value of the equipment} - \text{Depreciation}$$
$$= 60,000 - 24,000 = 36,000$$
$$\text{Depreciation third year} = 0.4 \times (\text{Value of the equipment})$$
$$= 0.4 \times 36,000 = 14,400$$
$$\text{Value after third year} = \text{Value of the equipment} - \text{Depreciation}$$
$$= 36,000 - 14,400 = 21,600$$

$$
\begin{aligned}
\text{Depreciation fourth year} &= 0.4 \times (\text{Value of the equipment}) \\
&= 0.4 \times 21,600 = 8640 \\
\text{Value after fourth year} &= \text{Value of the equipment} - \text{Depreciation} \\
&= 21,600 - 8640 = 12,960 \\
\text{Depreciation fifth year} &= 0.4 \times (\text{Value of the equipment}) \\
&= 0.4 \times 12,960 = 5184 \\
\text{Value after fifth year} &= \text{Value of the equipment} - \text{Depreciation} \\
&= 12,960 - 5184 = 7,776 \\
\text{Book value of the equipment after 5 years} &= 7776
\end{aligned}
$$

CHAPTER 14

Engineering Analysis

Important note on economic problems: Economic problems can be solved using either interest tables or equations. It is easier to use interest tables to solve economic problems. However, interest tables can be used only for very simple problems. To solve more difficult problems equations need to be used. For this reason, the problems in this book are solved using both interest tables and equations.

14.1 PRESENT WORTH AND FUTURE WORTH

It is not a secret that the value of money goes down with time. One thousand dollars today has more value than it will have a year from now.

If the interest rate is "i," present worth is "P," number of years is "n," and future worth is "F," the following equation can be used.

Representation:

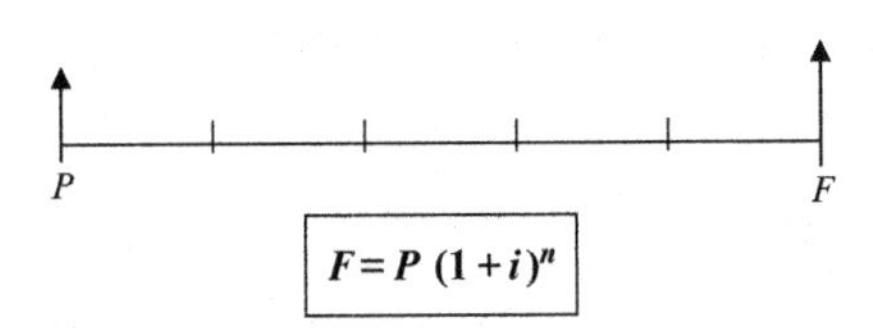

$$F = P\,(1+i)^n$$

Practice Problem 14.1

Find the value of $1200 in a savings account that generates 9% interest per year after 5 years.

Solution Using Interest Tables

P = Present value = 1200

i = 9%, n = 5

F needs to be found.

Go to interest tables and locate n = 5 and i = 9%.

Continued

http://dx.doi.org/10.1016/B978-0-12-809244-6.00014-7

Then look for the F/P column. This means value of F when P is given.
From the tables $F/P = 1.5386$
Hence, $F = 1.5386 \times P = 1.5386 \times 1200 = \$1846.32$
Please note that interest tables are not provided in this book, however, they are available in most economic books.

Solution Using Equations

$P = \$1200$
$i = 9\%,\ n = 5$
$F = P(1+i)^n$
$F = 1200(1+0.09)^5$
$F = 1846.4$

14.2 FINDING INTEREST RATE (*I*) WHEN *F* AND *P* ARE GIVEN

When interest rate i need to be found using given F and P values, the interest table method may be difficult. In such situations, the following equation can be used:

$$F = P(1+i)^n$$
$$F/P = (1+i)^n$$
$$(F/P)^{1/n} = (1+i)$$
$$\boxed{i = (F/P)^{1/n} - 1}$$

Practice Problem 14.2

An investor invests \$7000 dollars and receives \$11,000 after 7 years. What is the interest rate?

Answers: (A) 5%, (B) 4.3%, (C) 6.6%, (D) 7.2%

Solution Using Interest Tables

$F = \text{Future value} = 11{,}000$
$n = 7$
$i = $ Interest rate need to be found
$F/P = 11{,}000/7000 = 1.571$
Locate $n = 7$ and $F/P = 1.571$
Now you could see this is not easy to do since i (interest rate) is not given.
If you look at $n = 7$ and $i = 6\%$, $F/P = 1.5036$

However, we are looking for $F/P = 1.571$. Hence, we have to look for a higher interest rate.
Now look at $n = 7$ and $i = 7\%$, and $F/P = 1.6057$.
Now you can extrapolate.
$n = 7$ and $i = 6\%$, $F/P = 1.5036$
$n = 7$ and $i = 7\%$, $F/P = 1.6057$
Find i for $F/P = 1.571$
Use the following extrapolation equation:

$$(i - 6)/(1.571 - 1.5036) = (7 - 6)/(1.6057 - 1.5036)$$

If you look at the previous equation, i matches with 1.571, 6% matches with 1.5036 and 7% matches with 1.6057.
Solve the previous equation to find i.
$i = 6.660$

Solution Using Equations

$F = P(1 + i)^n$
$F = 11,000$
$P = 7000$, $n = 7$
$i = (F/P)^{1/n} - 1$
$i = (11,000/7000)^{1/7} - 1$
$i = 0.067 = 6.7\%$

14.3 SERIES PAYMENTS (*A*)

Business loans are paid using series payments. Assume a contractor obtained a loan and agreed to pay the loan on a yearly basis. This could be represented in an arrow diagram. Money coming into the contractor is represented with an upward arrow and money paid is represented with a downward arrow.

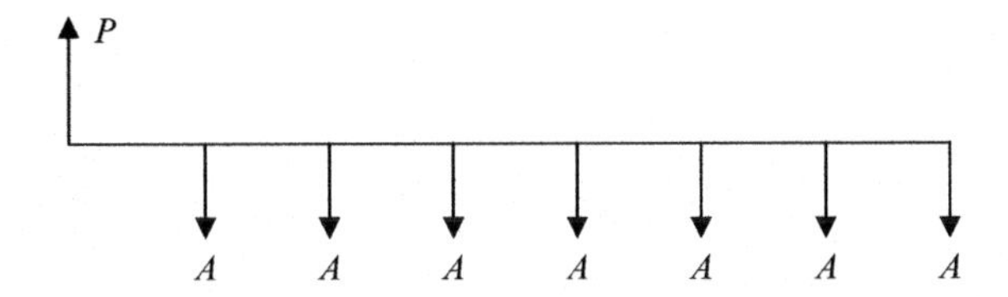

P = present worth (shown with an upward arrow to indicate money coming in)

A = series payments (shown with a downward arrow to indicate money going out)

$$P = A[(1+i)^n - 1]/[i(1+i)^n]$$

Practice Problem 14.3

A contractor obtained a 3 million dollar loan from a bank at an interest rate of 4% and agreed to make yearly payments for 5 years. What is his yearly payment?

Solution

P = 3,000,000, i = 4%, n = 5

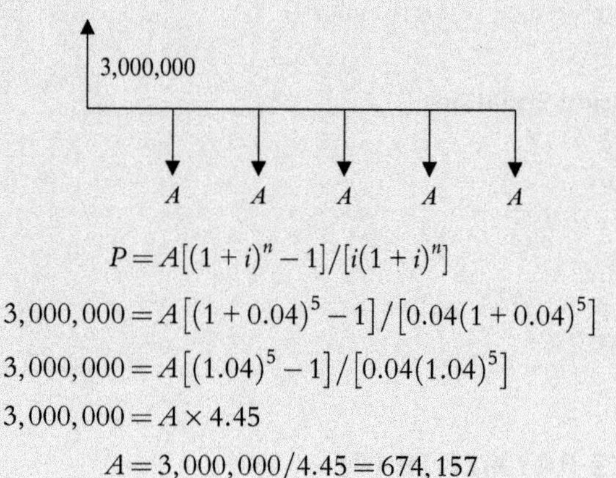

$$P = A[(1+i)^n - 1]/[i(1+i)^n]$$

$$3,000,000 = A\left[(1+0.04)^5 - 1\right]/\left[0.04(1+0.04)^5\right]$$

$$3,000,000 = A\left[(1.04)^5 - 1\right]/\left[0.04(1.04)^5\right]$$

$$3,000,000 = A \times 4.45$$

$$A = 3,000,000/4.45 = 674,157$$

14.4 SERIES PAYMENTS AND FUTURE VALUE

An employee can make series payments to a pension fund and expect to obtain a lump sum in the future.

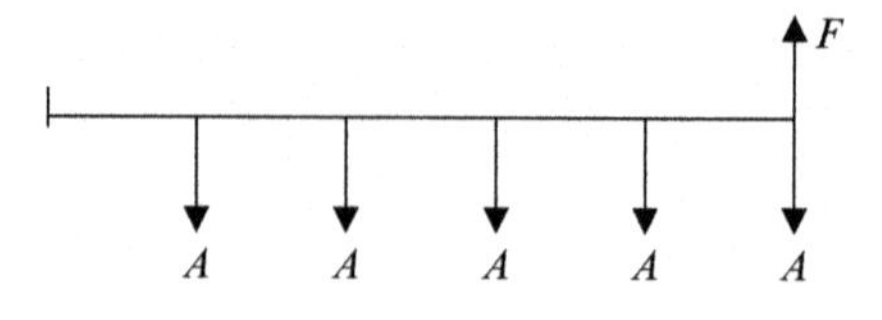

$$F = A[(1+i)^n - 1]/[i]$$

Practice Problem 14.4

An employee makes yearly payments for 20 years for a pension fund. His yearly payment is \$10,000 and interest rate is 7%. How much money he would be able to collect at the end of the 20-year period.

Solution

$F = A[(1+i)^n - 1]/[i]$

$A = 10,000$

$i = 0.07,\ n = 20$

$F = 10,000\left[(1+0.07)^{20} - 1\right]/[0.07]$

$F = 10,000\left[(1.07)^{20} - 1\right]/[0.07]$

$F = 409,954$

14.5 FUTURE VALUE AND ARITHMETIC GRADIENT: (*F* AND *G*)

Payments can increase after every payment. Such schemes are known as arithmetic gradients (*G*).

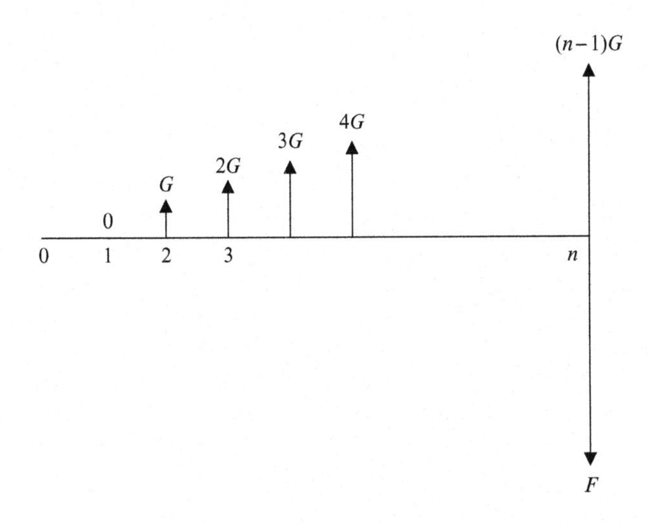

Note that first year increment is zero. Second year it is G and third year it is 2G and so forth. In the earlier figure, an individual is making constantly increasing payments to a bank and eventually obtain a future one-time sum of *F*.

The series in earlier figure is equal to one final payment of F. F is given as

$$F = \frac{G}{i}\left[\frac{(1+i)^n - 1}{i} - n\right]$$

If F is known, P or A can be calculated from equations given before.

14.6 UNIFORM SERIES (*A*) AND ARITHMETIC GRADIENT (*G*): (*A* AND *G*)

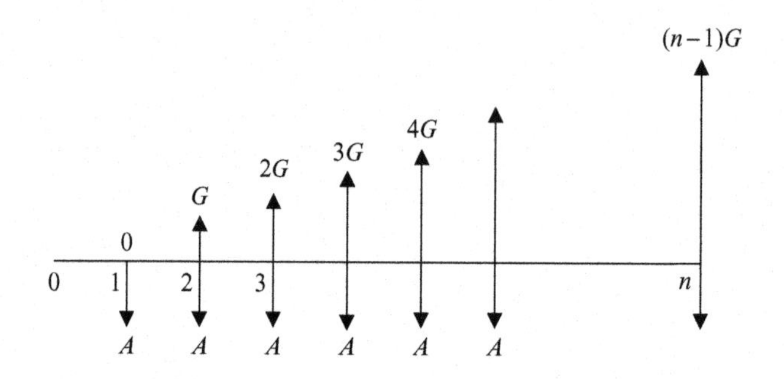

In the series in earlier figure, a man is making a constantly increasing sum of money and in return obtains uniform series payments (A). Please note the earlier figure is not drawn to a scale.

$$A = G\left[\frac{1}{i} - \frac{n}{(1+i)^n - 1}\right]$$

Practice Problem 14.5

A man is making monthly payments of 0, 500, 1000, 1500, etc., for 11 years. After 11 years, he wishes to obtain a lump sum payment. What is his lump sum payment if the interest rate is 5%?

Solution

$G=500$, $i=5\%$, $n=11$, $F=?$

Use the interest tables. Since F needs to be found, look for F/G.

From interest tables $F/G=64.135$

Hence $F=64.135 \times G=64.135 \times 500=\$32,067$

14.7 FINDING THE INTEREST RATE WHEN *G*, *N*, AND *F* ARE GIVEN

Practice Problem 14.6

A man is making monthly payments of 0, 500, 1000, 1500, etc., for 11 years. After 11 years, he wishes to obtain a lump sum payment of $40,000. What is the interest rate?

It is easier to use equations to solve such problems than using interest table.

Solution

$$G = 500, \quad i = ?, \quad n = 11, \quad F = 40{,}000$$

$$F = \frac{G}{i}\left[\frac{(1+i)^n - 1}{i} - n\right]$$

$$40{,}000 = \frac{500}{i}\left[\frac{(1+i)^{11} - 1}{i} - 11\right]$$

In the exam, four answers will be given. The student can insert the values and see which one works. In the real world, the problem has to be solved through iteration.

Use an interest rate of 5% and check the value in the right-hand side of the equation.

For interest rate of 5%, the value on the right-hand side is 32,067.

Use a higher interest rate and check the value again.

For interest rate of 6%, the value on the right-hand side is 33,097.

For interest rate of 12%, the value on the right-hand side is 40,227.

This value may be close enough for many situations.

14.8 UNIFORM SERIES (*A*) AND ARITHMETIC GRADIENT (*G*): (*A* AND *G*)

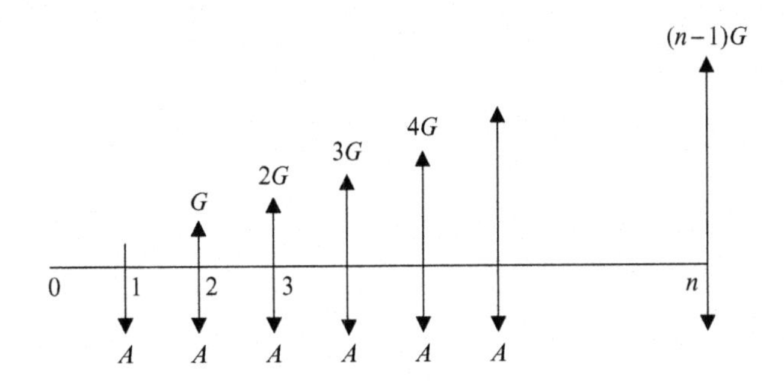

In the series in earlier figure, a man is making a constantly increasing sum of money and in return obtains uniform series payments (A). Please note the earlier figure is not drawn to a scale.

$$A = G\left[\frac{1}{i} - \frac{n}{(1+i)^n - 1}\right]$$

Practice Problem 14.7

A man is making monthly payments of 0, 600, 1200, 1800, etc., for 7 years. During this time, he likes to obtain a uniform payment every year. If the interest rate is 8% what is the uniform payment.

Solution

$G = 600$, $i = 8\%$, $n = 7$, $A = ?$

Since A needs to be found, from the interest tables find A/G.

$A/G = 2.6936$ (from interest tables)

$A = 2.6936 \times 600 = \$1616.2$

14.9 COST–BENEFIT ANALYSIS

All projects have costs and benefits. It is important to investigate the costs and benefits involved in a project.

Practice Problem 14.8

A developer is considering two development projects. The first project is a housing complex and the other one is a shopping mall. The following information is available.

Initial construction cost for the housing complex = 12 million dollars

Initial construction cost for the shopping mall = 17 million dollars

Yearly income from rental payments from the housing complex = 1.5 million dollars

Yearly income from rental payments from the shopping mall = 2.2 million dollars

Interest rate is at 6% and both projects have a lifetime of 20 years.

What is the better investment?

Solution

Arrow diagram for the housing complex:

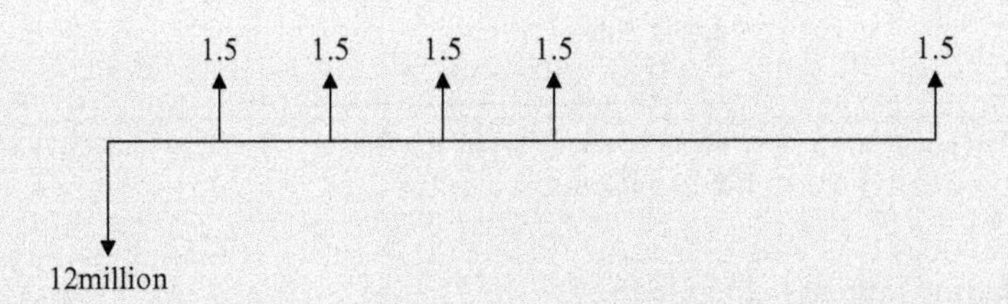

Housing complex has an initial cost of 12 million dollars. However, the developer will get yearly payments of 1.5 million dollars for next 20 years.

It is important to convert yearly payments to present worth.

What is the present worth of the yearly payments?

$P = A[(1+i)^n - 1]/[i(1+i)^n]$

$A = 1.5, \quad n = 20, \quad i = 6\%$

$P = 1.5\left[(1+0.06)^{20} - 1\right]/\left[0.06(1+0.06)^{20}\right]$

$P = 1.5\left[(1.06)^{20} - 1\right]/\left[0.06(1.06)^{20}\right]$

$P = 17.2$ million dollars

An arrow diagram can be used to represent this information:

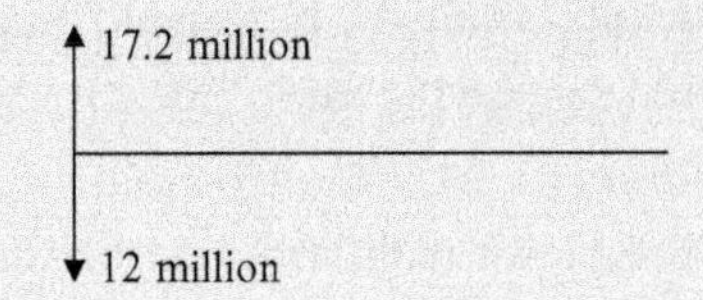

Initial investment = 12 million

Present value of return = 17.2 million

Benefit = 5.2 million

Practice Problem 14.9

A shopping mall has an initial cost of 17 million dollars. However, the developer will get yearly payments of 2.2 million dollars for next 20 years. What is the profit of the buyer based on present worth?

Solution

$P = A[(1+i)^n - 1]/[i(1+i)^n]$

$A = 22, \quad n = 20, \quad i = 6\%$

$P = 2.2\left[(1+0.06)^{20} - 1\right]/\left[0.06(1+0.06)^{20}\right]$

$P = 2.2\left[(1.06)^{20} - 1\right]/\left[0.06(1.06)^{20}\right]$

$P = 25.2$ million dollars

Continued

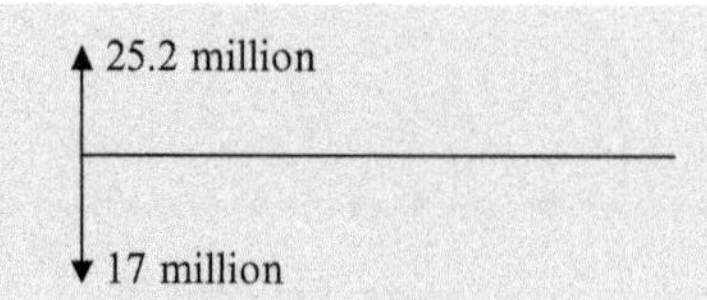

Initial investment = 17 million
Present value of return = 25.2 million
Benefit = 8.2 million
The shopping mall will bring a profit of 8.2 million.

Practice Problem 14.10

A city is considering whether to build a new power plant or to upgrade the existing power plant. The new power plant costs 2.5 million dollars. It has a maintenance cost of $50,000 per year for the first year and expected to increase by $5000 per year. The existing plant can be upgraded by spending $300,000 per year for the next 5 years. The existing plant has a maintenance cost of $60,000 per year and is expected to increase by $6000 per year. Which alternative cost is smaller if you consider the costs for next 10 years? Assume an interest rate of 5%.

Solution

All costs and benefits need to be transformed to present value.

Alternative A:
Initial cost = 2,500,000
Annual maintenance cost (A) for next 10 years = 50,000
Gradient (G) = 5000

Alternative B:
Initial cost = 0
Annual cost for upgrading ($A1$) for next 5 years = 300,000
Annual maintenance cost ($A2$) for next 10 years = 60,000
Gradient (G) = 6000

Alternative A: Arrow diagram;

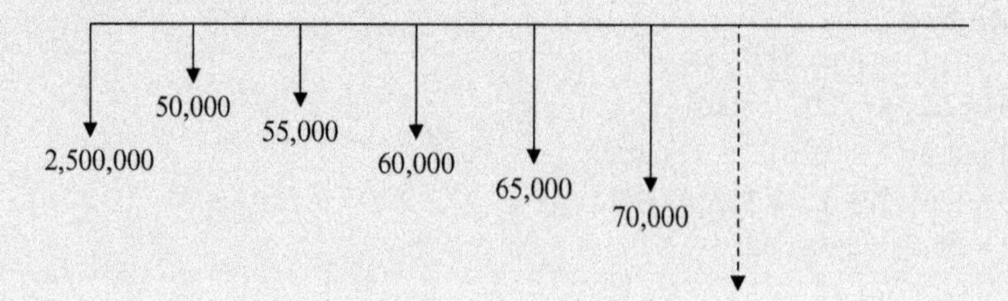

Convert all costs and benefits to present value.
Maintenance cost is divided into a uniform series and a gradient.
Maintenance cost = Uniform series (A) + Gradient (G)

$$A = 50,000,\ G = 5000$$

Uniform series (A) of \$50,000 and a gradient (G) of 5000.
Present value of uniform series of 50,000.
Use the interest table to find P/A value for $n = 10$ and $i = 5\%$.
$P/A = 7.722$
Hence $P = 7.722 \times 50,000 = -\$386,100$
(Negative value is used since it is a cost).
Convert the gradient ($G = 5000$) to present value.
Find using interest tables P/G for $i = 5\%$ and $n = 10$
$P/G = 31.652$
Hence $P = 31.652 \times 5000 = -158,260$

$$\text{Total maintenance cost} = -\$386,100 - \$158,260 = -\$544,360$$

$$\text{Total cost} = \text{Cost to build the new plant} + \text{Total maintenance cost}$$
$$= -\$2,500,000 - \$544,360 = -\$3,044,360$$

Alternative 2: The alternative cost is divided into two arrow diagrams. The first arrow diagram shows the plant upgrade cost. The second arrow diagram shows the maintenance cost.

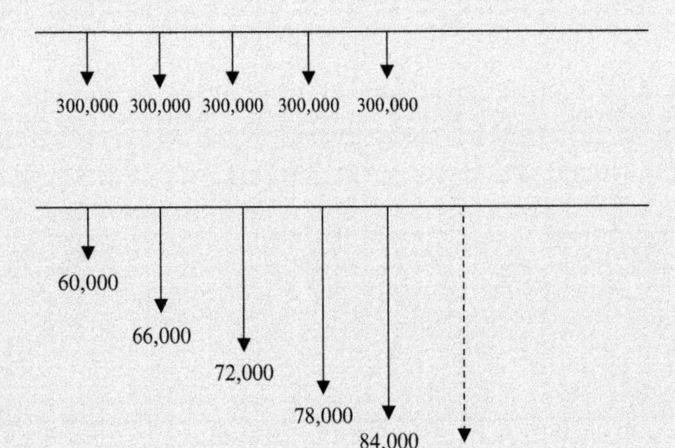

Upgrade cost:

Upgrade cost is 300,000 per year for 5 years.
$(P/A,\ i = 5,\ n = 5) = 4.329$ (from interest tables)
Present cost of the upgrade cost $= 4.329 \times 300,000 = -\$1,298,700$

Continued

Maintenance cost:

Maintenance cost is divided into a uniform series and a gradient.
Maintenane cost = Uniform series (A) + Gradient (G)

$$A = 60,000, \quad G = 5000$$

Uniform series (A) of \$60,000 and a gradient (G) of 6000.
Present value of uniform series of 60,000.
Use the interest table to find P/A value for $n = 10$ and $i = 5\%$.
$P/A = 7.722$
Hence $P = 7.722 \times 60,000 = -\$463,320$
(Negative value is used since it is a cost).
Convert the gradient ($G = 6000$) to present value.
Find using interest tables P/G for $i = 5\%$ and $n = 10$
$P/G = 31.652$
Hence $P = 31,652 \times 6000 = -189,912$

$$\text{Total maintenance cost} = -\$463,320 - \$189,912 = -\$653,232$$

$$\text{Total cost} = \text{Cost to upgrade the power plant} + \text{Total maintenance cost}$$
$$= -\$1,298,700 - \$653,232 = -\$1,951,932$$

Alternative 2 is better.

CHAPTER 15

Earned Value Management

The earned value (EV) management section is one of the easiest sections in the examination. It is highly likely that there will be questions from this section on the exam.

Consider the following scenario. A contractor receives a lump sum contract to construct a 1000 ft long pipeline for \$500 per linear foot. The contractor is required to complete the project in 10 months. The contractor will be paid monthly based on the work completed.

$$\text{Budgeted cost} = 1000 \times 500 = \$500,000$$

Assume that the contractor has completed 80 ft of pipe after 1 month and he has expended \$55,000. This is contractor's actual cost (AC).

15.1 COST IMPLICATIONS

Since the contractor has completed only 80 ft, he is entitled to \$40,000 (80 ft × 500). The contractor's entitlement is known as earned value (EV).

$$\text{Contractor's actual cost after 1 month (AC)} = \$55,000$$

$$\text{Contractor's earnings (EV)} = \$40,000$$

$$\text{Contractor's loss after 1 month} = \$15,000$$

It is clear that the contractor has to cut costs; otherwise, he will end up with a loss at the end of the project.

15.2 SCHEDULE IMPLICATIONS

A contractor is supposed to complete 1000 ft of pipe in 10 months. Hence, he is scheduled to complete 100 ft every month. His planned value (PV) after 1 month is \$50,000 (100 ft × 500).

However, he has completed only 80 ft of pipe after 1 month. It is clear that the contractor is behind schedule.

Scheduled cost or planned value $(\text{PV}) = 100 \times 500 = 50,000$.

Construction Engineering Design Calculations and Rules of Thumb
http://dx.doi.org/10.1016/B978-0-12-809244-6.00015-9

Table 15.1 Planned value, earned value, and actual cost

	Month 1	Month 2	Month 3	Month 4	Month 5	Month 6	Month 7	Month 8	Month 9	Month 10
PV	50,000	100,000	150,000	200,000	250,000	300,000	350,000	400,000	450,000	500,000
EV	40,000									
AC	55,000									

Table 15.2 Planned value, earned value, and actual cost

	Month 1	Month 2	Month 3	Month 4	Month 5	Month 6	Month 7	Month 8	Month 9	Month 10
PV	50,000	100,000	150,000	200,000	250,000	300,000	350,000	400,000	450,000	500,000
EV	40,000	95,000								
AC	55,000	102,000								

Hence, for this project, the following information can be formulated after 1 month (Table 15.1).

Assume that after the second month he has completed a total of 190 ft of pipes (including the 80 ft he did in the first month) and his total expenses (including the first month) is $102,000.

Find EV, PV, and AC (Table 15.2).

$$\text{Actual cost (AC)} = \$102,000$$

$$\text{Planned value (PV)} = 100,000 \text{ (as per schedule)}$$

$$\text{Earned value (EV)} = 190\text{ft} \times 500 = \$95,000$$

Practice Problem 15.1

Mr. Drake of ABC contracting has obtained a project to drive 600 ft of piles in 4 months. The contractor will be paid $200 per each foot of piling. After 2 months, the contractor has completed 350 ft of piling and his AC is $82,000.

As a construction engineer what is your advice to the contractor?

Solution

$$\text{Total budget} = 600\text{ft} \times 200 = \$120,000$$

After 2 months as per schedule, 50% of the project should be completed.

$$\text{Planned value (PV)} = 50\% \times 120,000 = \$60,000$$

After 2 months contractor has completed 350 ft of piles.

$$\text{Earned value (EV)} = 350\text{ft} \times 200 = \$70,000$$

$$\text{Actual cost for the contractor (AC)} = \$82,000$$

Analysis

The contractor has earned $70,000 but expended $82,000. Hence, contractor has lost $12,000 after 2 months.

As per schedule, the contractor should have earned $60,000 (PV).

However, the contractor has earned 70,000. Hence, contractor is ahead of schedule.

The contractor is moving faster than the original schedule.

Your advice to the contractor: "Mr. Drake, you are spending more than you earn. Hence, you are losing money. If you want to be profitable, you have to cut costs. As per progress, you are moving faster than the schedule. You can slow down a little bit if that helps in cutting costs."

15.3 FANCY TERMS FOR EV, PV, AND AC

Some prefer to use terms such as BCWP, ACWP, BCWS for EV, AC, and PV.

$$\text{Earned value (EV)} = \text{BCWP (Budgeted cost for work performed)}$$

$$\text{Actual cost (AC)} = \text{ACWP (Actual cost for work performed)}$$

$$\text{Planned value (PV)} = \text{BCWS (Budgeted cost for work scheduled)}$$

Unfortunately, you have to remember these terms as well.

15.4 COST PERFORMANCE INDEX

Cost Performance Index (CPI) is defined as follows:

$$\boxed{\text{CPI} = \text{EV}/\text{AC} = \text{BCWP}/\text{ACWP}}$$

CPI = 1.0: If CPI is equal to 1.0, the EV is equal to the AC. In other words, a contractor is spending exactly what he is earning. Contractors are in the business to make money. If he is spending exactly what he is earning, that is not productive.

CPI < 1.0: In this case, the EV is less than the AC. In other words, he is earning less than what he is spending. The contractor is losing money.

CPI > 1.0: In this case, the EV is greater than the AC. In other words, he is earning more than what he is spending. The contractor is making a profit.

15.5 SCHEDULE PERFORMANCE INDEX

Schedule Performance Index (SPI) is defined as follows:

$$\boxed{\text{SPI} = \text{EV}/\text{PV} = \text{BCWP}/\text{BCWS}}$$

SPI = 1.0: If SPI is equal to 1.0, the EV is equal to the PV. In other words, the contractor is earning exactly what he planned as per schedule. The contractor is moving along as per schedule.

SPI < 1.0: In this case, the EV is less than the PV. In other words, he is earning less than what he planned as per schedule. The contractor is falling behind the schedule.

SPI > 1.0: In this case, the EV is greater than the PV. In other words, he is earning more than what he planned as per schedule. The contractor is ahead of schedule.

Some questions and answers

Question: SPI of a project is 1.5. Is the contractor making a profit?
Answer: It is not possible to tell whether the contractor is making a profit by looking at the SPI.
Question: If SPI is 1.5, is the contractor ahead of the schedule?
Answer: If SPI is 1.5, then the contractor is earning more than what he planned. Hence, he is ahead of schedule.
Question: If CPI is 0.8, is the contractor ahead of the schedule?
Answer: CPI cannot be used to answer questions about the schedule.
Question: If CPI is 0.8, is the contractor making a profit.
Answer: If CPI is 0.8, EV is less than AC. He is earning less than what he is spending. The contractor is losing money.

15.6 COST VARIANCE

Cost variance (CV) is defined as follows:

$$\boxed{\text{CV (cost variance)} = \text{EV} - \text{AC} = \text{BCWP} - \text{ACWP}}$$

If CV is a positive value, then the contractor is earning more than what he is spending. That is a positive development. If CV is negative then the contractor is losing money.

15.7 SCHEDULE VARIANCE

Schedule variance (SV) is defined as follows:

$$\boxed{\text{SV (schedule variance)} = \text{EV} - \text{PV} = \text{BCWP} - \text{BCWS}}$$

If SV is a positive value, then the contractor is earning more than what he planned at the start of the project. That means the contractor is ahead of schedule. If SV is negative, then the contractor is earning less than what he planned at the beginning. The contractor is behind schedule.

Summary

$\text{CPI} = \text{EV}/\text{AC} = \text{BCWP}/\text{ACWP}$

$\text{CPI} > 1$: Contractor is making a profit
$\text{CPI} < 1$: Contractor is losing money
CPI cannot be used to tell whether a contractor is ahead of schedule or behind schedule

SPI = EV/PV = BCWP/BCWS

SPI > 1: Contractor is ahead of schedule

SPI < 1: Contractor is behind schedule

SPI cannot be used to tell whether contractor is making a profit or not

CV (cost variance) = EV – AC = BCWP – ACWP

If CV is positive, the contractor is making a profit

If CV is negative, the contractor is losing money

SV (schedule variance) = EV – PV = BCWP – BCWS

If SV is positive, the contractor is ahead of schedule

If SV is negative, the contractor is behind schedule

15.8 EV GRAPHS

In Fig. 15.1, the contractor agreed with the client to get paid \$1 per each square foot of painting.

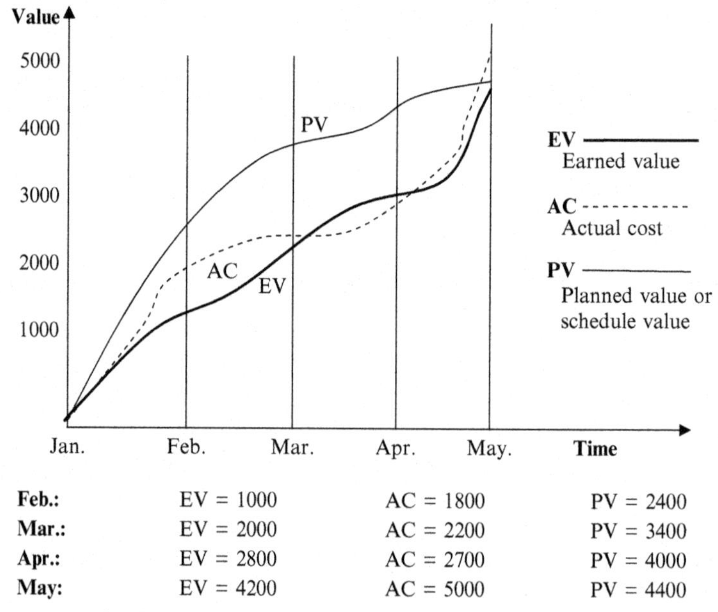

Fig. 15.1 Earned value graph.

February

- *Earned value:* By Feb., the contractor has completed 1000 sq ft of painting and client pays him \$1000 (EV) as per contract.

- *Actual cost:* But the contractor found out that he has spent \$1800 by Feb. The owner was furious and after investigating, he was able to find the reason. He found out that some workers were inefficient and highly paid. To cut costs, he decides to replace highly paid workers with low paid workers. Further, he takes steps to increase the productivity.

$$\text{CPI} = \text{EV}/\text{AC} = 1000/1800 = 0.56$$

In other words, to earn \$1, he has to spend ~\$1.8. This is not a healthy situation.

$$\begin{aligned}\text{CV (cost variance)} &= \text{EV} - \text{AC} = 1000 - 1800 \\ &= -800 \text{ (Cash flow is negative)}\end{aligned}$$

- *Planned value:* If the contractor had completed everything as per schedule, he would be paid the PV. It is very rare for any project to go exactly as the schedule. As per schedule, the contractor is supposed to finish 2400 sq ft of painting by Feb. If he was able to stick to the schedule, he would have earned \$2400 by Feb. However, the contractor had painted only 1000 sq. ft; hence, the contractor is behind schedule.

$$\text{SPI} = \text{EV}/\text{PV} = 1000/2400 = 0.41$$

$$\text{SV (schedule variance)} = \text{EV} - \text{PV} = 1000 - 2400 = -1400$$

The contractor is way behind the schedule. He should have completed ~\$2400 worth of work. However, he has completed only 1000 sq ft.

March

- *Earned value:* By Mar., the contractor has completed 2000 sq ft of painting and client pays him \$2000 (EV) as per contract.
- *Actual cost:* Due to hiring low salary workers, the AC of the contractor has reduced. Now the AC is \$2200.
- $\text{CPI} = \text{EV}/\text{AC} = 2000/2200 = 0.91$
  Still his ACs are higher than his earnings. However, the situation has significantly improved.
  CV (cost variance) = EV − AC = 2000 − 2200 = −200 (Cash flow is still negative. Things have vastly improved.)
- *Planned value:* As per schedule, the contractor is supposed to finish 3400 sq ft of painting by Mar. If he was able to stick to the schedule, he would have earned \$3400 by Mar. However, the contractor had painted only 2000 sq ft, hence, the contractor is behind schedule.
- $\text{SPI} = \text{EV}/\text{PV} = 2000/3400 = 0.59$

$$\text{SV (schedule variance)} = \text{EV} - \text{PV} = 2000 - 3400 = -1400$$

April

- *Earned value:* By Apr., the contractor has completed 2800 sq ft of painting and client pays him $2800 (EV) as per contract.
- *Actual cost:* Due to hiring low salary workers, the AC of the contractor has reduced. Now the AC is $2700.
- CPI = EV/AC = 2800/2700 = 1.03

  His CPI has gone above 1.0. His ACs are lower than his earnings.

  CV (cost variance) = EV − AC = 2800 − 2700 = 100 (Cash flow is now positive. Things are looking good for the contractor).
- *Planned value:* As per schedule, the contractor is supposed to finish 4000 sq ft of painting by Apr. If he was able to stick to the schedule, he would have earned $4000 by Apr. However, the contractor had painted only 2800 sq ft; hence, the contractor is still behind schedule.
- SPI = EV/PV = 2800/4000 = 0.70

$$\text{SV (schedule variance)} = \text{EV} - \text{PV} = 2800 - 4000 = -1200$$

May

- *Earned value:* By May, contractor has completed 4200 sq. ft of painting and client pays him $4200 (EV) as per contract.
- *Actual cost:* Now the AC is $5000.
- CPI = EV/AC = 4200/5000 = 0.84

  The contractor is losing money.

$$\text{CV (cost variance)} = \text{EV} - \text{AC} = 4200 - 5000 = -800$$

- *Planned value:* As per schedule, the contractor is supposed to finish 4400 sq ft of painting by May. If he was able to stick to the schedule, he would have earned $4400 by May. However, the contractor had painted only 4200 sq ft, hence, the contractor is still behind schedule.
- SPI = EV/PV = 4200/4400 = 0.95

$$\text{SV (schedule variance)} = \text{EV} - \text{PV} = 4200 - 4400 = -200$$

The contractor is behind schedule; he will exceed the contract duration.

CHAPTER 16

Construction Operations and Methods

16.1 INTRODUCTION

Construction work requires many operations. Lifting and rigging, dewatering, concreting, bracing of temporary structures, underpinning, jacking, grouting, and pumping are some common construction operations.

16.2 LIFTING AND RIGGING

Construction cannot be done without the lifting and rigging of material and equipment to the proper location. "Rigging" is a word that comes from the early sailing days. Rigging meant moving a ship, which involved many operations. In the construction industry, the word "rigging" is used to indicate the moving and lifting of material and equipment.

You cannot build if you do not move it to the correct location. Moving involves lifting, rolling, pulling, and pushing. Ancient Egyptians used wooden cylinders to move large stones. Today many other forms of equipment are used. As a construction engineer, you need to be familiar with the equipment and methods used to lift and move material and equipment.

Typically, in the construction industry, steel beams and steel columns have to be lifted. In high-rise building construction, concrete has to be lifted. Construction equipment such as jacks, tools, concrete mixers, and testing equipment also have to be moved and lifted to the place where they are needed.

16.3 SHEAVES (PULLEYS) AND BLOCKS

"Sheave" and "pulley" means the same thing.

Sheaves (Pulleys): A sheave is same as a pulley. A block is a set of pulleys connected together (Fig. 16.1).

In a sheave, the mechanical advantage is one. In other words, $T = W$.

Construction Engineering Design Calculations and Rules of Thumb
http://dx.doi.org/10.1016/B978-0-12-809244-6.00016-0

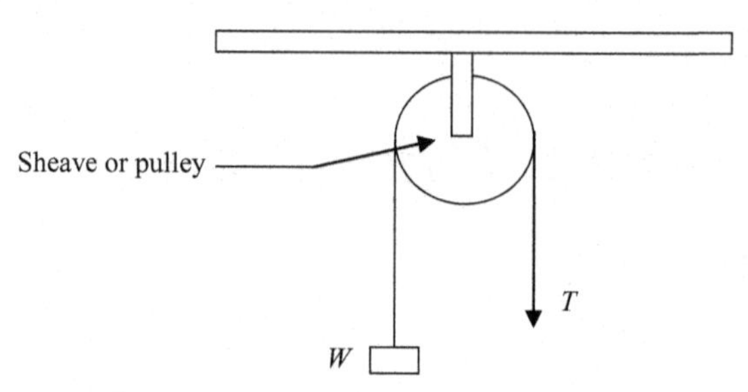

Fig. 16.1 Sheave or pulley.

This is assuming that the pulley and rope have no friction. Later, we will discuss how to calculate the force when friction is present.

Blocks: When more than one sheave is connected together, it is known as a *block*.

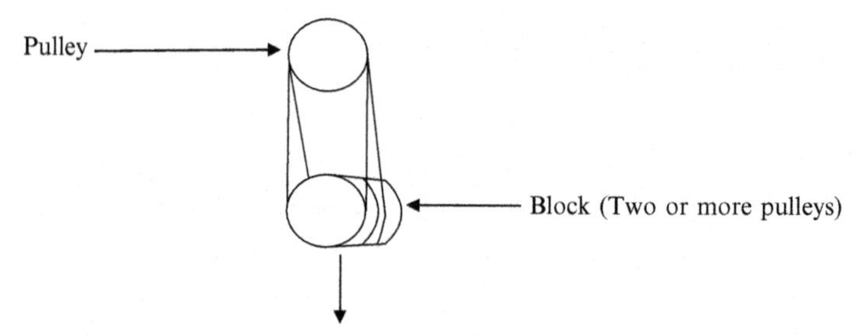

Fig. 16.2 Pulley and a block.

Look at Fig. 16.2. It shows a pulley or a sheave on the upper level. The lower level has two pulleys. The two pulley system is known as a block.

16.4 BLOCK AND TACKLE

When two or more blocks are combined together, it is known as a block and tackle.

16.5 SINGLE WHIP

Lifting with one sheave is known as single whip (Fig. 16.3).

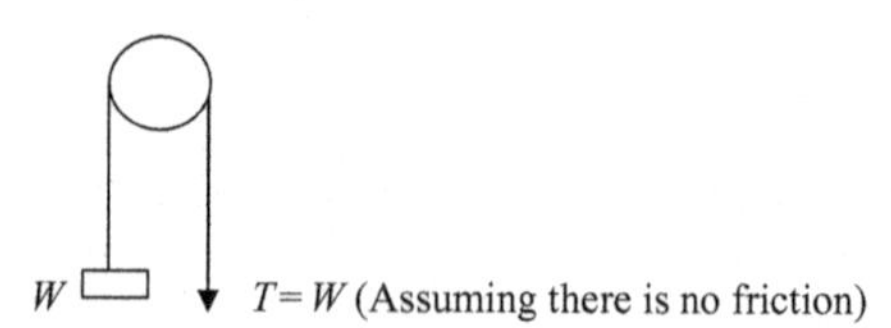

Fig. 16.3 Single whip.

The mechanical advantage of a single whip is 1.0. In other words, there is no advantage. The force required to lift weight W is W.

Parts of line: Parts of line is defined as the number of ropes attached to the weight. In this case, only one rope is attached to the weight. Hence, the parts of line is 1.0.

Movement of single whip: When weight is moved 1 ft upwards, the effort moves 1 ft downward.

16.6 GUN TACKLE

Two sheaves assembled one below the other is known as a gun tackle (Fig. 16.4).

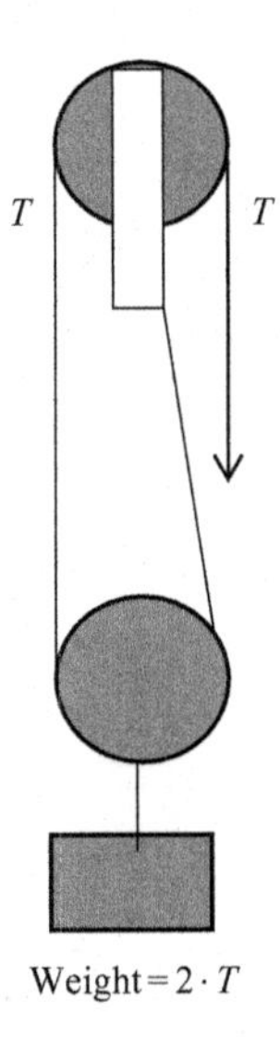

Fig. 16.4 Gun tackle.

Let us assume the applied force is "T." Then this tension travels through the ropes as shown if there is no friction. Now if you look at the lower block, it has two ropes pulling up. Hence, the block is been pulled by a force of $2T$.

$$\text{The force applied} = 100\text{lbs}$$

$$\text{The weight that can be lifted} = 200\text{lbs}$$

$$\text{Mechanical advantage} = 2.0$$

16.6.1 Parts of Line

In this case, two ropes are attached to the weight. Hence, the parts of line for a gun tackle is 2.0. Note that mechanical advantage and parts of line are the same.

16.6.2 Movement of Gun Tackle

Let us see how much weight moves when the effort moves by 1 ft (Fig. 16.5).

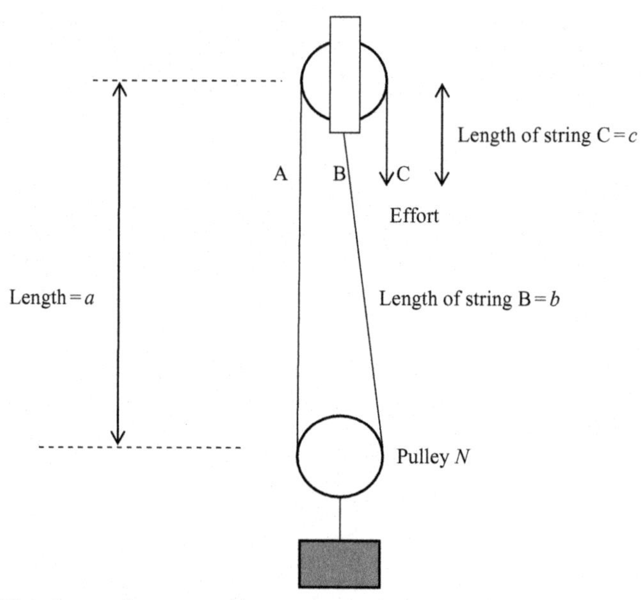

Fig. 16.5 Movement of a gun tackle.

Strings are named A, B, and C.

$$\text{Length of string A} = \text{“}a\text{”}$$

$$\text{Length of string B} = \text{“}b\text{”}$$

$$\text{Length of string C} = \text{“}c\text{”}$$

$$\text{Total length of strings} = a + b + c$$

Now let us assume that effort moves by X ft downward (Fig. 16.6).
Let us say string C is pulled by distance X downwards.
Then let us assume pulley N would go up by distance Y.

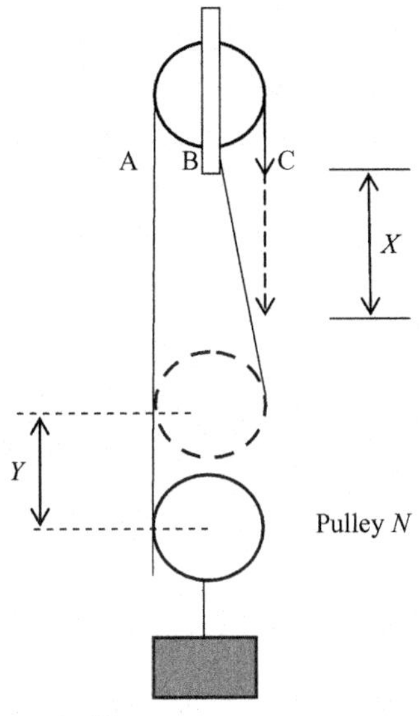

Fig. 16.6 Movement of a gun tackle.

New lengths of strings:

$$\text{Length of string A} = a - Y$$

$$\text{Length of string B} = b - Y$$

$$\text{Length of string C} = c + X$$

$$\text{Total length of strings} = (a - Y) + (b - Y) + c + X$$

The total lengths of strings do not change:
Hence,

$$a + b + c = (a - Y) + (b - Y) + c + X$$

$$0 = -2Y + X$$

$$Y = X/2$$

Hence, we can see that when the string is pulled by a distance of X, the pulley "N" would go up by a distance of $X/2$.

When the effort moves by a distance of 1 ft, the weight moves up by a distance of 1/2 ft.

In addition, we can say that when weight moves 1 ft, effort has to move 2 ft.

16.7 PULLEYS WITH FRICTION

So far, we have considered only pulleys without friction. In the real world, friction can be significant. Let us look at the simple pulley. Assume the weight is W and the force required to pull the weight is T. Note that T is NOT equal to W when friction is present. Also assume that the friction coefficient is μ (Fig. 16.7).

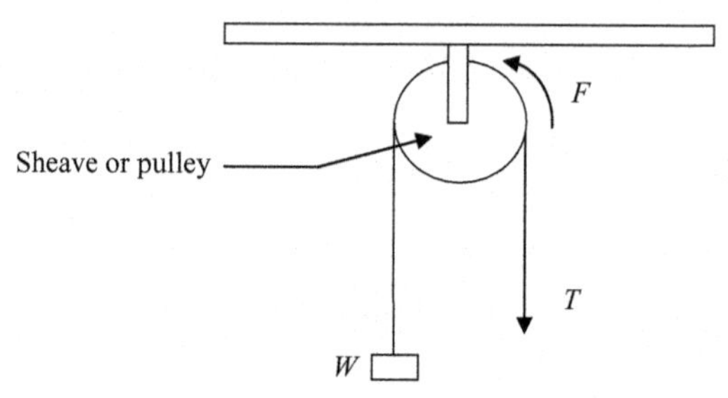

Fig. 16.7 Pulleys with friction.

When the rope is pulled down, friction (F) is acting against the movement of the rope.

Hence, we can write:

$$T = W + F \tag{16.1}$$

W and F are acting in the same direction.

$$\text{Now } F = \mu \cdot T \tag{16.2}$$

One may think why F is not equal to $\mu \cdot W$.

In reality, a portion of the rope has a tension of T and other portion has a tension of $W \cdot F = \mu \cdot T$ is larger than $F = \mu \cdot W$. Hence, it is conservative.

$$T = W + \mu \cdot T$$

$$W = T - \mu \cdot T = T(1 - \mu)$$

$$T = W/(1 - \mu)$$

Practice Problem 16.1

Assume the friction factor of a rope and pulley is 0.24. A load of 150 lbs needs to be lifted with a simple pulley. What is the force required?

Solution

$$T = W/(1-\mu)$$

$$T = 150/(1-0.24) = 197.37\,\text{lbs}$$

Gun Tackle With Friction: Fig. 16.8.

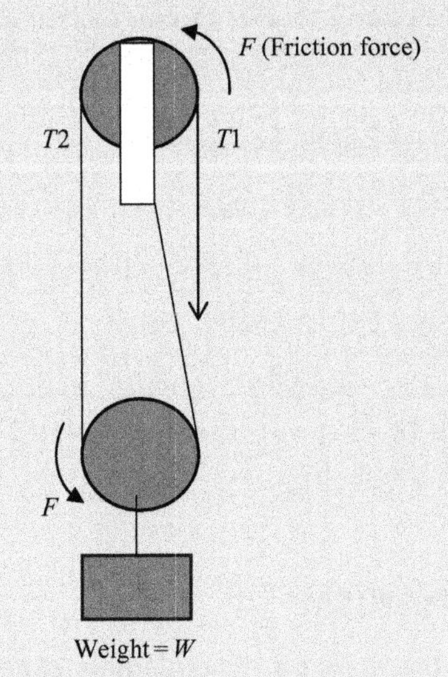

Fig. 16.8 Gun tackle with friction.

$$T1 = T2 + F$$

$T2$ and F acts on the same direction.

$$T1 = T2 + \mu \cdot T1$$
$$T2 = T1 \times (1-\mu) \quad (16.3)$$

Similarly;

$$T3 + F = T2$$

$T3$ and F acts in the same direction.

$$T3 + \mu \cdot T2 = T2$$
$$T3 = T2 \times (1-\mu) \quad (16.4)$$

From (16.3): $T2 = T1 \times (1-\mu)$

Continued

Hence $T3 = T1 \times (1-\mu) \times (1-\mu) = T1 \times (1-\mu)^2$

$W = T3 + T2$

$W = T1 \times (1-\mu)^2 + T1 \times (1-\mu) = T1 \times \left[(1-\mu)^2 + (1-\mu)\right]$

Practice Problem 16.2

A gun tackle is used to lift a weight of 230 lbs. The friction coefficient of two pulleys is 0.13. Find the force required to lift the weight.

Solution

$$W = 230\text{lbs} \ \text{ and } \ \mu = 0.13$$

$$W = T1 \times \left[(1-\mu)^2 + (1-\mu)\right]$$

$$230 = T1\left[(1-0.13)^2 + (1-0.13)\right]$$

$$230 = T1 \times 1.6269$$

$$T1 = 141.37\text{lbs}$$

A 230 lbs load can be lifted with an effort of 141.37 lbs. If there was no friction, the same load could have lifted with 115 lbs.

16.8 CRANE MECHANISM

Now we have acquired enough knowledge to tackle the crane mechanism. A simple crane consists of a winch, an upper pulley, and a moving block. A winch provides power to the rope through an internal combustion engine. Let us look at the figure below. You can see the upper pulley, moving block, and the winch.

The winch provides power to the ropes.

Practice Problem 16.3

A typical hoisting mechanism for a crane is shown in the Fig. The load lifted is 900 lbs. The friction coefficient of ropes and pulleys is 0.15. What is the effort required to lift the weight (Fig. 16.9)?

Solution

*F*1, *F*2, and *F*3 are frictional forces. Assume the effort to be *T* and the rope tensions as shown. Friction always acts against the movement of the rope.

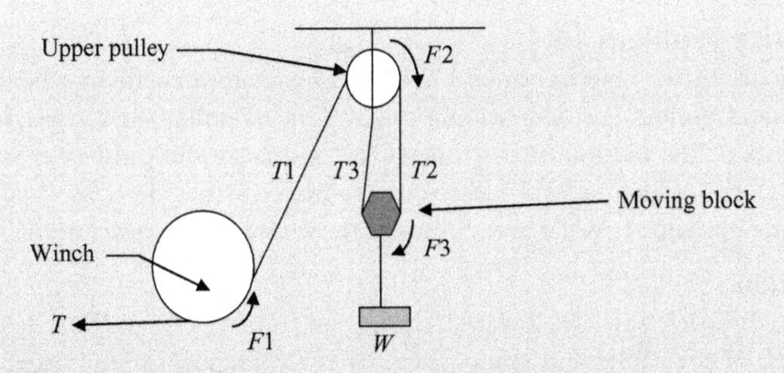

Fig. 16.9 Crane mechanism.

$$T = T1 + F1 = T1 + \mu \cdot T$$

Friction force and $T1$ act in the same direction.

$$T1 = T \times (1 - \mu) \quad (16.5)$$

Similarly at upper pulley:

$$T1 = T2 + F2 = T2 + \mu \cdot T1$$

$$T2 = T1 \times (1 - \mu) \quad (16.6)$$

From (16.5)

$$T2 = T \times (1 - \mu)^2 \quad (16.7)$$

Also:

$$T2 = T3 + F3 = T3 + \mu \cdot T2$$

$$T3 = T2 \times (1 - \mu)$$

From above (16.7)

$$T3 = T \times (1 - \mu)^2 \times (1 - \mu) = T \times (1 - \mu)^3$$

$$W = T3 + T2$$

$$W = T \times (1 - \mu)^3 + T \times (1 - \mu)^2$$

$$W = T \times [1 - \mu)^3 + (1 - \mu)^2]$$

$$T = W/[1 - \mu)^3 + (1 - \mu)^2]$$

$$T = 900/[1 - 0.15)^3 + (1 - 0.15)^2] = 673.33\text{lbs}$$

In this case, a 900 lb load is lifted with an effort of 673.33 lbs.

If there is no friction, the load can be lifted with an effort of 450 lbs since the mechanical advantage is 2.0.

When the hook has to be lowered, the moving block should move down. It should have enough weight to overcome the friction.

Practice Problem 16.4

In Fig. 16.9, the load has to be lowered. The friction coefficient between ropes and pulleys is 0.15. Assume the weight of pulleys and ropes to be negligible. The tension of the rope at the winch should not be zero. The rope in the winch should have some tension so that the rope will not become entangled. What is the minimum weight of the moving block?

Solution

When the hook has to be lowered, the moving block needs to overcome the friction of the cables and come down by itself. Hence, the moving block should be heavy enough to overcome the friction in the ropes. The weight of the moving block allows the hook to be brought down when there is no weight attached (Fig. 16.10).

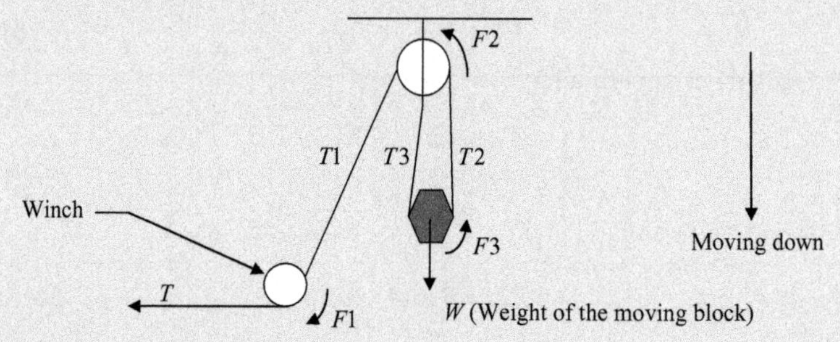

Fig. 16.10 Crane mechanism. The weight is moving down.

When the moving block comes down, friction is the acting against the movement of ropes.

$$T1 = F1 + T$$

$$F1 = \mu \cdot T1$$

$$T1 = \mu \cdot T1 + T \tag{16.8}$$

$$T = T1 \cdot (1 - \mu)$$

$$T1 = T/(1 - \mu)$$

Also:

$$T1 + F2 = T2$$

$$T1 + \mu \cdot T2 = T2 \tag{16.9}$$

$$T1 = T2 \cdot (1 - \mu)$$

From (16.8)

$$T1 = T/(1-\mu)$$

$$T/(1-\mu) = T2(1-\mu) \quad (16.10)$$

$$T2 = T/(1-\mu)^2$$

At the moving block:

$$T3 = T2 + F3$$

$$F3 = \mu \cdot T3$$

$$T3 = T2 + \mu \cdot T3$$

$$T3 = T2/(1-\mu)$$

From (16.10) $T2 = T/(1-\mu)^2$

$$T3 = T/(1-\mu)^3 \quad (16.11)$$

$$W = T2 + T3$$

Hence:

$$W = T/(1-\mu)^2 + T/(1-\mu)^3$$

$$W = T \cdot \left[1/(1-\mu)^2 + 1/(1-\mu)^3\right]$$

$$\mu = 0.15 \text{ and } T = 100$$

$$W = 100 \times 3.012 = 301.2\text{lbs}$$

Headache ball: In some instances, a headache ball is used instead of a moving block to keep the ropes tight when a load is not attached.

16.9 CHAIN HOISTS

Chain hoists are different from block and tackles. In the case of block and tackles, the radius of sheaves did not come into the equation. On the other hand, the radius of sheaves is important for the mechanism of chain hoists. Chain hoists are largely used to lift heavy weights. The main advantage of chain hoists is that a large load can be lifted with little force. The mechanism of a chain hoist is different from the mechanism of a block and tackle.

16.9.1 Chain Hoist Mechanism

Let us look at an example to understand the mechanism of chain hoists.

A chain hoist is shown in Fig. 16.11. Let us assume the radii of pulleys are 6.0, 5.0, and 5.5 in. as shown. Assume a force of 130 lbs is applied (Fig. 16.12).

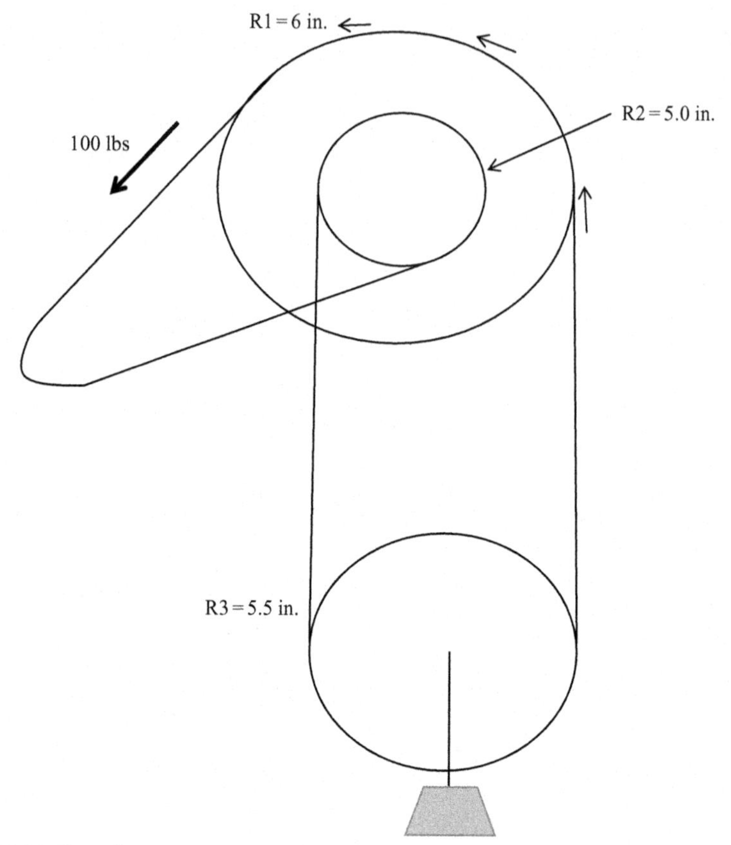

Fig. 16.11 Chain hoist.

Wheels R1 and R2 are rigidly attached. When the wheel R1 makes one revolution, wheel R2 also makes one revolution. Though it is the same chain that goes around, the chain segments are named A, B, C, and D.

Now let us look at the mechanism.

STEP 1: Find the movement of the weight:

A worker pulls the chain and the wheel R1 makes one revolution. When wheel R1 makes one revolution, wheel R2 also makes one revolution since two wheels are rigidly attached.

The chain is pulled by a distance of $2 \cdot \pi \cdot \mathrm{R1}$.

The chain segment D would go *up* by $2 \cdot \pi \cdot \mathrm{R1}$.

When the chain segment D goes up by $2 \cdot \pi \cdot \mathrm{R1}$, the weight "*W*" will go up by $(2 \cdot \pi \cdot \mathrm{R1})/2$.

When wheel R2 makes one revolution, the chain segment C goes *down* by $2 \cdot \pi \cdot \mathrm{R2}$

Net movement of the weight is $(2 \cdot \pi \cdot \mathrm{R1} - 2 \cdot \pi \cdot \mathrm{R2})/2 = \pi \cdot (\mathrm{R1} - \mathrm{R2})$

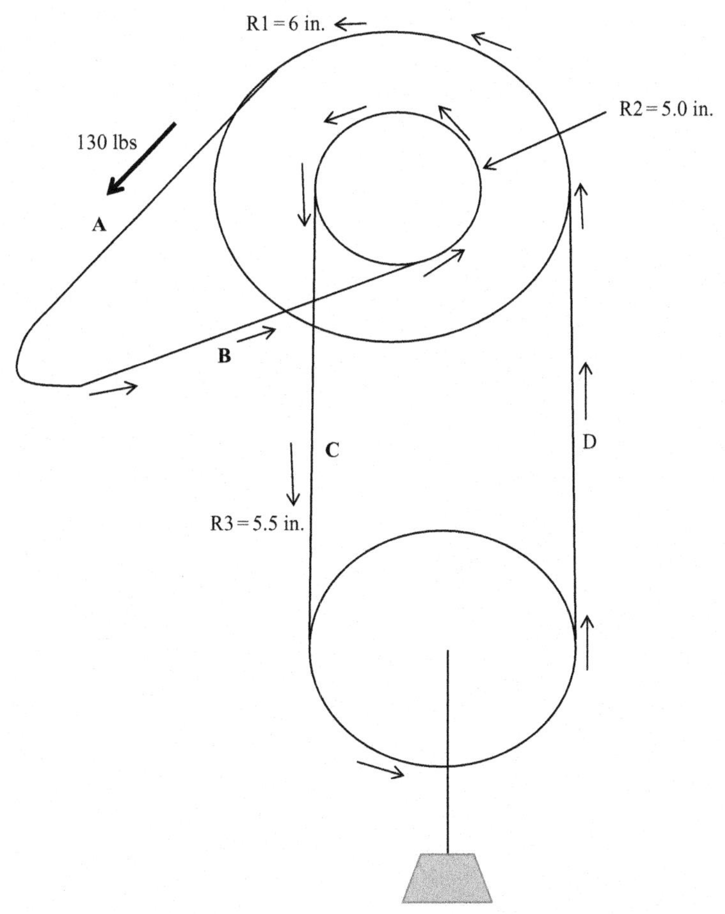

Fig. 16.12 Chain hoist movement.

STEP 2: Work done by weight and effort is the same if there is no friction.

The effort is given to be 130 lbs. The weight that can be lifted is to be found.

$$\text{Work done by effort} = \text{Effort} \times \text{Distance travelled by effort}$$

$$\text{Work done by weight} = \text{Weight} \times \text{Distance travelled by the weight}$$

$$\text{Work done by effort} = 130 \times 2 \cdot \pi \cdot \text{R1}$$

$$\text{Work done by weight} = W \times \pi \cdot (\text{R1} - \text{R2})$$

$$130 \times 2 \cdot \pi \cdot \text{R1} = W \times \pi \cdot (\text{R1} - \text{R2})$$

$$130 \times 2 \times 6 = W(6 - 5) \text{ lbs.in.}$$

Weight is in lbs and distance is in inches on both sides of the equation.

$$W = 1560 \text{ lbs}$$

CHAPTER 17

Dewatering in Construction

17.1 DEWATERING

Construction work is difficult or sometimes impossible when water is present. Dewatering is conducted to facilitate construction work in excavations. Excavations are needed for shallow foundations, retaining walls, and basements.

Construction engineers need to investigate the elevation of groundwater prior to a dewatering plan. Stability of the bottom may be affected by the presence of groundwater. In such situations dewatering method needs to be planned. Maintaining sidewall stability is another problem that needs to be addressed (Fig. 17.1).

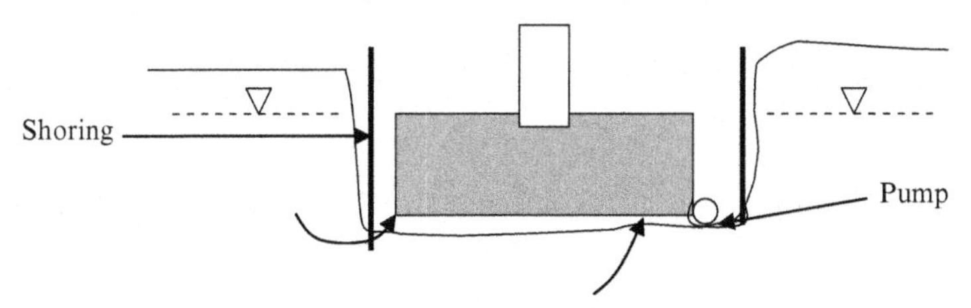

Fig. 17.1 Dewatering for a shallow foundation.

Concreting work for shallow foundations is conducted by installing a pump. Continuous pumping can be used to maintain a dry bottom. Stable side slopes can be maintained by providing shoring supports.

17.1.1 Well Points

In some cases it is not possible to pump enough water to maintain a dry bottom for concrete shallow foundations. In such situations well points are constructed to lower the groundwater table.

Use of well points to lower the groundwater table.

Small scale dewatering for column footings: Pump water from the excavation.

Construction Engineering Design Calculations and Rules of Thumb
http://dx.doi.org/10.1016/B978-0-12-809244-6.00017-2

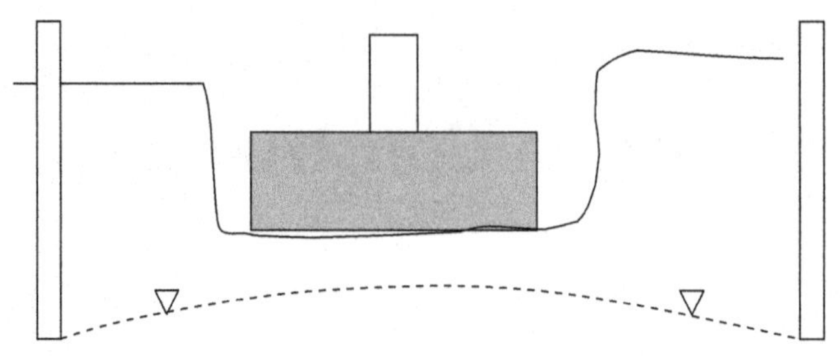

Fig. 17.2 Well points.

The groundwater level is lowered by constructing a small hole or a trench inside the excavation (as shown in Fig. 17.2) and placing a pump (or several pumps) inside the excavation. For most column footing construction work, this method will be sufficient (Fig. 17.3).

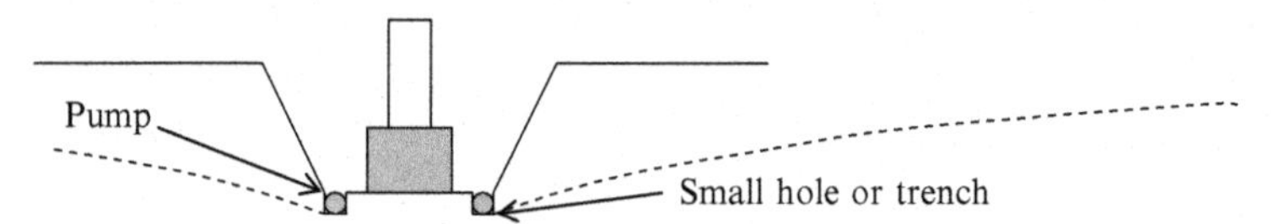

Fig. 17.3 Pumping water from an excavation.

17.1.2 Dewatering of a Column Footing

Medium scale dewatering for basements or deep excavations: Pump water from trenches or wells. For medium scale dewatering projects, trenches or well points can be used (Fig. 17.4).

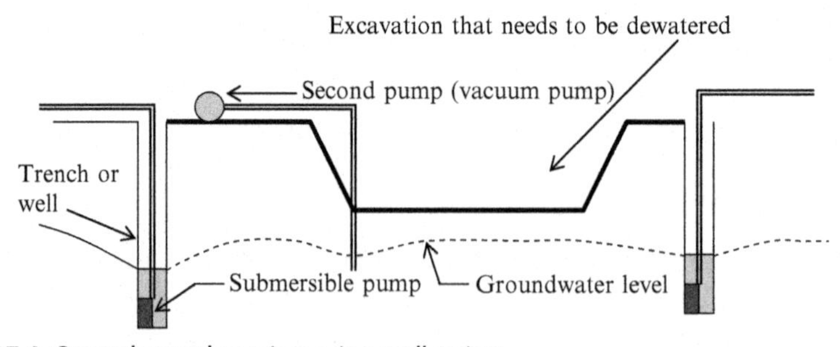

Fig. 17.4 Groundwater lowering using well points.

Add more pumps as necessary to keep the excavation dry. A combination of submersible pumps and vacuum pumps can be used. Large scale dewatering for basements or deep excavations:

Alternative 1: Well points or trenches are constructed. The main artery pipe is connected to each pump as shown. A strong, high-capacity pump will suck the water out of all the wells as shown (Fig. 17.5).

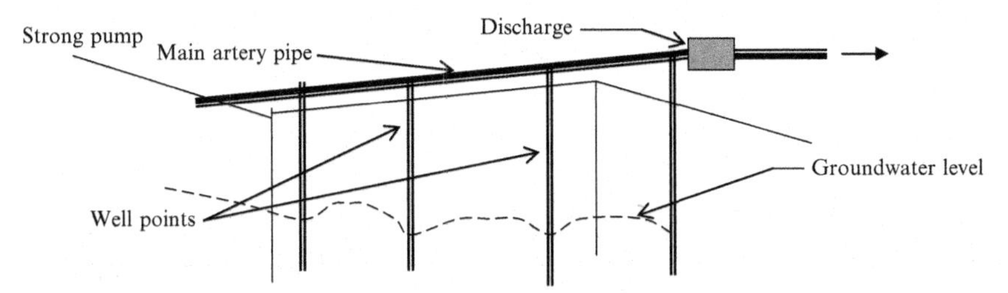

Fig. 17.5 Well points.

17.1.3 Well Points in Series With One Large Pump

Alternative 2: A similar dewatering system can be designed using submersible pumps. In this case, instead of one pump, each well would get a submersible pump as shown in Fig. 17.6.

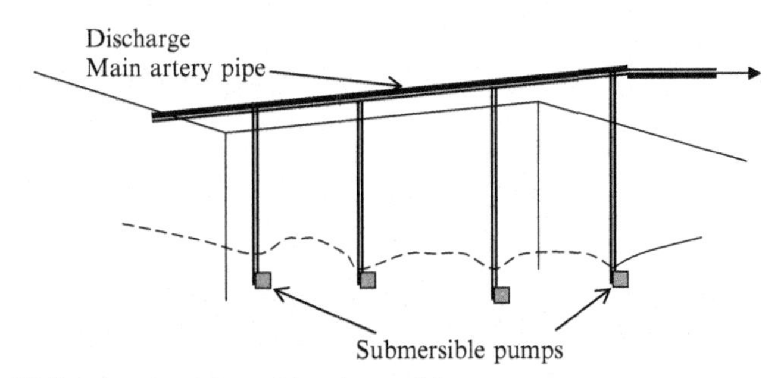

Fig. 17.6 Well points in series with submersible pumps.

This method is more effective than the alternative one. The main disadvantage is that it requires high maintenance. Since there is more than one pump, more maintenance work will occur compared to alternative one. On the other hand alternative two can be modified to include less well points since pumping effort can be increased significantly.

For most construction work, groundwater should be lowered at least 2 ft below the excavation.

If water needs to be treated prior to discharge, dewatering may become extremely expensive. In that case, "cutoff walls" or "ground freezing" options may be cheaper.

17.2 DESIGN OF DEWATERING SYSTEMS

Initial study: Study the surrounding area and locate nearby rivers, lakes, and other water bodies. Groundwater normally flows toward surface water bodies such as rivers and lakes. If the site is adjacent to a major water body (such as a river or lake), the groundwater elevation will be the same as the water level in the river.

Construct borings and piezometers: Soil conditions: Create soil profiles based on borings. Assess the permeability of the existing soil stratums. Sandy soils will transmit more water than clayey soils.

Seasonal variations: Groundwater elevation readings should be taken at regular intervals. In some sites, groundwater elevation could be very sensitive to seasonal changes. The groundwater elevation during the summer could be drastically different from the winter.

Tidal effects: Groundwater elevation changes with respect to tidal flow. In some sites groundwater elevation may show a very high sensitivity to high and low tides.

Artesian conditions: Check for artesian conditions. Groundwater could be under pressure and well pumping or any other dewatering scheme could be a costly procedure.

Groundwater contamination: If contaminated groundwater is found, then groundwater cutoff methods should be studied.

17.3 MONITORING WELLS

Monitoring wells are installed to obtain the groundwater elevation. Monitoring wells are typically constructed using PVC pipes (Fig. 17.7).

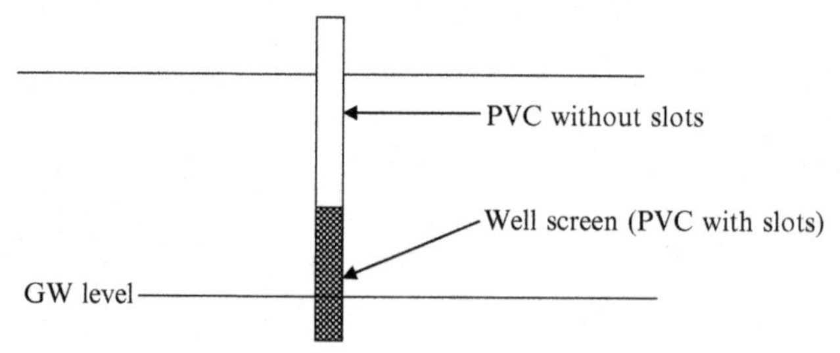

Fig. 17.7 Groundwater monitoring well.

A slotted section of the PVC is known as the well screen and allows water to flow into the well. If there is no pressure, the water level in the well indicates the groundwater level.

17.4 AQUIFERS WITH ARTESIAN PRESSURE

The groundwater in some aquifers can be under pressure. The monitoring wells will register a higher water level than the groundwater level. In some cases, water will spill out from the well due to artesian pressure (Fig. 17.8).

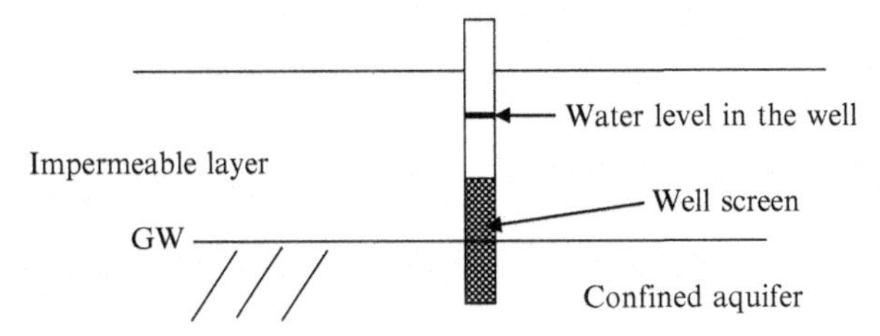

Fig. 17.8 Water level in a confined aquifer.

Monitoring well in a confined aquifer: The water level in the well is higher than the actual water level in the aquifer since the aquifer is under pressure.

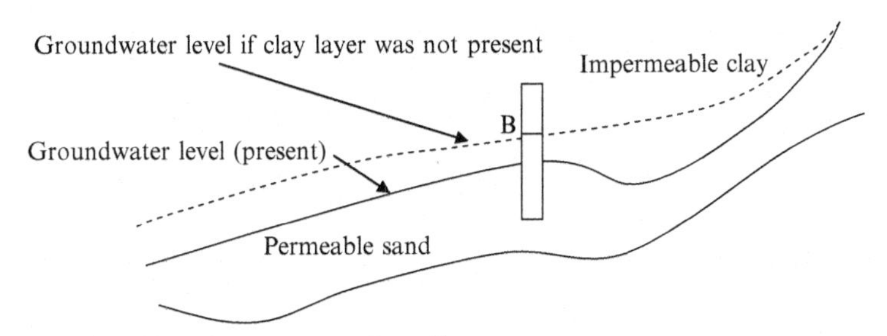

Fig. 17.9 Groundwater in a confined aquifer.

Artesian conditions: In Fig. 17.9, the impermeable clay layer is shown lying above the permeable sand layer. The dotted line shows the groundwater level if the clay layer is absent. Due to the impermeable clay layer, the groundwater cannot reach the level shown by the dotted line. When a well is installed, the water level will rise to point B, higher than the initial water level due to artesian pressure.

CHAPTER 18

Pumps and Pump Curves

Pumps are required in construction to pump water from excavations and trenches. On a larger scale, pumps are also required to dewater cofferdams.

Head Added by a Pump: Let us look at the pump shown. Assume the head loss due to friction is f (Fig. 18.1).

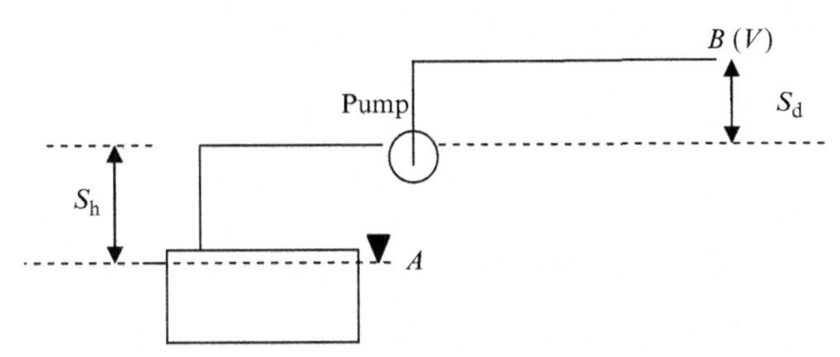

Fig. 18.1 Head added by a pump.

The water level in the excavation is at A. Water needs to be sucked into the pump. S_h is known as the suction head. Once the water is sucked into the pump, it has to be discharged. S_d is known as the discharge head. In addition, the pump has to overcome friction in the pipes.

Head added by the pump $(h_A) = S_h + S_d + f + V^2/2g + P_b/\gamma$

P_b = Pressure at point B

V = Velocity at point B

f = Head loss due to friction.

The equation above can be obtained using the energy equation.

18.1 ENERGY EQUATION

$$\text{Energy of water in the excavation} + \text{Head added by the pump} = \text{Energy at point B} + \text{Head loss due to friction} \quad (18.1)$$

There is energy in the water of the excavation. Energy has to be added to the water in the excavation to pump out.

Construction Engineering Design Calculations and Rules of Thumb
http://dx.doi.org/10.1016/B978-0-12-809244-6.00018-4

Energy of water has three terms. They are:

$$\text{Energy due to pressure} = \text{Pressure}/\text{Density} = P/\gamma$$

$$\text{Energy due to velocity} = V^2/2g$$

$$\text{Energy due to datum} = h$$

$$\text{Energy of water in the excavation} = 0$$

In the excavation, water is exposed to atmosphere. Hence, pressure energy is zero.

Water in the excavation is not moving. Hence, kinetic energy is zero.

If we assume the datum to be the water level in the excavation, then datum head is also zero.

Hence, from energy equation:

$$\text{Energy at point A} + \text{Head added by the pump}$$
$$= \text{Energy at point B} + \text{Friction head loss}$$
$$0 + h_A = S_h + S_d + f + V^2/2g + P_b/\gamma$$

18.2 HORSEPOWER OF PUMPS

Pump horsepower is given by the following equation.

$$\boxed{\text{Pump horsepower (HP)} = h \times m'/550}$$

h = Head added by the pump

m' = Flow in lbs/s

Note that in pump horsepower calculations, the unit of flow is NOT cu ft/s but lbs/s.

Practice Problem 18.1

The suction head of a pump is 6 ft and the discharge head is 7 ft. The flow is 300 gallons/min (gpm). Find the horsepower of the pump. Head loss due to friction is 3 ft. Ignore the velocity head and pressure head in pipes.

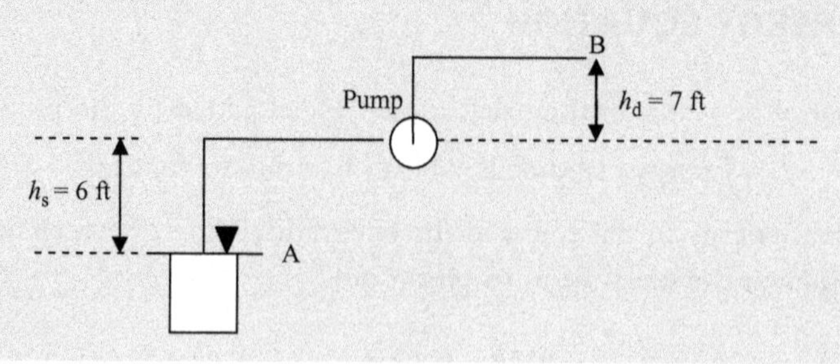

Solution

Pump horsepower is given by the following equation.

$$\boxed{\text{Pump horsepower (HP)} = h \times m'/550}$$

h = Head added by the pump

m' = Flow in lbs/s

Note that in pump horsepower calculations, the unit of flow is NOT cu ft/s but lbs/s.

Typically, flow is given in cu ft/s or gpm. This has to be converted to lbs/s.

STEP 1: Find the head added by the pump:

$$\text{Head added by the pump} = S_h + S_d + f + V^2/2g + P_b/\gamma$$

Ignore the velocity head and pressure head.

Hence;

$$\text{Head added by the pump} = S_h + S_d + f = 6 + 7 + 3 = 16\,\text{ft}$$

STEP 2: Find the flow in lbs/s:

Flow is given in gpm. This has to be converted to lbs/s.

$$\text{Flow} = 300\,\text{gpm} = 300/60\,\text{gal/s} = 5\,\text{gal/s}$$

$$1\,\text{gallon} = 8.342\,\text{lbs};\quad \text{Hence}\ 5\,\text{gal/s} = 5 \times 8.342\,\text{lbs/s} = 41.71\,\text{lbs/s}$$

STEP 3: Find the pump horsepower:

$$\text{Pump horsepower (HP)} = h \times m'/550$$

$$h = 16\,\text{ft};\ m' = 41.71\,\text{lbs/s}$$

$$\text{Pump horsepower (HP)} = h \times m'/550 = 16 \times 41.71/550 = 1.21\,\text{HP}$$

18.3 PUMP PERFORMANCE CURVE, PUMP EFFICIENCY CURVE, AND SYSTEM CURVE

As a construction engineer, you may have to pick the correct pump for the job. Let us say in one project you have to pump water to an elevation of 25 ft at a rate of 100 gpm. In another project, you need to pump water to an elevation of 15 ft at a rate of 200 gpm. The same pump may not be the best choice for both cases (Fig. 18.2).

When a pump is operating at its highest efficiency, the energy bill would be lowest. Hence, you need to find a pump that will operate at its highest efficiency. Highest efficiency of pump A is attained when pumping at

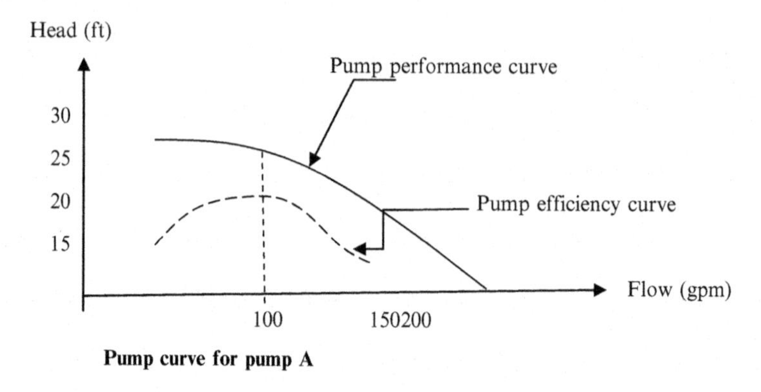

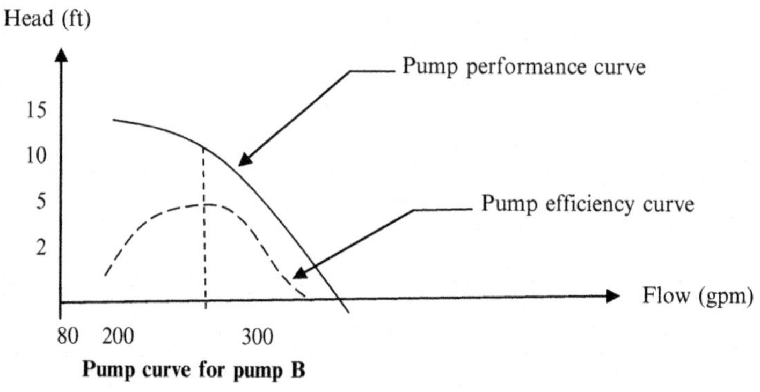

Fig. 18.2 Pump curves.

100 gpm. The highest efficiency of pump B is attained when pumping at 200 gpm.

Let us say you need to pump at a rate of 100 gpm to an elevation of 25 ft, then pump A should be your best choice. However, if you want to pump at a rate of 150 gpm to an elevation of 20 ft, pump A is not very efficient. Yet pump A can do the job.

Now if you want to pump at a rate of 150 gpm to an elevation of 50 ft, pump A may not be able to do the job. (See the pump performance curve).

If you look at pumps A and B, pump A can pump to higher elevations, but at a slower rate. On the other hand, pump B can pump at a faster rate, but it cannot lift the water to higher elevations.

18.3.1 System Curve

A system curve has nothing to do with pump curve. A system curve depends on the friction head loss in the pipes in the system. Let us assume the pipe line has certain characteristics. When water is flowing at a certain velocity, it has a certain friction. Friction head loss is given by:

$$\text{Friction head loss} = H_f = f \cdot L \cdot V^2/(2g \cdot d)$$

where f is the friction coefficient; L is the length of pipe; V is the velocity of water; and d is the diameter of the pipe.

The length of pipes, friction coefficient, and diameter of pipe are dependent on the specific system you have.

In addition, you will notice that the higher the flow, the higher the friction head loss.

Let us look at a typical dewatering scheme (Fig. 18.3).

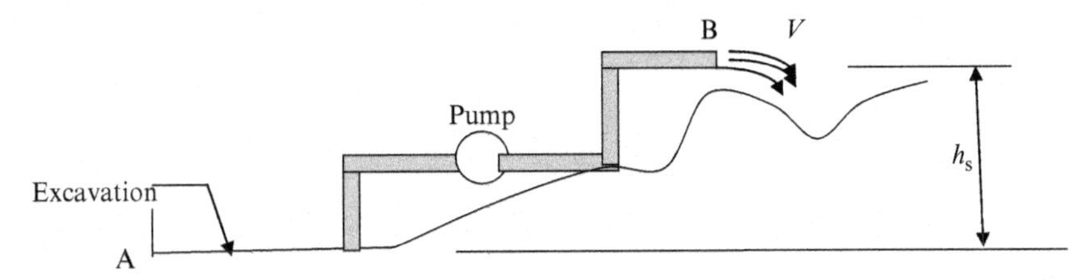

Fig. 18.3 Pumping water.

Water in the excavation is exposed to atmosphere. Hence, the pressure head is zero. In addition, water does not move. Hence, the kinetic head is zero. We are required to pump water at velocity V.

$$\text{Energy at point A} = 0$$

$$\text{Energy at point B} = h_s + V^2/2g + P_B/\gamma$$

where h_s is the static head; V is the velocity; and P_B is the pressure at point B.

In this case, water at point B is exposed to atmosphere. Hence, P_B is zero.

$$H_R = \text{Head added by the pump} = H_f + h_s + V^2/2g$$

H_f is the friction head loss.

H_f and $V^2/2g$ depend on velocity of water. The higher the velocity, the higher the head loss due to friction (Fig. 18.4).

The higher the flow, the higher the head required. The above curve depends on the length of pipes, friction coefficient, and static head. All these parameters are dependent on the system (Fig. 18.5).

Now let us look at some pump curves drawn alongside the system curve.

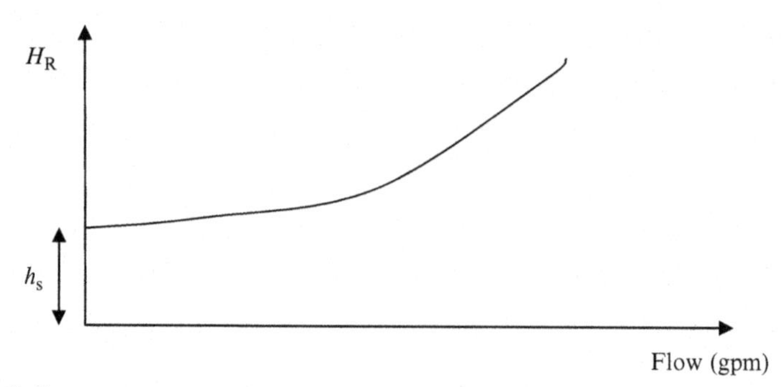

Fig. 18.4 System curve.

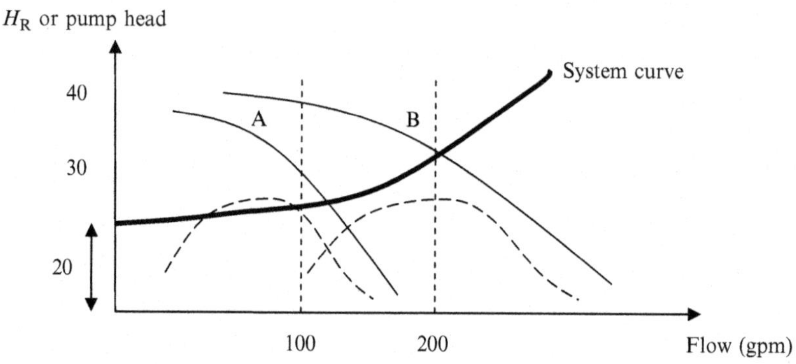

Fig. 18.5 System curve and pump curves.

Two pump performance curves (A and B) and their efficiency curves are shown.

Let us assume that we need to pump at 100 gpm. To pump at 100 gpm, we need a pump that can deliver a head of around 22 ft as per the system curve given. The system requires at least 22 ft of head to pump at 100 gpm to provide the static head required and to overcome the friction. Both pumps A and B are capable of doing so. However, pump A is better suited for the job since it has a higher efficiency than pump B at 100 gpm. Now if we want to pump at 200 gpm, pump A is not suitable. Pump A is not capable of providing a flow of 200 gpm. Then we find pump B to be a better candidate.

18.3.2 System Curve When Pumping Horizontally

Water is pumped from one reservoir to the other. Both are at the same elevation. The system curve for this situation is shown below (Figs. 18.6 and 18.7).

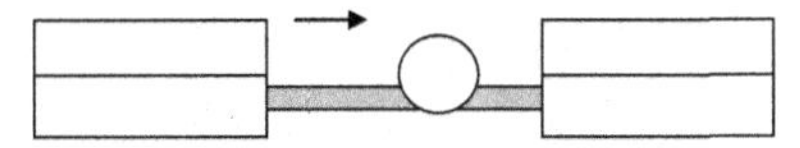

Fig. 18.6 System curve when pumping horizontally.

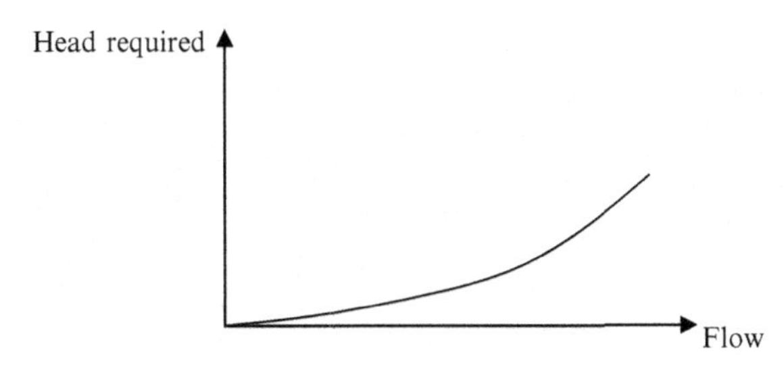

Fig. 18.7 Flow versus head required.

When the flow is increased, the head required goes up since the friction head loss and the velocity head depends on flow.

18.4 NET POSITIVE SUCTION HEAD

When selecting pumps, one has to look for the net positive suction head (NPSH) of the pump as well. Now let us look what this means (Fig. 18.8).

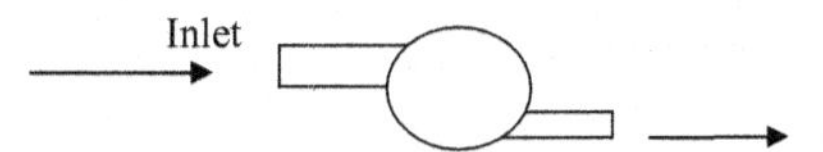

Fig. 18.8 Net positive suction head.

In the above pump, water comes in from the inlet. For the impellers to work properly, a certain amount of head is required. If the head available at the inlet is very low, there would be air pockets inside the pump. This is known as cavitation. Cavitation means cavities of air have developed. In such situations, a pump will not operate as designed. Therefore every pump would indicate the NPSH required for the pump to operate properly. Some pumps may require a larger NPSH while other pumps may require a smaller NPSH.

On the other hand, the head available in the system has nothing to do with the pump. Available NPSH at the inlet of the pump depends on the suction head, friction head loss, and vapor pressure.

If water is pumped from an open reservoir, the available NPSH is given by the following equation.

$$\text{NPSH}_{\text{Available}} = \text{Atmospheric head} - \text{Suction head} - \text{Friction loss} \\ - \text{Vapor pressure}$$

Let us look at an example.

Practice Problem 18.2

Water is pumped from an excavation 12 ft deep. Friction head loss in the suction pipes is 10 ft. The vapor pressure of water is 0.6 ft. The atmospheric pressure is 35 ft. What is the available NPSH at the inlet of the pump?

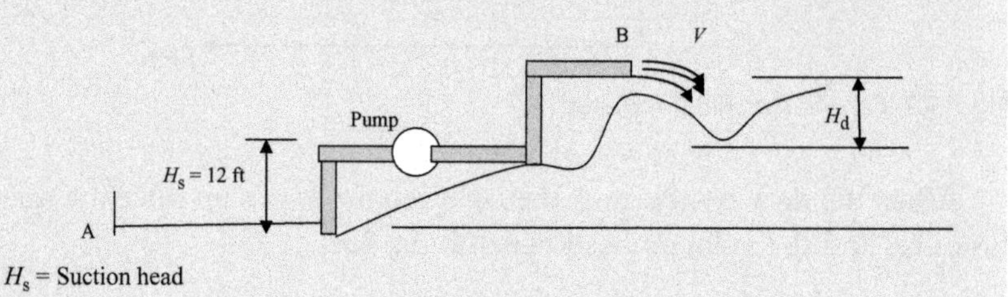

Solution

$$\text{NPSH}_{\text{Available}} = \text{Atmospheric head} - \text{Suction head} - \text{Friction loss} \\ - \text{Vapor pressure}$$

$$\text{NPSH}_{\text{Available}} = 35 - 12 - 10 - 0.6 = 12.4$$

$\text{NPSH}_{\text{Available}}$ is dependent on the friction head loss and the suction head. These two parameters have nothing to do with the pump.

Once you have calculated the $\text{NPSH}_{\text{Available}}$, you need to find a pump that requires a NPSH less than what is available. If $\text{NPSH}_{\text{required}}$ of the pump is greater than the $\text{NPSH}_{\text{Available}}$, then the pump will not work.

For the pump to work:

$$\text{NPSH}_{\text{Available}} > \text{NPSH}_{\text{required}}$$

Practice Problem 18.3

Two pumps are available at a job site.

$$\text{Pump A} \rightarrow \text{NPSH}_{\text{required}} = 15\,\text{ft}$$

$$\text{Pump B} \rightarrow \text{NPSH}_{\text{required}} = 8\,\text{ft}$$

Which pump is best suited for the previous example?

Solution

$$NPSH_{Available} = 12.4 (\text{Computed earlier})$$

Pump A requires a NPSH of 15 ft. What is available is 12.4. This pump will not perform properly. When the available NPSH is less than the required NPSH, water vapor and air pockets will form inside the pump. This is known as cavitation. This is not a desirable situation.

On the other hand, the NPSH required for pump B is only 8 ft of water. For this application, pump B should be selected.

Practice Problem 18.4

Dewatering from an excavation is shown below.

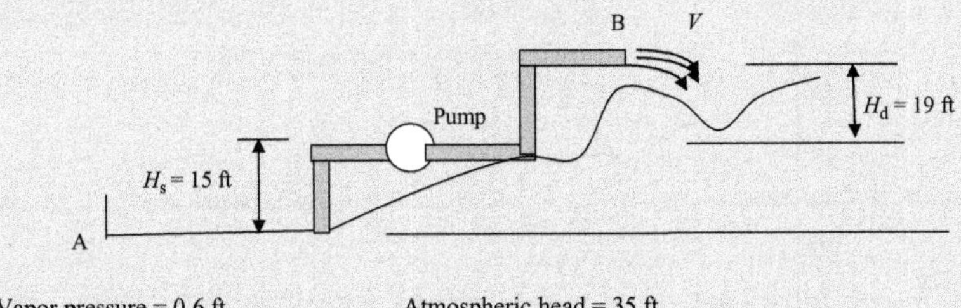

Vapor pressure = 0.6 ft
Pump rate required = 400 gpm
Ignore the velocity head

Atmospheric head = 35 ft

System curve is shown below:

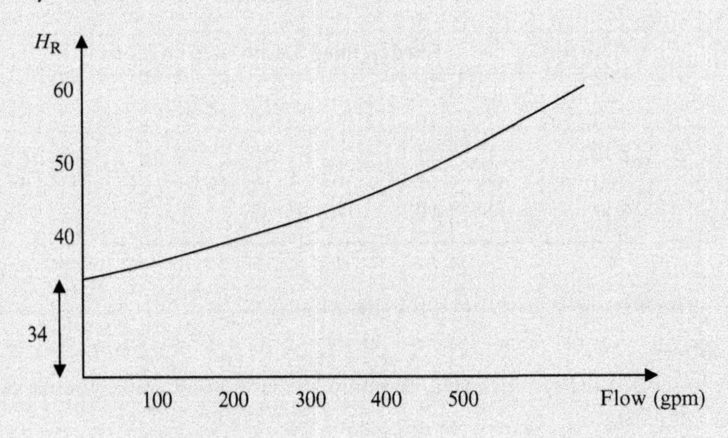

Performance curves for two pumps are given below.

Pump A: (NPSH = 7 ft)

Continued

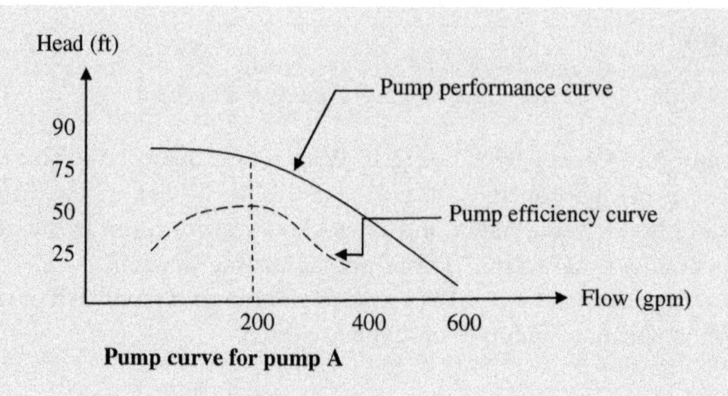

Pump curve for pump A

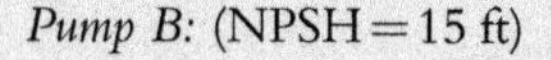
Pump B: (NPSH = 15 ft)

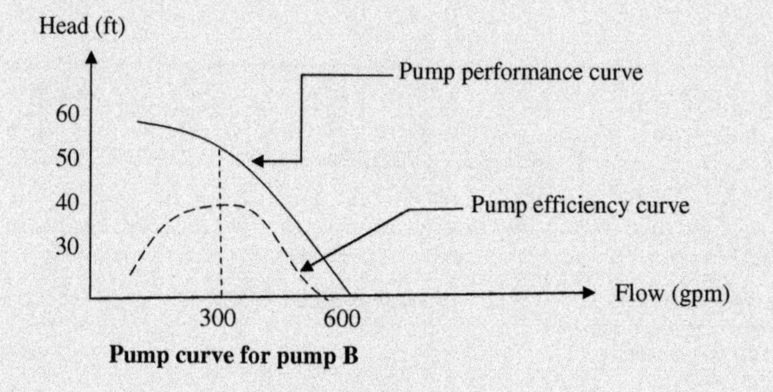

Pump curve for pump B

Friction head loss in suction pipes is given by the following table:

Flow (gpm)	Friction head loss in suction pipes
100	8
150	10
200	11
400	12
500	19

Which pump is best suited for the project?

Solution

If you look at the system curve, to pump at a rate of 400 gpm, you need a head of 45 ft.

Now let us look at the pump curve of pump A.

Can pump A pump at a rate of 400 gpm and at the same time provide a head of 45 ft?

The answer is yes. However, there is a problem. The efficiency of the pump at 400 gpm is very low.

Now let us look at pump B. We will ask the same question.

Can pump B pump at a rate of 400 gpm and at the same time provide a head of 45 ft?

The answer is yes. The efficiency of the pump B at 400 gpm seems to be reasonable.

Now let us look whether the pumps would operate properly.

Find available NPSH:

$$\begin{aligned} NPSH_{Available} &= \text{Atmospheric head} - \text{Suction head} \\ &\quad - \text{Friction head in suction pipes} - \text{Vapor pressure} \end{aligned}$$

$$NPSH_{Available} = 35 - 15 - 12 - 0.6 = 7.4\,\text{ft}$$

A table has been provided to find the friction head loss in suction pipes. For a rate of 400 gpm, friction head loss in suction pipes is 12 ft.

The NPSH required by pump A is 7 ft. The available NPSH is larger than required by NPSH.

On the other hand, the NPSH required for pump B is 15 ft. The required NPSH is much larger than what is available. Hence pump B cannot be used.

Pump A has to be selected. This pump has a low efficiency level and high-energy bill to be expected. However, there is no choice. Pump B cannot be used since the available NPSH is not enough for this pump to operate.

Practice Problem 18.5

What can be done to increase the $NPSH_{avialable}$?

Solution

$$\begin{aligned} NPSH_{Available} &= \text{Atmospheric head} - \text{Suction head} \\ &\quad - \text{Friction head in suction pipes} - \text{Vapor pressure} \end{aligned}$$

One can replace the old piping with new piping. This would decrease the friction head. In addition, the suction head can be reduced. This can be done by placing the pump at a lower elevation. This option may not be feasible depending on the conditions of the site.

CHAPTER 19

Scheduling

19.1 CONSTRUCTION SEQUENCING

It is important to sequence construction activities. Walls cannot be built without constructing the footings. Footings cannot be built before the earthwork is finished. The contractor needs to identify activities that need to be completed and sequence them.

Site clearing → Excavation for footings → Formwork for footings → Footing construction

19.2 ACTIVITY ON NODE NETWORKS AND CPM NETWORK ANALYSIS

There are two types of networks. They are (Figs. 19.1 and 19.2):

- activity on node networks and
- activity on arrow networks

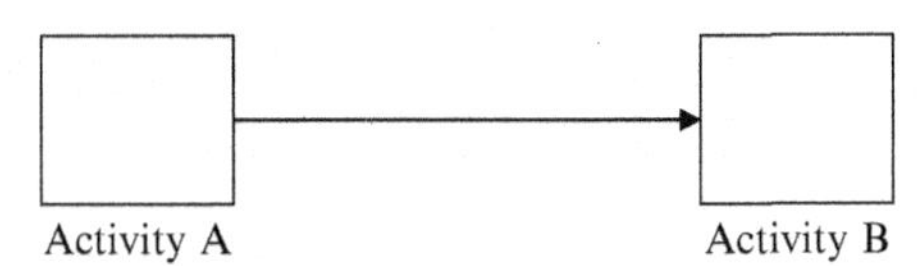

Fig. 19.1 Activity on node network.

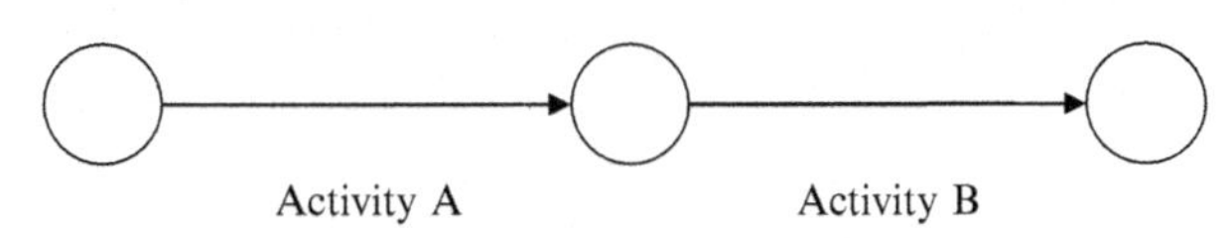

Fig. 19.2 Activity on arrow network.

Activity on node networks: Let us first look at activity on node networks. It is fair to say that activity on node networks are more widely used than activity on arrow networks.

The critical path method is the most popular technique adopted for scheduling using both networks.

Construction Engineering Design Calculations and Rules of Thumb
http://dx.doi.org/10.1016/B978-0-12-809244-6.00019-6

In the critical path method, all activities have the following attributes:

- start time,
- finish time, and
- duration.

Let us look at the following example:

Practice Problem 19.1

The contractor has forecasted that he will be able to complete site clearing in 10 days and construct footings in 9 days.

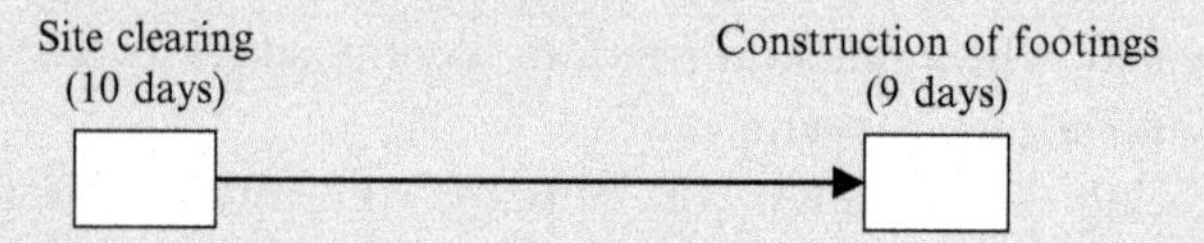

Site clearing is the first activity. It can be started on day 1. Some authors start the activity in day zero. However, NCEES examples book uses day 1 as the first day.

Site clearing: Early start (ES) = 1

Site clearing: Early finish (EF) = 1 + 10 = 11

Construction of footings: Early start (ES) = 11

Construction of footings: Early finish (EF) = 11 + 9 = 20

Activity name		
ES	Duration	LS
EF		LF

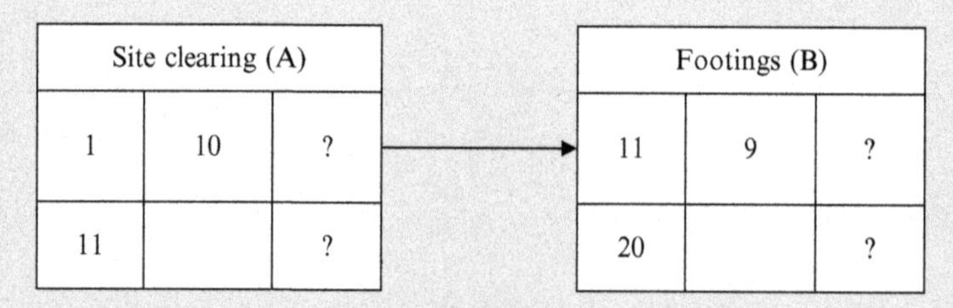

Late start (LS) and late finish (LF) has to be completed.

What is the late finish of activity B?

Early finish time of activity B is 20. Since there is no any other information available, the late finish also would be 20.

Late finish of footing construction (LF) = 20

Late start of activity B = LF − duration = 20 − 9 = 11
Now late finish of activity A can be found.
Late finish of activity A = 11
Late start of activity A = LF − duration = 11 − 10 = 1
These values can be included in the schedule.

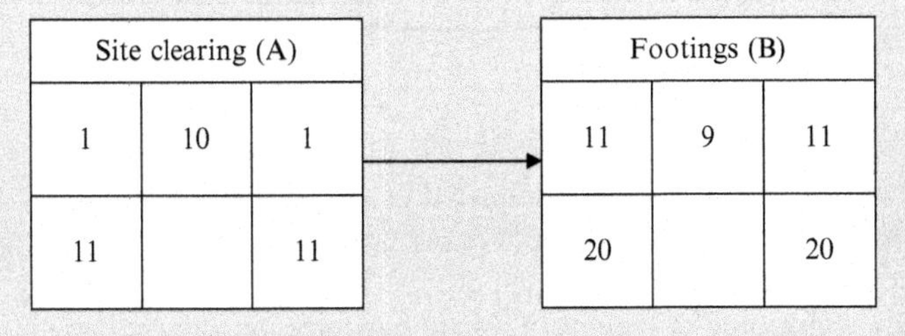

Practice Problem 19.2

The contractor can complete the construction of building walls in 20 days. He is planning to start the construction of the roof immediately after constructing the building walls. The contractor has estimated that he needs 12 days to construct the roof.

Activity (A) = Construction of walls

Activity (B) = Construction of roof

(A) Construction of walls (20 days) (B) Roof (12 days)

Forward pass: A forward pass means going forward, starting from the first activity. During the forward pass, early start time (ES) and early finish time (EF) of activities will be completed.

Late start time (LS) and late finish time (LF) of activities will NOT be completed during forward pass.

LS and LF are completed during backward pass.

Construction of walls (A): ES = 1, EF = 1 + 20 = 21

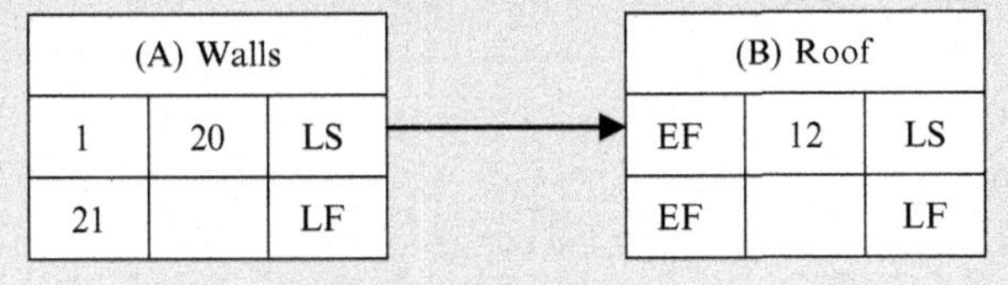

The contractor can start the roof construction immediately. Hence, the early start (ES) of roof construction is 21.

Continued

Construction of the roof "B": ES = 21, EF (Early finish) = 21 + 12 = 33
Early finish time (EF) of roof construction is 21 + 12 = 33
Now these numbers can be inputted into the CPM diagram.

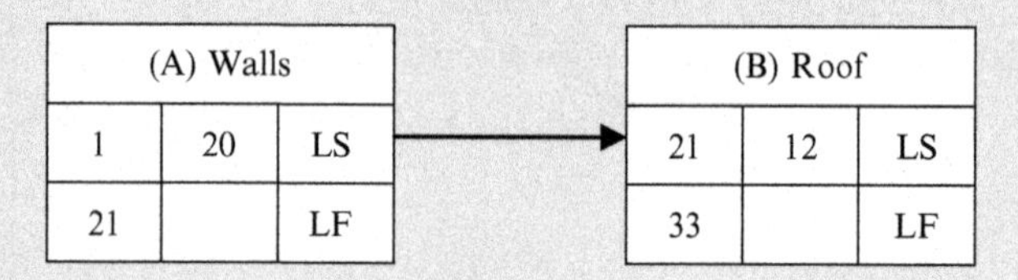

Now the forward pass is completed.
Backward pass: Late finish time of activity B = 33
Now you will be able to compute the late finish time of activity "A."
Late finish time of activity A = Late start time of activity B = 21
Late start time of activity A = Late finish time of activity A − duration of A = 21 − 20 = 1

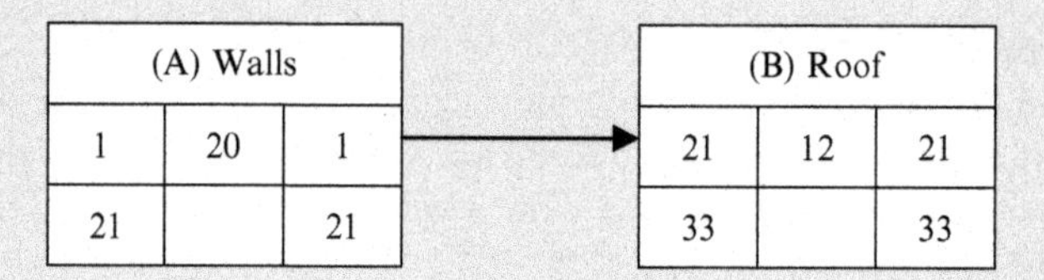

Predecessor and successor:

Activity A is the predecessor and activity B is the successor.
Dependence:

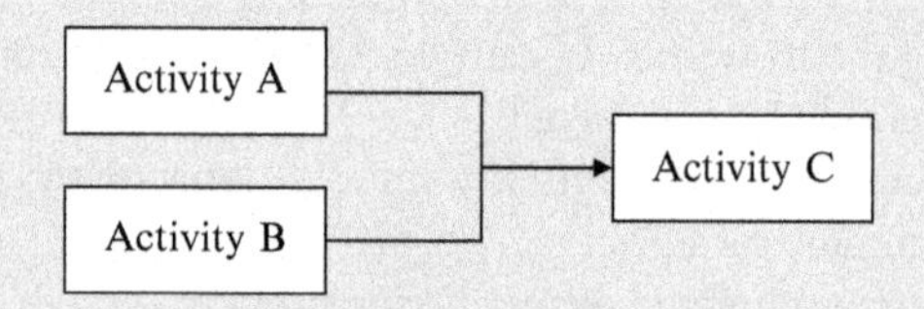

Activity C is dependent upon activity A and activity B.
Activity B does NOT depend on activity A.

Practice Problem 19.3

The contractor has to complete the construction of walls and fabrication of the roof truss before constructing the roof. The roof truss is constructed outside and brought in and installed.

$$\text{Activity A} = \text{Wall construction}(\text{Duration} = 20\,\text{days})$$

$$\text{Activity B} = \text{Fabrication of the roof truss (Duration} = 25\,\text{days})$$

$$\text{Activity C} = \text{Installation of the roof}(\text{Duration} = 15\,\text{days})$$

Activity A—No predecessors
Activity B—No predecessors
Activity C—Predecessors (A and B)
Conduct the forward and backward passes.

Solution

STEP 1: Draw the activity diagram (input the durations).

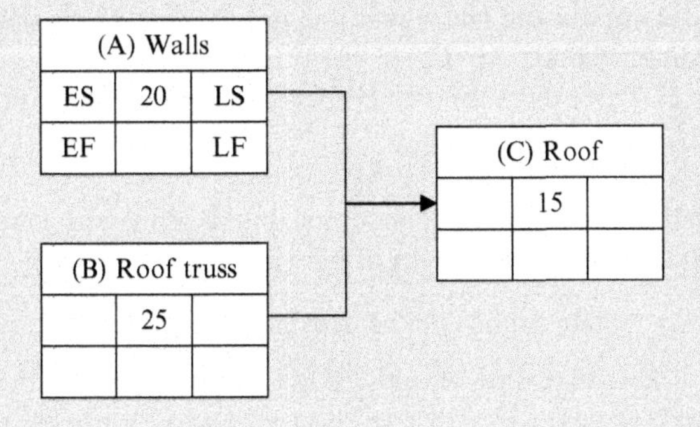

STEP 2: Forward pass: (Find early start time and early finish time of each activity).

Activity A:

$$\text{Early start time(ES)} = 1$$

$$\text{Early finish time(EF)} = 21$$

Activity B:

$$\text{Early start time(ES)} = 1$$

$$\text{Early finish time (EF)} = 26$$

Activity C:

Activity C cannot be started until both activities A and B are completed. Activity A is completed on 21 and B is completed on 26.

$$\text{Early start time of activity C(ES)} = 26$$

$$\text{Early finish time of activity C(EF)} = 26 + 15 = 41$$

Forward pass is completed. Input the above values in CPM diagram.

Continued

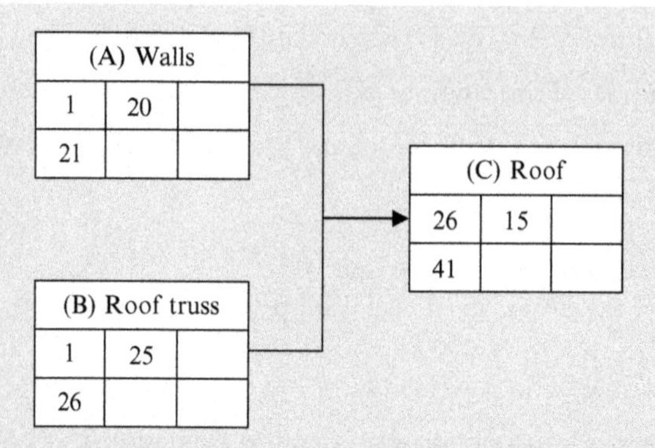

STEP 3: *Backward pass:* (Late start time and late finish time of each activity).

Start to compute the late start time and the late finish time of each activity starting from activity C.

Activity C: Since there is no any other information available, late finish time of activity C is 41.

Late start time of activity $C = 41 - 15 = 26$.

Activity B: Activity C cannot be started until B is finished. Late start time of activity C is 26. Hence, activity B has to be finished by 26.

$$\text{Late finish time of activity B(LF)} = 26$$

$$\text{Late start time of activity B(LS)} = 26 - 25 = 1$$

Activity A: Activity C cannot be started until A is finished as well. The late start time of activity C is 26. Hence, activity A has to be finished by 26.

$$\text{Late finish time of activity A(LF)} = 26$$

$$\text{Late start time of activity A} = 26 - 20 = 6$$

Look at the late start time of activity A.

Activity A *can* be started on day 6, without affecting the schedule.

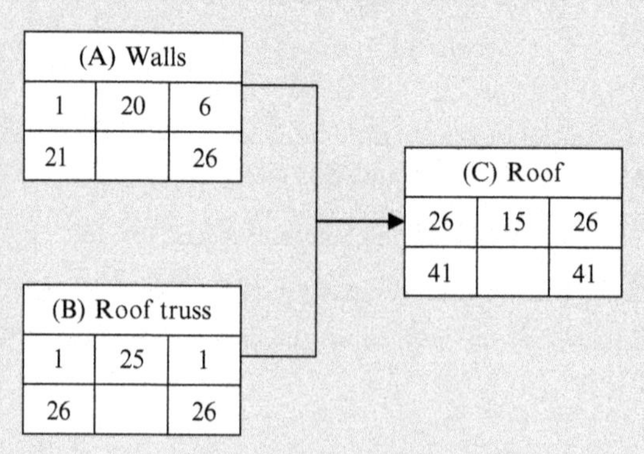

Practice Problem 19.4

The following information is given. Draw the CPM diagram.

Activity A = No Predecessors (Duration = 12 days)
Activity B = No Predecessors (Duration = 15 days)
Activity C = Predecessors (A and B) (Duration = 16 days)
Activity D = Predecessors (A and B) (Duration = 18 days)

(a) Draw the CPM diagram.
(b) Conduct the forward and backward passes.

Solution

STEP 1: Draw the activity diagram (input the durations).

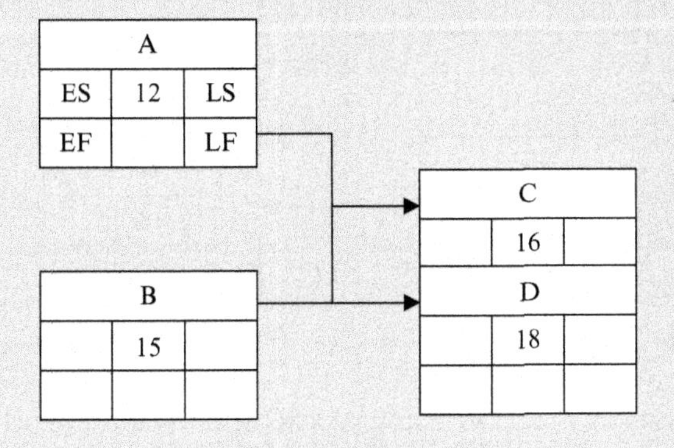

STEP 2: Forward pass: (Early start time and early finish time of each activity).

Activity A:

$$\text{Early start time(ES)} = 1$$

$$\text{Early finish time(EF)} = 1 + 12 = 13$$

Activity B:

$$\text{Early start time(ES)} = 1$$

$$\text{Early finish time(EF)} = 1 + 15 = 16$$

Activity C:

Activity C cannot be started until both activities A and B are completed. Activity A is completed on 13 and B is completed on 16.

$$\text{Early start time of activity C(ES)} = 16$$

$$\text{Early finish time of activity C(EF)} = 16 + 16 = 32$$

Continued

Activity D:

Activity D cannot be started until A and B are completed. Activity B is completed on 16 and activity A is completed on 13.

$$\text{Early start time of activity D(ES)} = 16$$

$$\text{Early finish time of activity D(EF)} = 16 + 18 = 34$$

Forward pass is completed. Input the above values in CPM diagram.

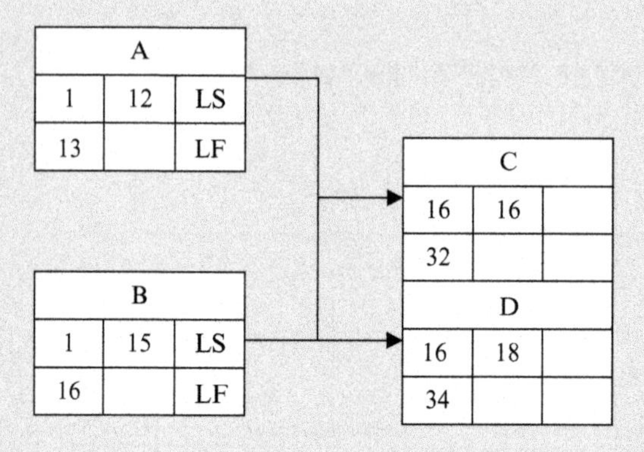

STEP 3: *Backward pass:* (Late start time and late finish time of each activity).

Start to compute the late start time and late finish time of each activity from activity D.

Activity D: Late finish time of activity D is 34.

$$\text{Late start time of activity D} = 34 - 18 = 16$$

Activity C: Activity C can be completed by 32. But it can be delayed till day 34 without affecting the project schedule.

Why?

Activity D will be completed by day 34 for the project to be completed. Hence, if required, activity C can be delayed till day 34.

$$\text{Late finish time of activity C(LF)} = 34$$

$$\text{Late start time of activity C(LS)} = 34 - 16 = 18$$

Let us input the numbers we acquired so far in the CPM diagram.

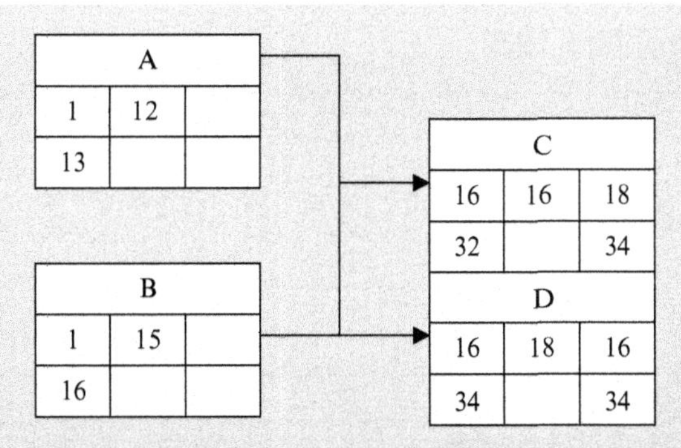

Activity B: Activity C and D cannot be started until B is finished. Late start time of activity C is 18 and late start time of activity D is 16. Hence, activity B has to be finished by 16.

If activity B is finished after 16, activity D cannot be started by 16.

$$\text{Late finish time of activity B(LF)} = 16$$

$$\text{Late start time of activity B} = 16 - 15 = 1$$

Activity A: Activity A has to be completed to start activities C and D.

Late start time of activity C is 18 and late start time of activity D is 16. Hence, activity A has to be finished by day 16.

$$\text{Late finish time of activity A(LF)} = 16$$

$$\text{Late start time of activity A(LS)} = 16 - 12 = 4$$

Now you can input the numbers in the CPM diagram.

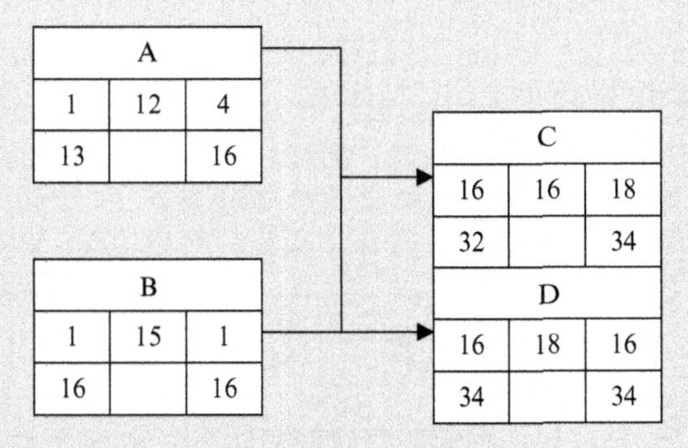

What is the critical path of the above project?

Activities A and C can be delayed without delaying the project. Activities B and D cannot be delayed.

Hence, the critical path of the project is

$$B \rightarrow D$$

Practice Problem 19.5

Complete the network shown. Durations are as shown.

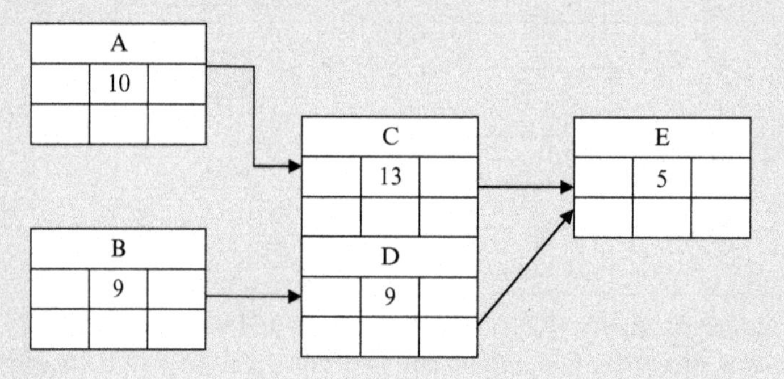

Solution

Complete the forward pass: During the forward pass early start times and early finish times are found.

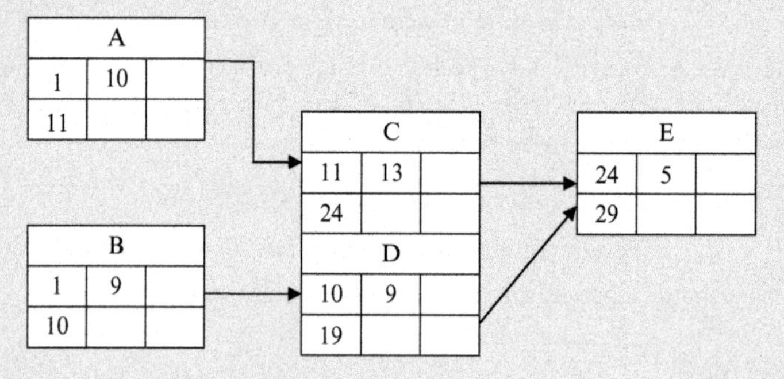

Complete the backward pass:

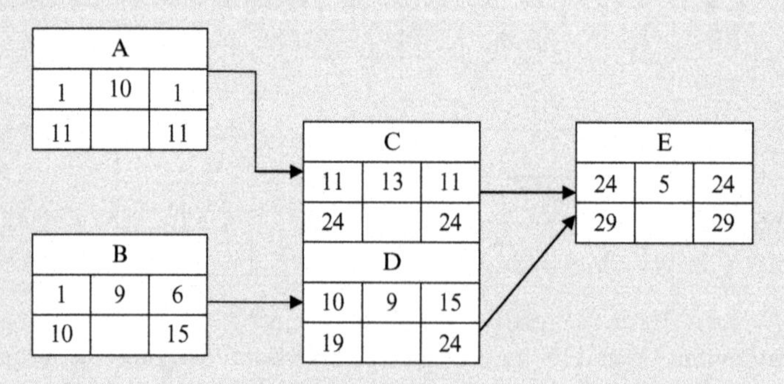

A critical path is the path with zero float.

$$\text{Critical path}: A \rightarrow C \rightarrow E$$

Finish to start lag time: So far, we have considered that the next activity would start just after the previous activity is finished. There are some situations where this is not possible. A concrete footing has to be constructed and a steel column has to be placed on the footing. After construction of the footing, there is a lag of 10 days for the concrete to cure.

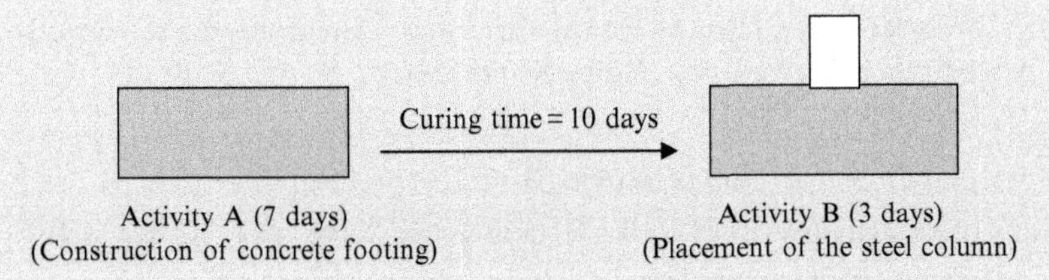

Finish to start lag = 10 days

In the above example, the steel column cannot be placed on the concrete footing until 10 days after the construction of the footing. This can be shown symbolically as below.

Activity A (7 days) —— F-S = 10 ——→ Activity B (3 days)

F − S = 10 indicates that finish to start there is a delay of 10 days. In this case it is for the curing of concrete.

It is important to note that this equation is valid for both early and late times.

$$F - S = 10$$

$$EF - ES = 10$$

(Early finish of activity A to Early start of activity B, there is a delay of 10 days).

And also

$$LF - LS = 10$$

(Late finish of activity A to Late start of activity B, there is a delay of 10 days).

Construction of footing (A)		
1 (ES)	7	LS
8 (EF)		LF

F-S = 10 →

Place the steel column (B)		
18 (ES)	3	LS
21 (EF)		LF

Continued

$$\text{Early start of activity A} = 1$$

$$\text{Early finish of activity A} = 1 + 7 = 8$$

$$\text{Early start of activity B} = 8 + 10 = 18 \quad \text{(10 is the finish to start delay)}$$

$$\text{Early finish of activity B} = 18 + 3 = 21$$

Now let us find LS and LF of two activities. Note that finish to start lag exists for both forward pass and backward pass.

LF of activity B is 21.

$$\text{LS of activity B} = 21 - 3 = 18$$

LF of activity $A = 18 - 10 = 8$ (remember there is a 10-day delay between activities A and B).

$$\text{LS of activity A} = 8 - 7 = 1$$

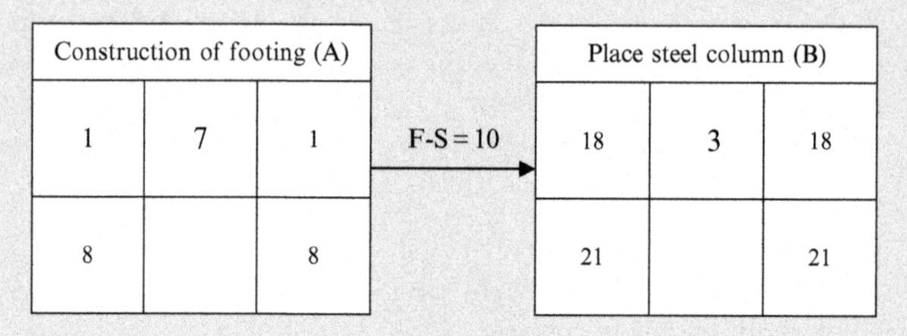

Start-to-start relationships: A contractor is erecting steel columns. He is supposed to erect 100 steel columns. He could deliver all 100 steel columns and start erecting steel columns.

However, the contractor is planning to deliver a portion of the steel columns and start erecting prior to the completion of delivery of all steel. He is planning to deliver a portion of the steel columns in the first two days and start erecting right away. While erecting, he will keep delivering steel to the site. Hence, two activities, delivering steel and erecting steel, will happen simultaneously.

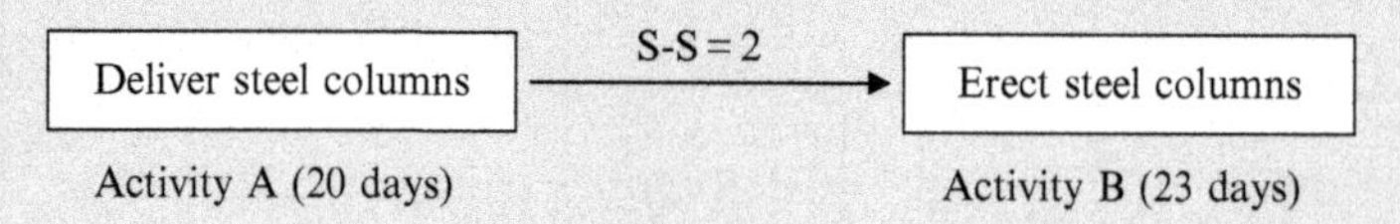

$S - S = 2$ means 2 days after start of activity A, activity B would start. After that, activity A and activity B would happen simultaneously.

S − S = 2 can be represented as follows.
Early start of activity B = Early start of activity A + 2
It will be true for late start as well.
Late start of activity B = Late start of activity A + 2
Now let us fill the boxes.

$$\text{Early start of activity A} = 1$$

$$\text{Early finish of activity A} = 1 + 20 = 21$$

$$\text{Early start of activity B} = 1 + 2 = 3$$

(Contractor is planning to start activity B, just 2 days after starting activity A).

Early finish of activity B = 3 + 23 = 26

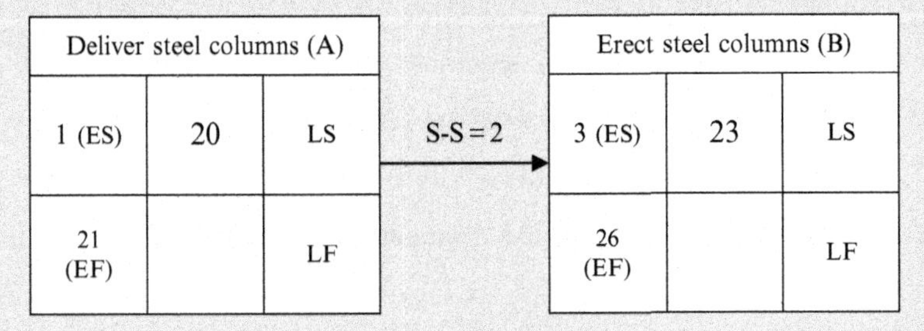

Now let us find LS and LF of two activities.
LF of activity B is 26.
LS of activity B = 26 − 23 = 3
LS of activity B = LS of activity A + 2
(Remember two activities are connected by S-S relationship.)
Hence LS of activity A = LS of activity B − 2 = 3 − 2 = 1.
LF of activity A = 1 + 20 = 21.

Deliver steel columns (A)		
1 (ES)	20	1
21 (EF)		21

S-S = 2 →

Erect steel columns (B)		
3 (ES)	23	3
26 (EF)		26

Finish-to-finish relationships: Let us consider the same example above. A contractor is planning to start delivering steel columns and start erecting steel columns while the steel been delivered. The contractor is planning to complete erecting all steel columns 12 days after completion of delivery of all steel. This can be represented as follows.

Continued

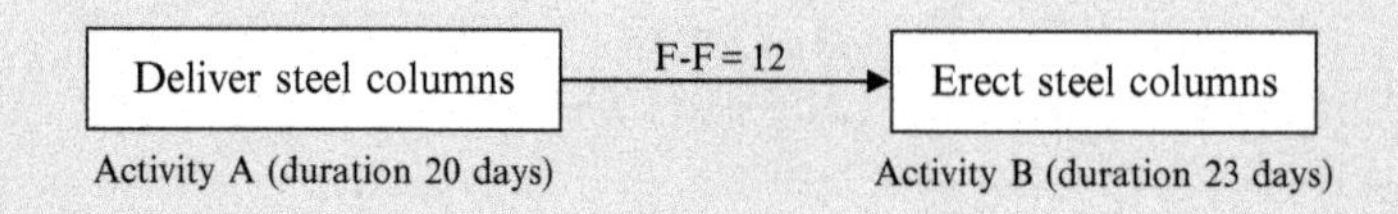

$F - F = 12$ means 12 days after activity A is *finished*, activity B would be *finished*.

$$\text{Early finish (EF) of activity B} = \text{EF of activity A} + 12$$

Similarly,

$$\text{Late finish (LF) of activity B} = \text{LF of activity A} + 12$$

Let us fill the boxes.

$$\text{Early start of activity A} = 1$$

$$\text{Early finish of activity A} = 1 + 20 = 21$$

$$\text{Early finish(EF) of activity B} = \text{EF of activity A} + 12$$

$$\text{Early finish(EF)of activity B} = 21 + 12 = 33$$

$$\text{ES of activity B} = 33 - \text{duration} = 33 - 23 = 10$$

Deliver steel columns (A)		
1 (ES)	20	LS
21 (EF)		LF

F-F = 12 →

Erect steel columns (B)		
10 (ES)	23	LS
33 (EF)		LF

Backward pass:

$$\text{LF of activity B} = 33$$

Now we can find the LF of activity A.
Since,

$$\text{Late finish (LF) of activity B} = \text{LF of activity A} + 12$$

$$\text{LF of activity A} = \text{LF of activity B} - 12 = 33 - 12 = 21$$

$$\text{LS of activity A} = 21 - 20 = 1$$

$$\text{LS of activity B} = \text{LF of activity B} - \text{duration} = 33 - 23 = 10$$

Now all boxes can be filled.

Deliver steel columns (A)		
1 (ES)	20	1
21 (EF)		21

F-F = 12 →

Erect steel columns (B)		
10 (ES)	23	10
33 (EF)		33

S-S and F-F relationships together: S-S and F-F relationships can be used together. Let us look at the same steel column example.

A contractor is planning to start delivering steel columns and start erecting steel columns while the steel is being delivered. The contractor is planning to start erecting steel columns 5 days after start delivering steel columns. At the same time the contractor is planning to finish erection of steel columns 15 days after completion of delivery of all steel. This can be represented as follows.

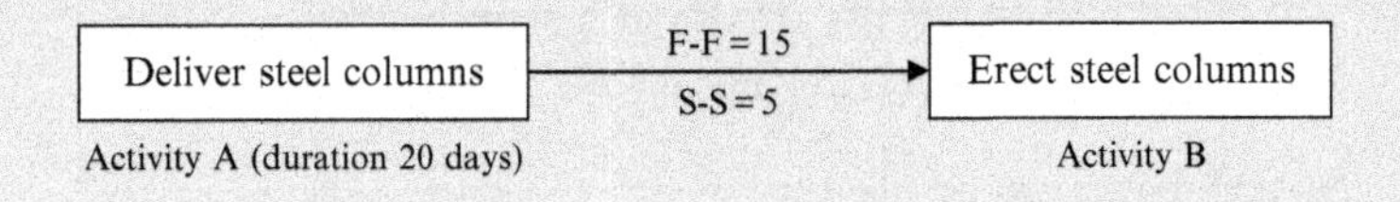

Note that duration of activity B is not given. If duration of activity B is given, there would be a conflict. Duration of activity B is not needed if two relationships are given.

We can write the following:

$$\text{ES of activity B} = \text{ES of activity A} + 5$$

$$\text{LS of activity B} = \text{LS of activity A} + 5$$

$$\text{EF of activity B} = \text{EF of activity A} + 15$$

$$\text{LF of activity B} = \text{LF of activity A} + 15$$

Forward pass:

$$\text{ES of activity A} = 1$$

$$\text{EF of activity A} = 1 + 20 = 21$$

$$\text{ES of activity B} = \text{ES of activity A} + 5 = 1 + 5 = 6$$

$$\text{EF of activity B} = \text{EF of activity A} + 15$$

$$\text{EF of activity B} = 21 + 15 = 36$$

Now these numbers can be represented in the boxes.

Continued

Deliver steel columns (A)		
1 (ES)	20	LS
21 (EF)		LF

F-F = 15
S-S = 5

Erect steel columns (B)		
6 (ES)		LS
36 (EF)		LF

Backward pass:

$$\text{LF of activity B} = 36$$

Now we can use the F-F relationship.

$$\text{LF of activity B} = \text{LF of activity A} + 15$$
$$\text{LF of activity A} = \text{LF of activity B} - 15$$
$$\text{LF of activity A} = 36 - 15 = 21$$
$$\text{LS of activity A} = 21 - 20 = 1$$

Now we can use the S-S relationship.

$$\text{LS of activity B} = \text{LS of activity A} + 5$$
$$\text{LS of activity B} = 1 + 5 = 6$$

Now all boxes can be filled.

Deliver steel columns (A)		
1 (ES)	20	1 (LS)
21 (EF)		21 (LF)

F-F = 15
S-S = 5

Erect steel columns (B)		
6 (ES)		6 (LS)
36 (EF)		36 (LF)

S-F relationships: Start to finish relationships rarely occur in the construction industry. Hence, many textbooks ignore start to finish relationships. Many software programs do not allow start to finish relationships. Start to finish relationship can be represented as follows.

$$\text{ES of activity A} + 5 = \text{EF of activity B}$$

Can you think of a practical application of such a situation?

Negative lags: In the paragraph above, we discussed positive lag between activities. A positive lag occurs due to curing of concrete or lead time of an item. Next, we consider negative lag.

Finish to start (negative lag): It is possible to have a negative lag between two activities. Let us assume that a contractor has to complete two activities, the erection of steel and painting. The contractor can start painting prior to finishing the steel erection. The contractor can erect 50% of the steel and then start painting while the steel erection is going on.

$$F - S = -5$$

The equation above means that the contractor is planning to erect steel. At the same time he is planning to start painting 5 days prior to the completion of the steel erection.

The finish to start difference is −5 days. It does not matter whether they are early or late start and finish times. Nevertheless, one has to be consistent. If it is an early start time, then finish should be an early finish time. Similarly, if F is late finish time then S should be late start time. Hence, two equations can be developed for early and late times.

$$EF - ES = -5$$

And

$$LF - LS = -5$$

The equation above means that the contractor is planning to start painting 5 days before finishing the steel erection.

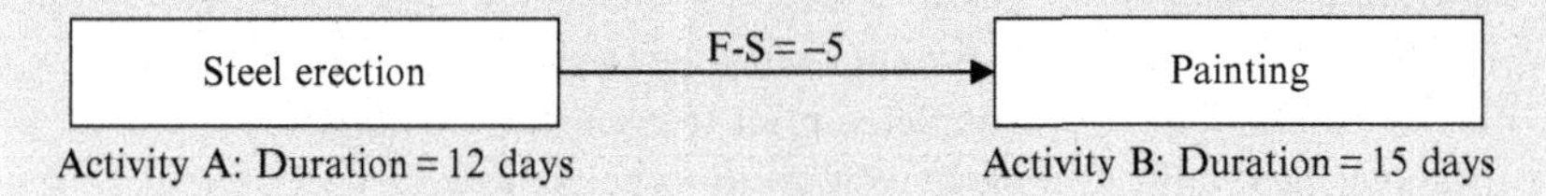

$$\text{ES of activity B} = \text{EF of activity A} - 5$$

(Activity B would start 5 days before activity A is finished).

And

$$\text{LS of activity B} = \text{LF of activity A} - 5$$

Let us do the forward pass:

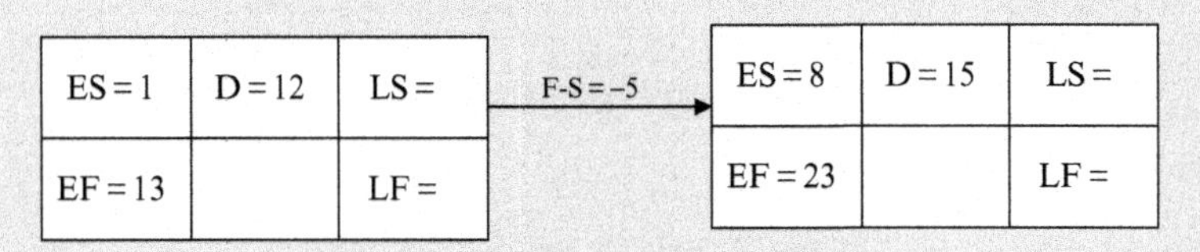

$$F - S = -5$$

$$\text{Early finish of activity A is } 13$$

$$\text{ES of activity B} = \text{EF of activity A} - 5$$

$$\text{ES of activity B} = 13 - 5 = 8$$

Continued

Next, we can do the backward pass:

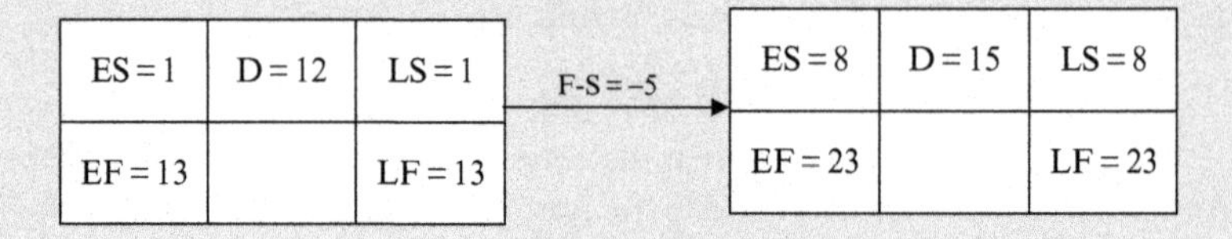

$$\text{LS of activity B} = \text{LF of activity A} - 5$$
$$\text{LS of activity B} = 8$$

Hence

$$8 = \text{LF of activity A} - 5$$
$$\text{LF of activity A} = 13$$

Finish-to-finish (negative lag): It is possible to have a negative lag between two activities with finish-to-finish relationship. Let us assume that a contractor has to build a retaining wall and paint it. The client does not want to paint the whole wall. The client wants only 75% of the wall to be painted. He believes that 25% of the wall is out of public view and need not be painted.

In this scenario, wall painting cannot be started until wall is built.

Wall painting *can be* completed prior to completion of the wall. (This is possible since only 75% of the wall need to be painted.)

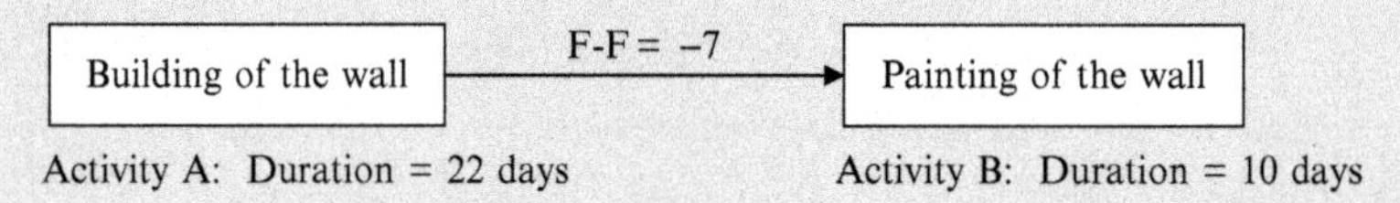

$$\text{EF of activity B} = \text{EF of activity A} - 7$$

(Activity B finishes 7 days before activity A).

And

$$\text{LF of activity B} = \text{LF of activity A} - 7$$

Let us do the forward pass:

$$\text{Activity A}: \text{ES} = 1 \text{ and EF} = 23$$
$$\text{EF of activity B} = \text{EF of activity A} - 7 = 23 - 7 = 16$$
$$\text{ES of activity B} = \text{EF of activity B} - \text{duration} = 16 - 10 = 6$$

ES = 1	D = 22	LS =
EF = 23		LF =

F-F = −7 →

ES = 6	D = 10	LS =
EF = 16		LF =

Backward pass:

LF of activity B = 16

LF of activity B = LF of activity A − 7

LF of activity A = LF of activity B + 7 = 6 + 7 = 23

LS of activity A = 23 − duration = 23 − 22 = 1

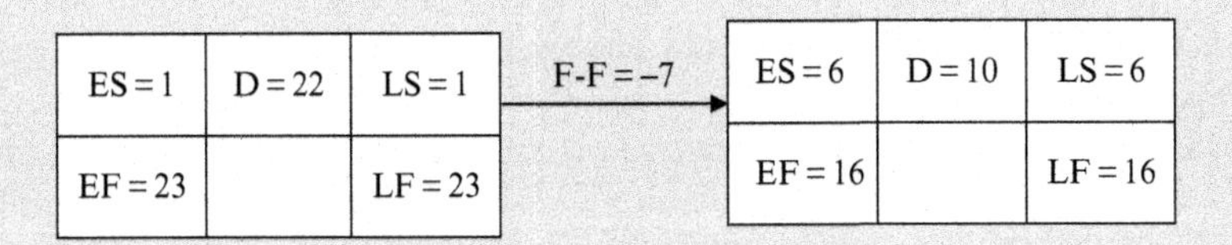

Practice Problem 19.6

Complete the network below. The relationship between activity C to E is F − F = − 4. Durations are as shown.

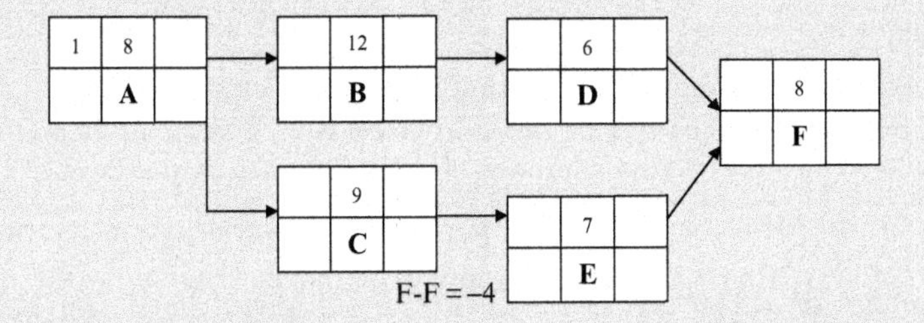

Solution

STEP 1: Complete the forward pass.

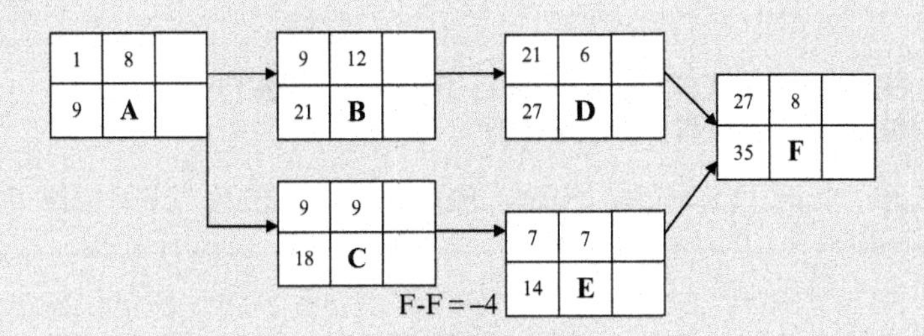

The forward pass of activities A, B, C, and D are straightforward.

Continued

Activity C to E: Activity C to E, one has to worry about the relationship given.

EF of activity C is 18.

Hence, EF of activity E should be 18 − 4 = 14. Note that the relationship is finish to finish and it is negative.

If EF of activity E is 14, then ES of activity E = 14 − duration = 14 − 7 = 7.

Activity F cannot be started until activity D and E are completed. Activity E is completed by 14 and activity D is completed by 27. Hence, ES of activity F is 27.

STEP 2: *Backward pass:*

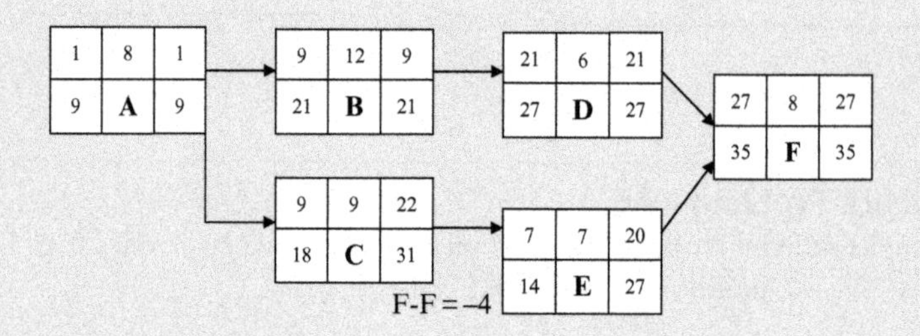

Backward Passes from F to D and F to E are standard.

Backward Pass from E to C: LF of activity E is 27. Hence, LF of activity C is 31. This may be little confusing. Look at EF of activities C and E.

F − F = −4

LF of activity E = LF of activity C − 4

27 = LF of activity C − 4

LF of activity C = 27 + 4 = 31

19.3 FLOATS (TOTAL FLOAT, FREE FLOAT, AND INDEPENDENT FLOAT):

ES = Early Start, LS = Late Start, EF = Early Finish, LF = Late Finish, D = Duration

Activity No:		
ES	D	LS
EF		LF

Floats: Three types of floats are identified.

19.3.1 Total Float

$$\boxed{\text{Total float} = \text{LF} - \text{EF}}$$

LF − EF is same as LS − ES.

This can be shown as follows. $\text{LF} = \text{LS} + \text{D}\,(\text{D} = \text{Duration})$

$$\text{EF} = \text{ES} + \text{D}$$
$$\text{LF} - \text{EF} = (\text{LS} + \text{D}) - (\text{ES} + \text{D}) = \text{LS} - \text{ES}$$

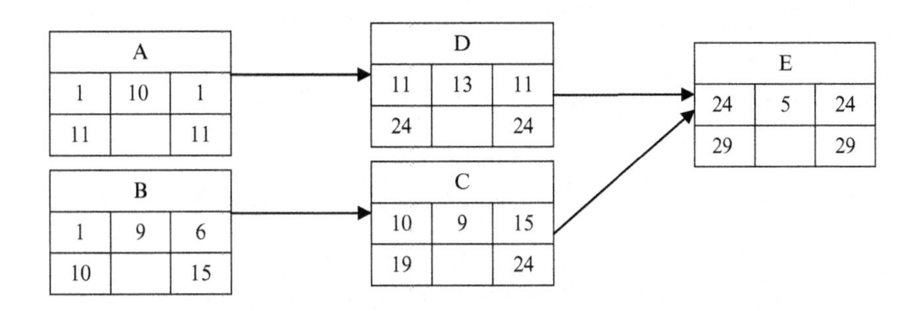

$$\text{Total float of activity A} = \text{LF} - \text{EF} = 11 - 11 = 0$$
$$\text{Total float of activity B} = \text{LF} - \text{EF} = 15 - 10 = 5$$
$$\text{Total float of activity C} = \text{LF} - \text{EF} = 24 - 19 = 5$$
$$\text{Total float of activity D} = \text{LF} - \text{EF} = 24 - 24 = 0$$
$$\text{Total float of activity E} = \text{LF} - \text{EF} = 29 - 29 = 0$$

Note: All floats are zero for critical path activities.

19.3.2 Free Float

$$\boxed{\text{Free float of activity A} = \text{ES}_{\text{successor}} - \text{EF of activity A}}$$

$$\text{Free float of A} = \text{ES}_{\text{successor}} - \text{EF of activity A}$$
$$\text{Successor of A is D. Hence } \text{ES}_{\text{successor}} = 11$$
$$\text{EF of activity A} = 11$$
$$\text{Free float} = 0$$

$$\text{Free float of B} = \text{ES}_{\text{successor}} - \text{EF of activity B}$$
$$\text{Successor of B is C. Hence } \text{ES}_{\text{successor}} = 10$$
$$\text{EF of activity B} = 10$$
$$\text{Free float} = 0$$

$$\text{Free float of C} = ES_{successor} - \text{EF of activity C}$$
$$\text{Successor of C is E. Hence } ES_{successor} = 24$$
$$\text{EF of activity C} = 19$$
$$\text{Free float} = 24 - 19 = 5$$

$$\text{Free float of D} = ES_{successor} - \text{EF of activity D}$$
$$\text{Successor of D is E. Hence } ES_{successor} = 24$$
$$\text{EF of activity D} = 24$$
$$\text{Free float} = 24 - 24 = 0$$

Free float and total float are zero along critical path.
Note that total float of B is five but free float is zero.
Total float and free float of C is 5.

19.3.3 Discussion of Total Float and Free Float

Let's assume that a certain project is of interest to a state senator and governor of the state. The governor tells the project manager that he does not want to move any of the early start times of major activities. On the other hand, the senator tells the project manager that he don't care about early start times of activities but cares only of the final completion date.

If the project manager was to work without changing the early start times of activities then he has to work with free floats.

If the project manager is interested only of the completion date then he can work with total floats.

For an example, if the project manager were to delay activity B by 5 days, he will delay the early start time of activity C, but will not delay the final completion date. The governor of the state would be a very angry man since activity C cannot be started at the early start time as scheduled. But on the other hand, the senator may not have a problem with delaying activity B by 5 days. The senator would ask the governor, "Why you care about early start time of activity C?" The governor would reply, "What if there is an unseen situation in activity C and it is delayed?" Then the whole project would be delayed.

Why do some people resist changes to the early start times of activities?

There is a fear that if all the available slack (total float) is utilized at the start of a project, there would not be any slack at the end. If any contingency were to occur at the end, the project would be delayed. Hence, some executives do not want to change early start times of activities.

19.3.4 Independent Float

$$\text{Independent float of activity A} = \text{ES}_{\text{successor}} - \text{LF}_{\text{predecessor}} - \text{Duration of A}$$

If you need to find the independent float of activity A, obtain the early start time of successor. Then obtain the LF time of predecessor. Use the above given equation to find the independent float.

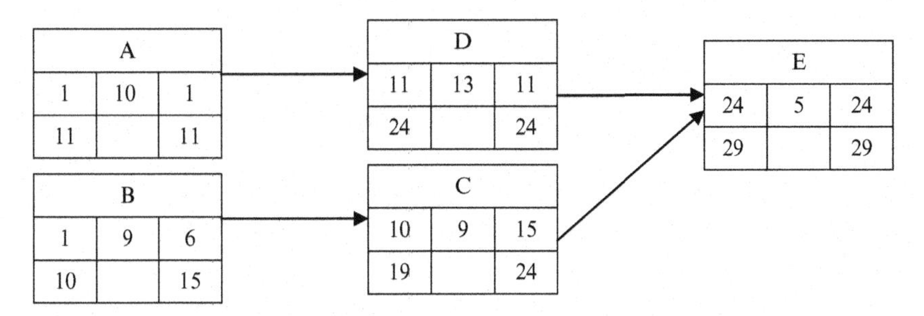

Independent float of activity A = There is no predecessor. Hence, no independent float

Independent float of activity B = There is no predecessor. Hence, no independent float.

Independent float of activity C = ES of activity E − LF of activity B − Duration of C = 24 − 15 − 9 = 0

Independent float of activity D = ES of activity E − LF of activity A − Duration of D = 24 − 11 − 13 = 0

Independent float of activity E = There is no successor. Hence, no independent float.

Discussion: An independent float indicates the slack that each activity has so that it will have absolutely no impact on preceding and successive activities. In other words the preceding activity can be finished at late finish time and the succeeding activity can start at early start time. In most cases, independent float is zero.

19.4 ACTIVITY ON ARROW NETWORKS

In activity on arrow diagrams, activities are represented in arrows. Nodes are considered to be events. The very first event is the "Start" event. Very last event is the "End" event. An activity cannot stat until the event prior to that activity is accomplished. For an event to be completed, all activities coming to that event (node) should be completed.

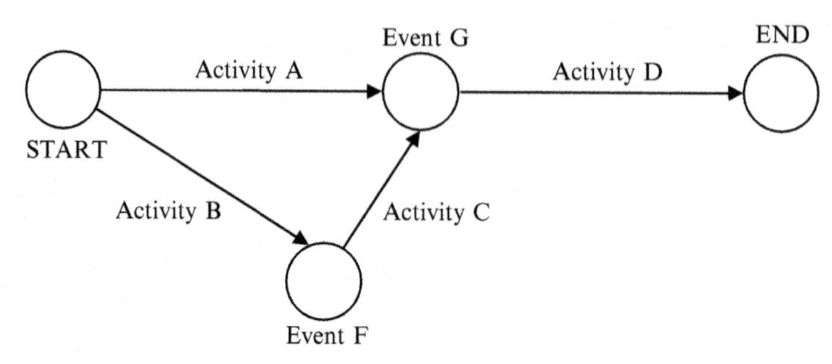

Fig. 19.3 Activity on arrow networks.

Fig. 19.3 shows an activity on arrow network. It has three activities. (activity A, B, and C). It also has three events (START event, Event F, Event G, and END event).

Activity D cannot start until event G is accomplished. Event G is accomplished when both activities A and A have been completed.

Practice Problem 19.7

Activity on arrow diagram is shown below. Find the duration of the project.

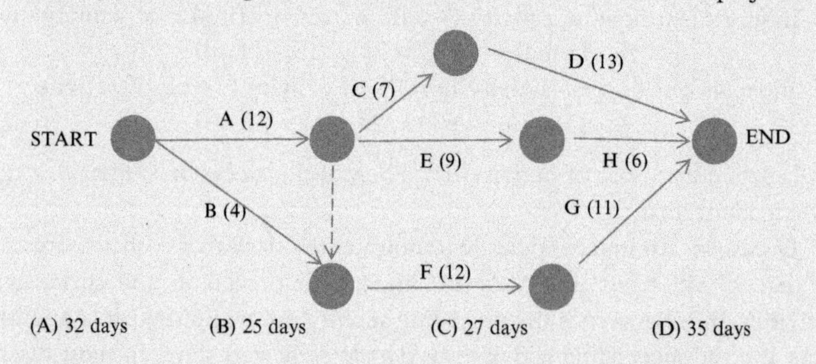

(A) 32 days (B) 25 days (C) 27 days (D) 35 days

Solution

Activities are shown on arrows in activity on arrow diagrams.

In activity on arrow diagrams, arrows are used to show activities.

In activity on arrow diagrams, nodes are called events.

First node is always the "START" event and last node is the "END" event.

On the other hand, in activity on node diagrams, activities are shown on the node.

19.4.1 Dummy Activities

Broken arrows are used to indicate dummy activities. Activity F cannot be started until activities A and B are completed. If the dummy activity is not shown, activity F can be started as soon as activity B is completed.

There are number of paths exist from start to end.

$$\text{Duration of path A, C, D} = 12 + 7 + 13 = 32$$

$$\text{Duration of path A, E, H} = 12 + 9 + 6 = 27$$

$$\text{Duration of path A, F, G} = 12 + 12 + 11 = 35$$

$$\text{Duration of path B, F, G} = 4 + 12 + 11 = 27$$

To complete the project, longest path needs to be completed.

$$\text{Duration of the project} = 35 \quad \text{(Ans D)}$$

Practice Problem 19.8

Find the early start and late start of activity D for the project given in the previous problem.

(A) ES = 12, LS = 15 (B) ES = 13, LS = 22 (C) ES = 19, LS = 22 (D) ES = 22, LS = 24

Solution

Activity D, cannot be started until activity C is completed.

$$\text{Early start of activity D} = 12 + 7 = 19 \text{ (ES} = 19)$$

Find the late start (LS) of activity D

Finding late start of an activity can be tricky.

We found that project duration to be 35 days.

Now we need to find what is the latest day that activity D can be started without delaying the project?

Latest day that activity D can be started without delaying the project = 35 − 13 = 22

35 is the duration of the project and 13 is the time need to complete activity D.

If activity D is started on day 22, there would not be any delay to the project. LS = 22

(Ans C)

Practice Problem 19.9

Find the total float of activity D.

(A) 1 (B) 3 (C) 9 (D) 2

Solution

Total float of an activity is given by the following equation:

$$\boxed{\text{Total float of an activity} = \text{LS} - \text{ES}}$$

$$\text{LS} = \text{Late start;}\quad \text{ES} = \text{Early start}$$

$$\text{LS of activity D} = 22$$

$$\text{ES of activity D} = 19$$

$$\text{Float} = \text{LS} - \text{ES} = 3$$

19.4.2 Activity Time Analysis

Activity times can be changed in order to change the critical path. The activity time of an activity can be either increased or decreased. The activity time can be decreased by increasing manpower. Similarly, the activity time can be increased by reducing resources to that particular activity.

19.4.3 Resource Leveling

We may develop a fast schedule that completes the project on time. But what about the resources? Does the contractor have resources (equipment and manpower) to do two or three activities at the same time? If the contractor does not have enough equipment and manpower to do two or three simultaneous activities, then resources should be increased. This can be done by renting new machines and hiring new personnel. Renting more machines will be an expensive thing to do. Hiring new people can also be a problem. It is not easy to find people who have suitable expertise.

Hence, the next option is to manipulate the activities.

Construction resources are labor, material, and equipment. In construction scheduling, conflicts can arise when activities compete for common resources that are available in limited quantities. After development of the critical path schedule, a resource utilization chart is developed.

Resources need to be allocated evenly during the lifetime of the project. In Fig. 19.4, when activities B and D are conducted, demand for resources increases. It is important to level the resources during the project duration. In many cases, it is not an easy task. In Fig. 19.4, it is possible to stretch activities B and D and shorten the duration of activities A and F.

Increase the duration of activities B and D. This can be done by reducing the daily quantity of resources.

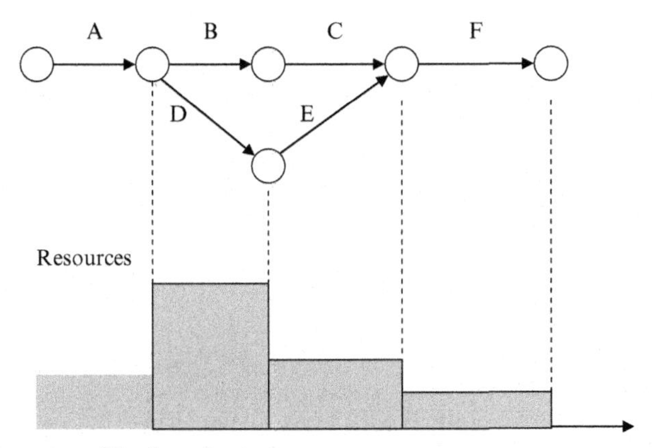

Fig. 19.4 Resource utilization chart.

Decrease the duration of activities A and F. This can be done by increasing the daily quantity of resources.

It is possible to level resources by manipulating duration of activities. In some cases it is possible to move activities around for the purpose of resource leveling.

Various computer algorithms are developed to level resources without affecting the schedule.

Competition for resources among activities:

Resource leveling procedure: Following procedure is normally adopted.

Construct the critical path schedule.

Develop the resource schedule.

Move around the activities to level the resources (Fig. 19.5).

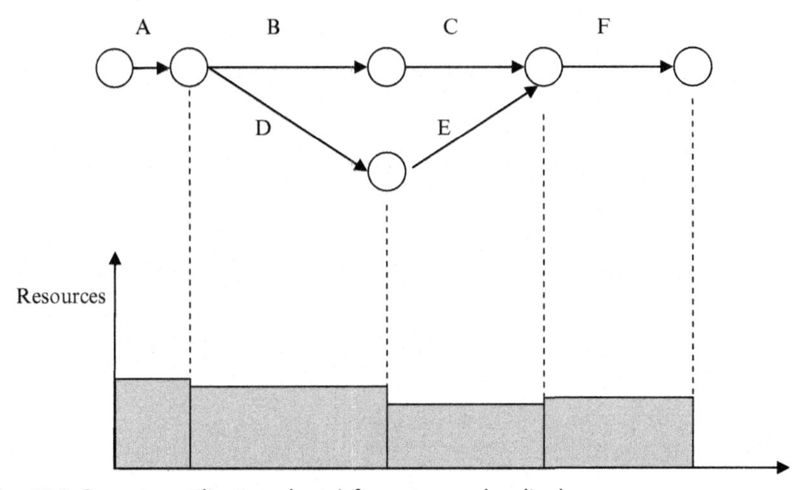

Fig. 19.5 Resource utilization chart (after resource leveling).

Practice Problem 19.10

Contractor has developed durations and resources required for construction of two slabs.

Resources required for each activity is given below:

Activity 1: Formwork slab in building A: 1 foreman, 8 carpenters, 6 laborers.

Activity 2: Rebar installation of slab in building A: 1 foreman, 5 iron workers, 3 laborers.

Activity 3: Concreting of slab in building A: 1 foreman, 8 concrete masons, 6 laborers.

Activity 4: Formwork of slab in building B: 1 foreman, 8 carpenters, 6 laborers.

Activity 5: Rebar installation of slab in building B: 1 foreman, 5 iron workers, 3 laborers.

Activity 6: Concreting of slab in building B: 1 foreman, 8 concrete masons, 6 laborers.

Activity 7: Opening ceremony.

Activity 1: Duration 7 days

Activity 2: Duration 10 days

Activity 3: Duration 7 days

Activity 4: Duration 12 days

Activity 5: Duration 11 days

Activity 6: Duration 7 days

Activity 7: Duration 1 day

Logic of activities:

Activities 1 and 4 can start at any time.

Predecessor of activity 2 is activity 1.

Predecessor of activity 3 is activity 2.

Predecessor of activity 5 is activity 4.

Predecessor of activity 6 is activity 5.

Predecessors of activity 7 are 3 and 6.

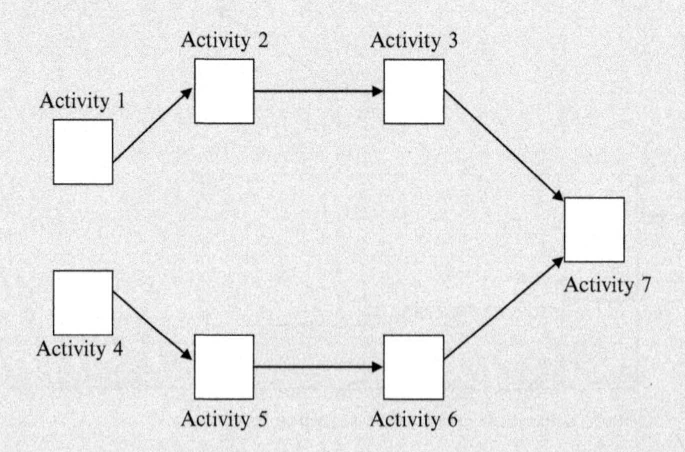

Maximum resources available at a given time:
2 foremen
8 carpenters
10 iron workers
16 concrete masons
12 laborers

What is the project duration?

Solution

There are only eight carpenters available. Hence, activities 1 and 4 cannot go parallel.

There are 10 ironworkers available. Hence, there are enough ironworkers to conduct activities 2 and 5 in parallel.

There are 16 concrete masons available. Hence, activities 3 and 6 can be done parallel.

Hence, new network can be drawn as shown.

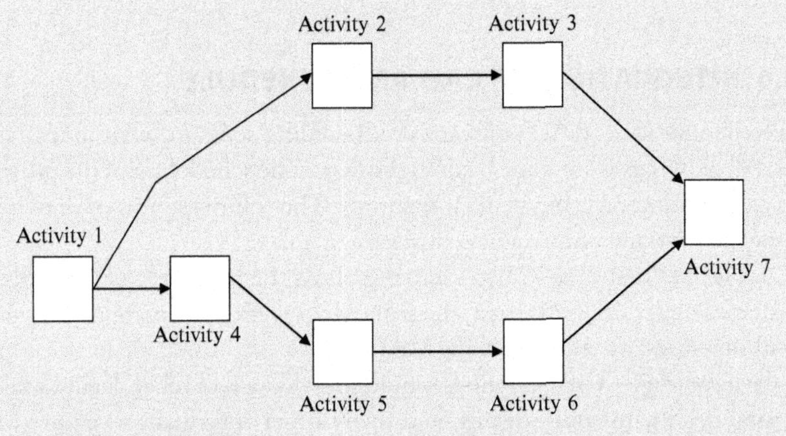

There are two paths.
Path 1, 2, 3, 7
Path 1, 4, 5, 6, 7

$$\text{Duration of path 1, 2, 3, 7} = 7 + 10 + 7 + 1 = 25\ \text{days}$$

$$\text{Duration of path 1, 4, 5, 6, 7} = 7 + 12 + 11 + 7 + 1 = 38\ \text{days}$$

$$\text{Project duration} = 38\ \text{days}$$

19.5 TIME-COST TRADEOFF

Acceleration of a project can be done by increasing the labor. Unfortunately, in some situations increasing manpower may bring diminishing returns. This could be due to number of reasons.

More management staff is needed to manage a bigger crew. The crew may not have enough space to work.

There may not be enough machinery to support a bigger crew. Delays would have a bigger cost impact due to the larger crew size (Fig. 19.6).

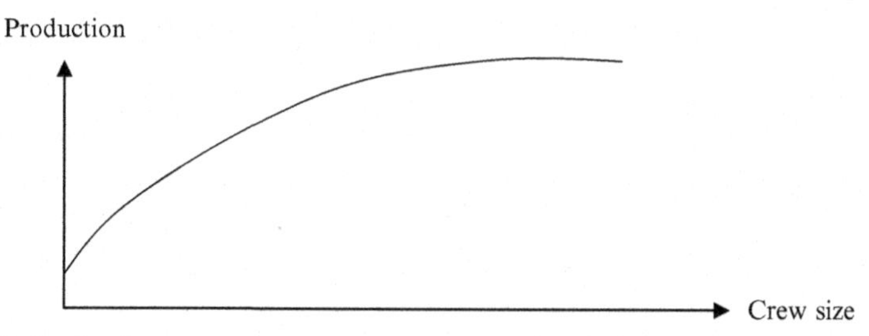

Fig. 19.6 Production versus crew size.

19.6 INTEGRATION OF CAD AND SCHEDULE

Three decades ago, CAD software or scheduling software were not available. All drawings were done by draughtsman. The scheduling of the project was also conducted using manual methods. The following sets of drawings are needed for any construction project:

- *Architectural drawings*—These drawings show the final view of the project after completion. In the past, these drawings were done manually. Today all drawings are done by using CAD.
- *Civil drawings*—Civil drawings would show concrete rebar details, structural details, masonry details, and many other information required to construct the project.
- *Mechanical drawings*—Mechanical drawings would show plumbing details, ducts, underground piping, piping inside walls, boilers, air conditioning units, and all other mechanical devices.
- *Electrical drawings*—Electrical drawings would show electrical wiring, electrical panels, transformers, and switches.
- *Communication drawings*—Communication network details such as communication cables, wiring, routers, telephone lines, and computers are shown in these drawings.

As far as scheduling is concerned, CPM technique is widely used for scheduling. Computer software such as Primevera is used for scheduling. Let us look at a simplified example of constructing small building (Fig. 19.7).

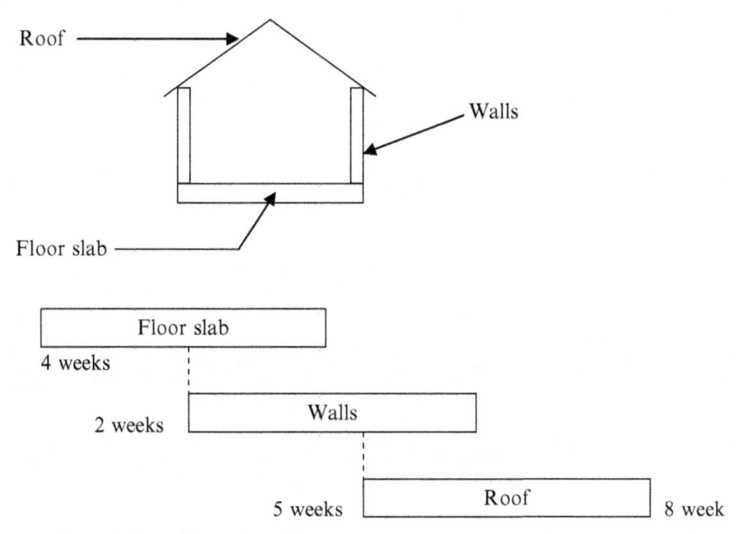

Fig. 19.7 Floor slab, walls and roof.

According to the schedule, 2 weeks after starting the project, wall construction would start. Floor slab will be completed in 4 weeks. Roof will be started in 5 weeks and completed in 8 weeks.

As mentioned earlier CAD drawings would show how the building would look at the end of the project. Integrated software such as *Rivet* can show how the site would look during construction. For an instance, it is possible for such software to show how the site would look after 3 weeks. After 3 weeks, most of the floor slab is completed and some of the walls also have been completed.

Advantages of integrated software: Integrated software can show conflicts that could occur during construction. For instance, it is possible to visualize an underground piping been done while a footing is been constructed. The resident engineer may decide to slow down the piping construction to facilitate the footing construction. Another main advantage is that owners (mostly laymen) can better visualize the construction process. Integrated software provides a better understanding between subcontractors.

CHAPTER 20

Material Quality Control

Material quality has to be determined during construction. Bad concrete may cause cracks in beams and slabs. Bad welds may create structural failures. Wood connections should be tested for quality.

20.1 MATERIAL TESTING

The material testing section of this text is one of the easiest sections and the student should take full advantage of it. Be thoroughly prepared to answer all of the questions in this section. This chapter will cover most of the common material testing procedures. The student is encouraged to conduct his own research and obtain a good knowledge of various testing methods and procedures.

Construction materials such as concrete, soil, asphalt, and steel need to be tested to assure the quality. This chapter is devoted to the testing of construction materials.

20.2 TESTING OF CONCRETE

It is necessary to make sure that concrete has achieved the specified strength requirements. Typically, concrete cylinders are obtained during construction and sent to a laboratory for testing.

Concrete testing procedure (concrete cylinders):

STEP 1: Obtain concrete test cylinders. Concrete is poured into the test cylinder in five layers and is compacted with a steel rod. Obtain a minimum of three cylinders (Fig. 20.1).

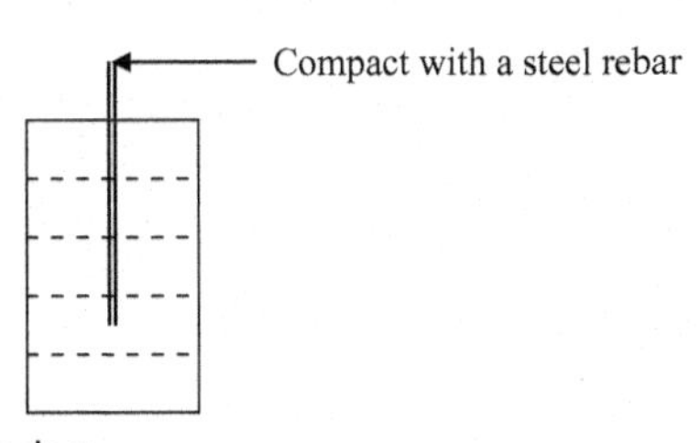

Fig. 20.1 Concrete cylinders.

Construction Engineering Design Calculations and Rules of Thumb
http://dx.doi.org/10.1016/B978-0-12-809244-6.00020-2

STEP 2: Leave the concrete cylinders close to the construction site for few days to emulate the curing conditions at the site. Then send the cylinders to a testing laboratory. The cylinders are typically broken after 7, 14, and 28 days.

Practice Problem 20.1

Assume 3000 psi concrete (which means that the strength expected after 28 days is 3000 psi) cylinders were taken and sent to the laboratory. The following test values were received. Is the concrete acceptable?
Seven-day strength: 1000 psi
Fourteen-day strength: 2200 psi
Twenty-eight day strength: 3100 psi
Required 28-day strength is 3000 psi. Achieved strength is 3100 psi. The concrete is acceptable.

Practice Problem 20.2

Assume 5000 psi concrete cylinders were taken and sent to the laboratory. The following test values were received. Is the concrete acceptable?
Seven-day strength: 2200 psi
Fourteen-day strength: 3700 psi
Twenty-eight day strength: 4700 psi
Required 28-day strength is 5000 psi

Solution

The achieved strength is 4700 psi. The concrete is not acceptable. In such situations the following options are available.

Inform the design engineer immediately. Ask him whether the concrete is used for a structural member. If the concrete is used for a structural member, can this low strength concrete support the loading?

If the low strength concrete is not adequate and does not comply with the design intent, the concrete has to be demolished and new concrete has to be placed.

If the concrete is used for a nonstructural members (such as steps, slab on grade, handicap ramp) then low strength concrete may be acceptable. In this situation the structural engineer who designed the structure needs to be involved.

Concrete slump test: A slump test is conducted by packing concrete to a standard cone. After compaction of concrete in three layers, the cone is removed. When the cone is removed, the concrete is left unsupported and will slump. The bigger the slump, the higher the water content.

Concrete with high water content is low in strength. To increase the strength, one has to decrease the water content.

On the other hand, concrete with low water content has low workability.

Slump test procedure:

STEP 1: Compact concrete to a standard cone. The standard cone is 12 in. (300 mm) in height. The concrete is placed in three layers and tamped with a steel rebar 25 times per layer.

STEP 2: The cone is removed and the concrete is left unsupported.

STEP 3: The slump is measured. Compare the slump obtained in the field to the slump specified by the engineer.

High slump means high water content. High water content means low strength.

Low slump means low water content. Low water content means high strength.

Hence, to achieve high strength, the water content needs to be kept low. However, when the water content is too low, it is difficult to pour concrete. Therefore the engineer needs to specify the range of slump that is acceptable.

Standard cone size for the slump test: 300 mm (12 in.) high cone, 200 mm (8 in.) wide at the bottom and 100 mm (4 in.) wide at the top (Fig. 20.2).

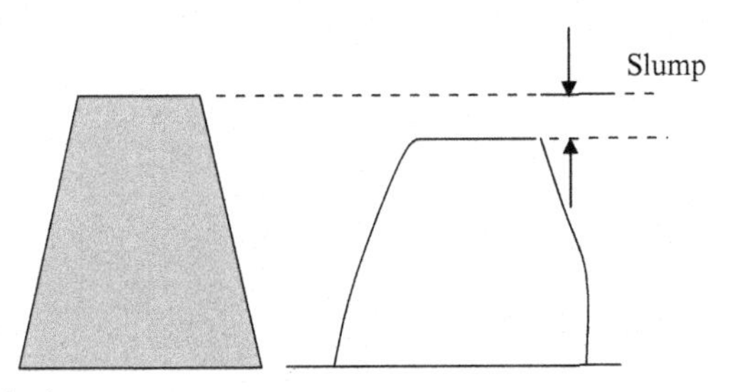

Fig. 20.2 Slump test.

20.3 SOIL TESTS

Laboratory testing: After completion of the boring program, laboratory tests are conducted on the soil samples. A laboratory test program is dependent on the project requirements. Some of the laboratory tests done on soil samples are given below:

- sieve analysis
- water content
- Atterberg limit tests (liquid limit and plastic limit)
- permeability test

- UU tests (undrained unconfined tests)
- density of soil
- consolidation test
- tri-axial tests
- direct shear test

20.3.1 Sieve Analysis

Sieve analysis is conducted to classify soil into sands, silts, and clays. Sieves are used to separate soil particles and group them based on their size. This test is used for the purpose of classifying the soil.

Standard sieve sizes are shown in Table 20.1.

Table 20.1 US sieve No. and mesh size

Sieve No.	Mesh size (mm)
No. 4	4.75
No. 6	3.35
No. 8	2.36
No. 10	2.00
No. 12	1.68
No. 16	1.18
No. 20	0.85
No. 30	0.60
No. 40	0.425
No. 50	0.30
No. 60	0.25
No. 80	0.18
No. 100	0.15
No. 200	0.075
No. 270	0.053

Gravel: Particles >#4 sieve are considered to be gravel.

Sands: Particles in the range #4–200 are considered to be sand.

Silts and clays: Particles smaller than #200 are considered to be silts and clays.

The differentiation of silts and clays cannot be done using sieve analysis. Clays are bound together due to chemical and electromagnetic forces. Silt particles are not bound together due to chemical forces.

A hypothetical sieve analysis test based on selected group of sieves are given below as an example.

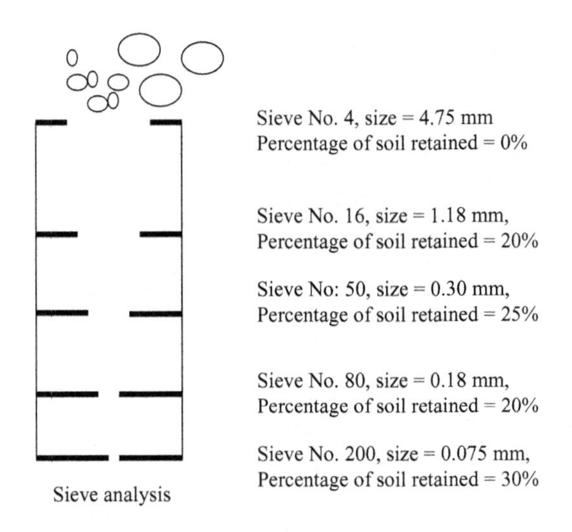

Sieve analysis

If we know the percentage of soil passed from a given sieve, we can find the percentage of soil retained in that sieve.

Sieve No. 4 (size 4.75 mm): All soil went passed sieve No. 4

Percent retained at this sieve is 0%

Percent passed = 100%

Sieve No. 16 (1.18 mm): 20% of soil was retained in sieve No. 16

Percent retained at sieve No. 16 = 20%

Percent passed = 100 − 20 = 80%

Sieve No. 50 (0.30 mm): 25% of soil was retained in sieve No. 50

Total retained so far = 20 + 25 = 45%

Percent passed = 100 − 45 = 55%

Sieve No. 80 (0.18 mm): 20% retained in sieve No. 80

Total retained so far = 20 + 25 + 20 = 65%

Percent passed = 100 − 65 = 35%

Sieve No. 200 (0.075 mm): 30% retained in sieve No. 200

Total retained so far = 20 + 25 + 20 + 30 = 95%

Percent passed through sieve No. 200 = 100 − 95 = 5%

Now it is possible to draw a graph indicating the percentage passing at each sieve (Fig. 20.3).

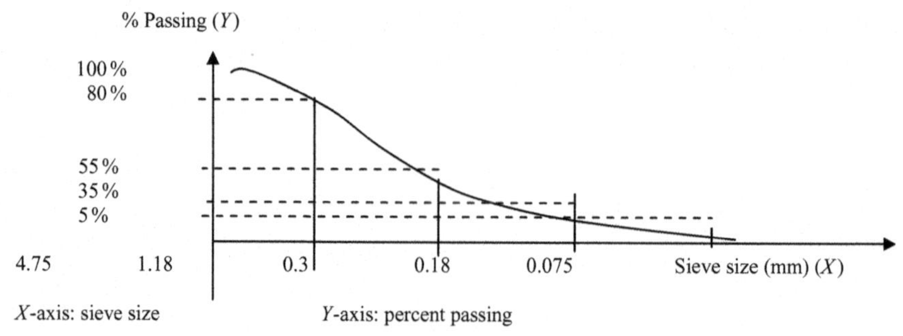

Fig. 20.3 Percent passing vs. particle size.

D_{60}: D_{60} is defined as the size of the sieve that allows 60% of the soil to pass. This value is used for soil classification purposes and frequently appears in geotechnical engineering correlations.

To find the D_{60} value, draw a line at 60% passing point. Then drop it down to obtain the D_{60} value (Fig. 20.4).

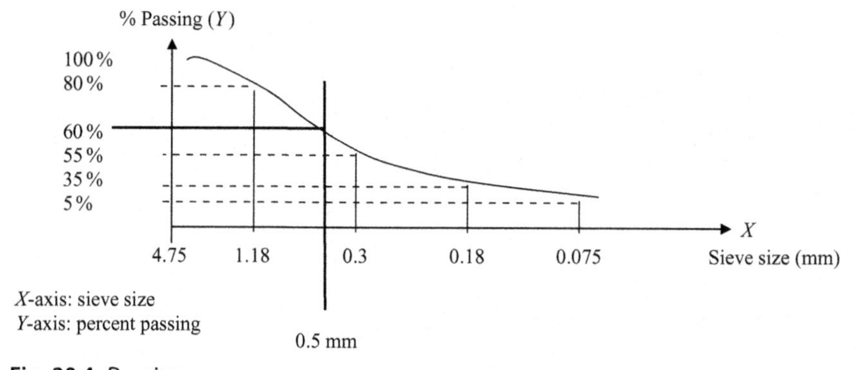

Fig. 20.4 D_{60} size.

Finding D_{60}: In this case D_{60} is closer to 0.5 mm.

Find D_{30}: As before, draw a line at 30% passing line. In this case, D_{30} happened to be ~0.1 mm.

Gradation: When a sand sample contains sand particles of all sizes, it is called a well-graded sand. When sand particles are more or less similar in size, the sand is considered to be poorly graded.

Well-graded sands can be compacted better than poorly graded sands (Fig. 20.5).

Fig. 20.5 Well-graded sands (various sizes) and poorly graded sands (similar sizes).

20.3.2 Soil Classification

Soils are classified into different categories.

Major categories of soils are:

- gravels
- sands
- silts
- organic clays
- inorganic clays
- peat

Silts and sands: Sand particles can be seen with the naked eye. Silt and clay particles cannot be seen with the naked eye. The difference between silts and clays is that clay particles bind together. Silt particles do not have cohesive properties.

In many cases, silt particles and clay particles are mixed together. Such soils are known as silty clay or clayey silt. When the predominant constituent is clay and silt is the secondary component, such soils are known as silty clay. If the predominant constituent is silt, such soils are known as clayey silt.

Symbols:

SP: Poorly graded sand

SW: Well-graded sand

Since symbol "S" was used for sands, another symbol was needed to represent silts. Symbol "M" is used to represent silts.

ML: Low plastic silts

MH: High plastic silts

Clays: Some clays contain plenty of organic matter. Organic matter is mainly decomposed tress and roots. Clays with a large quantity of organic matter is known as organic clays.

Clays could be high plastic to low plastic. Highly cohesive clays are known as high plastic clays and vice versa. High plastic clays are also known as fat clays.

CL: Low plastic inorganic clays
CH: High plastic inorganic clays
OL: Low plastic organic clays or silts
OH: High plastic organic clays or silts
PT: Predominantly organic soils, peat, muck, marsh soils

20.3.3 Soil Classification Procedure

Coarse grained soils and fine grained soils: Sands and silts are known as coarse grained soils and silts and clays are known as fine grained soils.

Particles larger than 0.075 mm (No. 200 sieve size) are classified as coarse grained soils or coarse fraction (sands and gravel).

Particles smaller than 0.075 mm (No. 200 sieve size) are classified as fine grained soils or fine fraction (silts and clays).

20.3.4 Classification of Gravels

If 50% or more of the coarse fraction is larger than 4.75 mm (No. 4 sieve) such soils are classified as gravels.

If 50% or more of the coarse fraction is smaller than 4.75 mm (No. 4 sieve) such soils are classified as sands.

How to differentiate between poorly graded gravel and well-graded gravel?

Conditions for well-graded gravels (GW): Two conditions have to be satisfied for well-graded gravel.

Condition 1: $D_{60}/D_{10} > 4$

Condition 2: $1 < D_{30}^2/(D_{10} \times D_{60}) < 3$

If any of these conditions are violated such soils are classified as GP.

If gravel contains silts, then it would be classified as GM.

Similarly, if the gravel contains clays, it would be classified as GC (Fig. 20.6).

20.3.5 Classification of Sands

If more than 50% of the soil sample is larger than 0.075 mm (No. 200 sieve) such soils are classified as gravels and sands.

If 50% or more of the coarse fraction is smaller than 4.75 mm (No. 4 sieve) such soils are classified as sands.

Conditions for well-graded sands (SW):

Condition 1: $D_{60}/D_{10} > 6$

Condition 2: $1 < D_{30}^2/(D_{10} \times D_{60}) < 3$

If any of these conditions are violated such soils are classified as poorly graded sands (SP).

If sand contains silts, then it would be classified as SM.
Similarly, if the sand contains clays, it would be classified as SC.

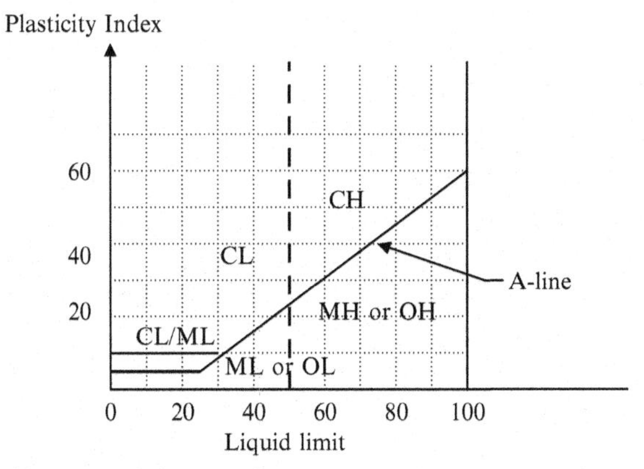

Fig. 20.6 Liquid limit vs. plasticity index.

20.3.6 Soil Compaction

Shallow foundations can be rested on controlled fill, also known as engineered fill or structural fill. Typically, such fill materials are carefully selected and compacted to 95% of the modified Proctor density.

A Modified Proctor test is conducted by placing soil in a standard mold and compacted with a standard ram.

Modified Proctor Test Procedure:

STEP 1: Soil that needs to be compacted is placed in a standard mold and compacted (Fig. 20.7).

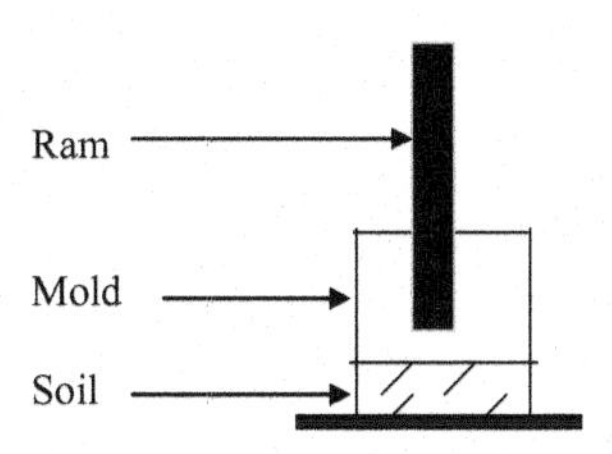

Fig. 20.7 Standard mold.

STEP 2: Compaction of soil is done by dropping a standard ram 25 times for each layer of soil from a standard distance. Typically, soil is placed in five layers and compacted.

STEP 3: After compaction of all five layers, the weight of the soil is obtained. The soil contains solids and water. Solids are basically soil particles.

$$M = M_s + M_w$$

M = total mass of soil including water
M_s = mass of solid portion of soil
M_w = mass of water

STEP 4: Find the moisture content of the soil.
Moisture content is defined as M_w/M_s.
Small sample of soil is taken and placed in the oven and measured.

STEP 5: Find the dry density of soil.
Dry density of soil is given by M_s/V.
M_s is the dry weight of soil and "V" is the total volume.

STEP 6: Repeat the test few times with different moisture content and plot a graph between dry density and moisture content (Fig. 20.8).

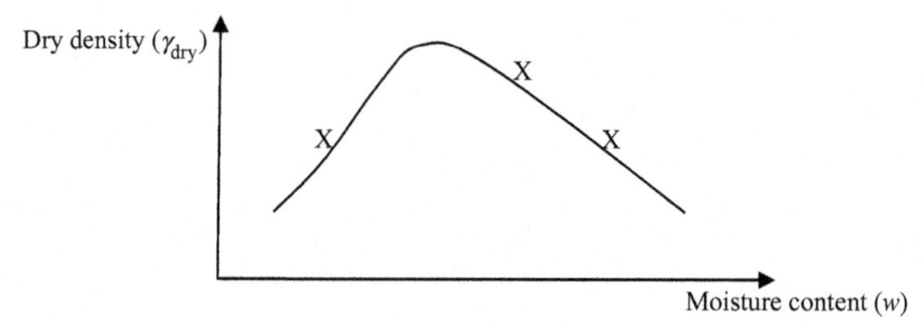

Fig. 20.8 Dry density and moisture content.

STEP 7: Obtain the maximum dry density and the optimum moisture content (Fig. 20.9).

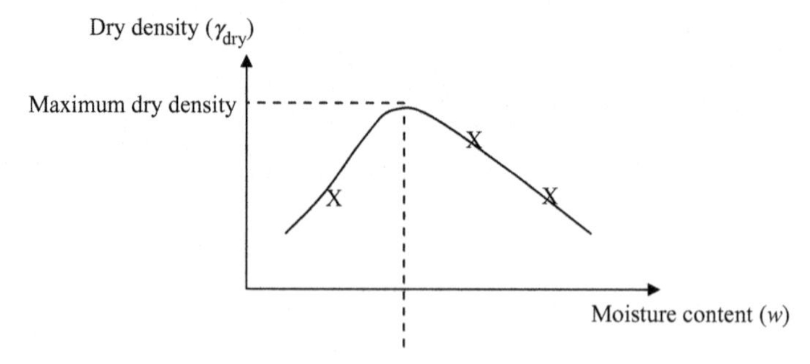

Fig. 20.9 Optimum moisture content.

For a given soil, there is an optimum moisture content that would provide the maximum dry density.

It is not easy to attain the optimum moisture content in the field. Usually soil that is too wet is not compacted. If the soil is too dry, water is added to increase the moisture content.

Liquid limit: Liquid limit is the water content where soil start to behave as a liquid.

Liquid limit is measured by placing a clay sample in a standard cup and making a separation (groove) using a spatula. The cup is dropped till the separation vanishes. Water content of the soil is obtained at this sample. The test is performed again by increasing the water content. Soil with low water content would yield more blows and soil with high water content would yield less blows.

A graph is drawn between number of blows and the water content (Fig. 20.10).

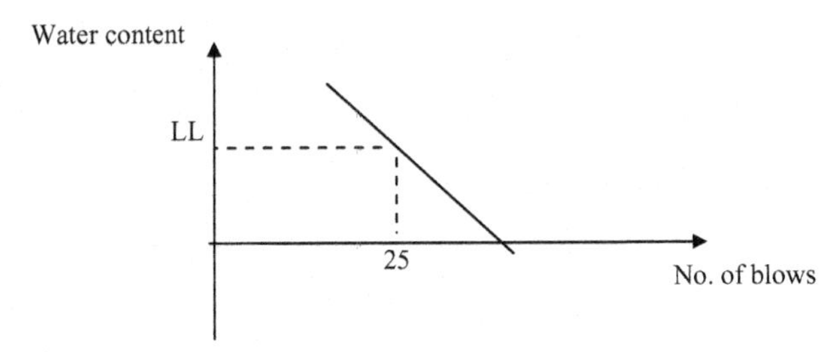

Fig. 20.10 Graph for liquid limit test.

Plastic limit: The plastic limit is measured by rolling a clay sample to a 3 mm diameter cylindrical shape. During continuous rolling at this size, the clay sample tends to lose moisture and cracks start to appear. Water content where cracks start to appear is defined as the plastic limit (Fig. 20.11).

20.3.7 Permeability Test

Transport of water through soil media depends on the pressure head, velocity head, and the potential head due to elevation. In most cases the most important parameter is the potential head due to elevation (Fig. 20.12).

Water travels from "A" to "B" due to high potential head. The velocity of traveling water is given by the Darcy equation.

$$v = k \cdot i \quad \text{(Darcys Equation)}$$

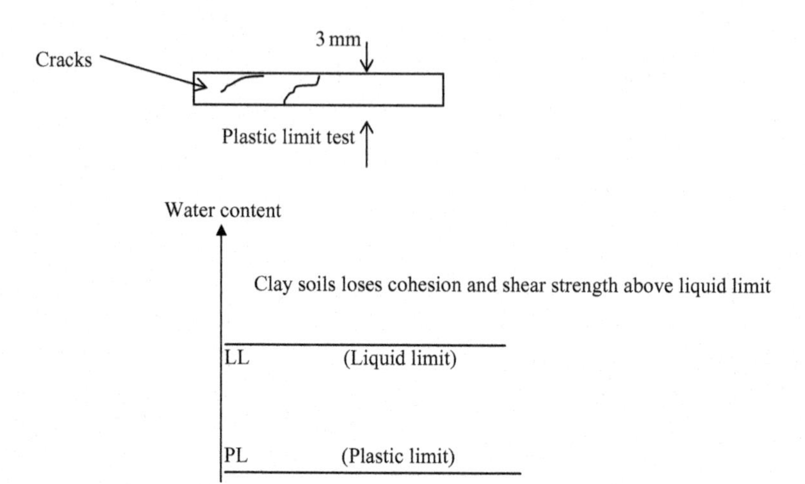

Fig. 20.11 Clay soils loses cohesion and shear strength below plastic limit.

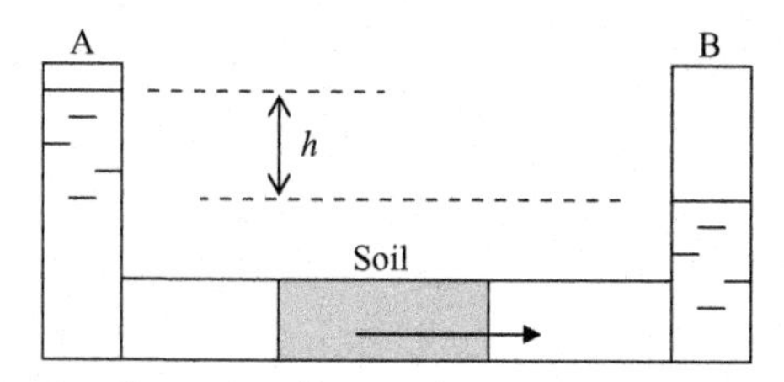

Fig. 20.12 Water flowing through soil.

v = velocity
k = coefficient of permeability (cm/s or in./s)
i = hydraulic gradient = h/L
L = length of soil

$$\text{Volume of water flow} = Q = A \times v$$

A = area
v = velocity

Practice Problem 20.3

Find the volume of water flowing in the pipe shown. Soil permeability is 10^{-5} cm/s. Area of the pipe is 5 cm^2. Length of soil plug is 50 cm.

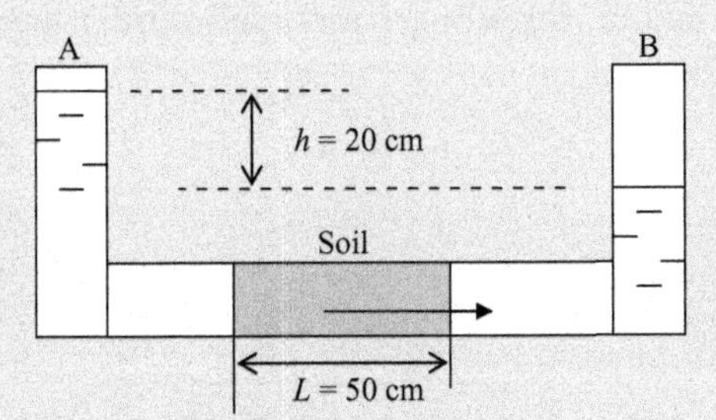

Water flow due to 20 cm gravity head

Solution

Apply the Darcy equation:

$$v = k \cdot i \text{ (Darcys equation)}$$

v = velocity
k = coefficient of permeability (cm/s or in./s)
i = hydraulic gradient = h/L
L = length of soil
A = area of the pipe = 5 cm^2

$$v = k \times (h/L)$$

$$v = 10^{-5} \times 20/5 = 4 \times 10^{-6}\,\text{cm}/s$$

$$\text{Volume of water flow} = A \times v = 5 \times 4 \times 10^{-6}\,\text{cm}^3/s = 2 \times 10^{-5}\,\text{cm}^3/s$$

20.3.8 Unconfined Undrained Compressive Strength Tests (UU Tests)

Unconfined compressive strength test is designed to measure the shear strength of clay soils. This is the easiest and most common test done to measure the shear strength.

Since the test is done with the sample in an unconfined state and the load is applied fast so that there is no possibility of draining, the test is known as unconfined–undrained test (UU test).

UU test procedure: Soil sample is placed in a compression machine and compressed until failure. Stresses are recorded during the test and plotted (Figs. 20.13 and 20.14).

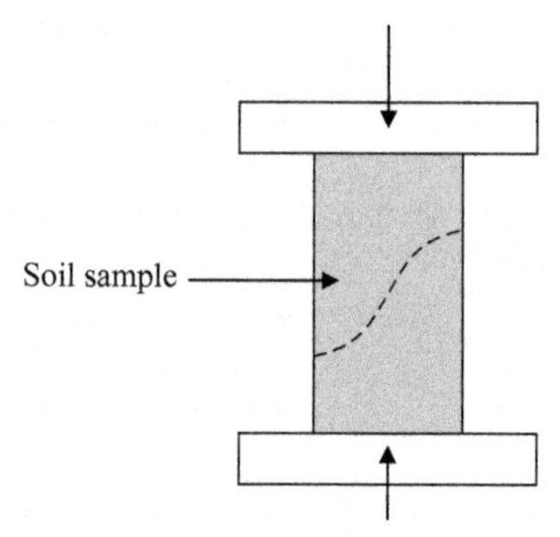

Fig. 20.13 UU test apparatus.

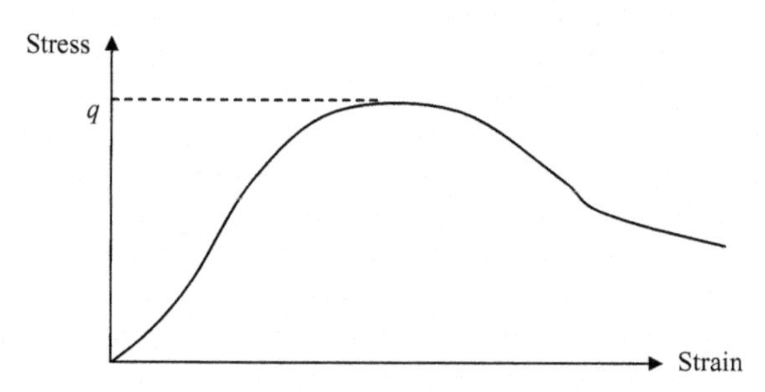

Fig. 20.14 UU apparatus and stress–strain curve.

20.3.9 Tensile Failure

When a material is subjected to a tensile stress, it would undergo tensile failure.

Fig. 20.15 shows material failure under tension. Tensile failure of soil is not as common as shear failure and tensile tests rarely conducted. On the other hand, tensile failure is common in tunnels.

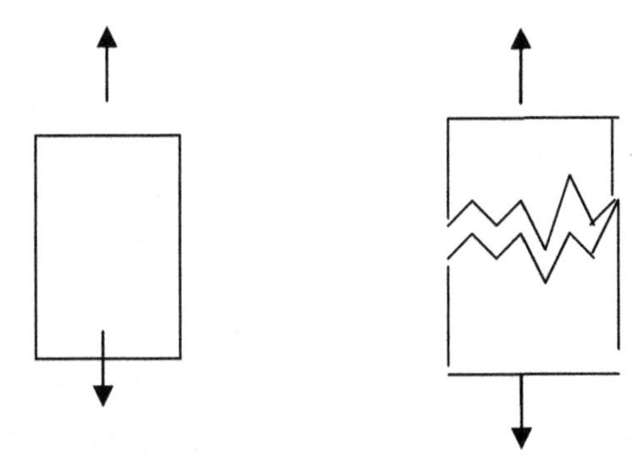

Fig. 20.15 Tensile strength test.

20.4 ASPHALT

Asphalt is a bituminous material manufactured using petroleum by-products. Asphalt is mixed with aggregates and the base course is manufactured. The base course is laid on top of compacted soil and rolled. Another layer of asphalt, known as top, is laid on top of the base course (Fig. 20.16).

Both base course and top course are manufactured by mixing asphalt with aggregates. The main difference is that base course is mixed with coarse aggregates while top course is mixed with fine aggregates.

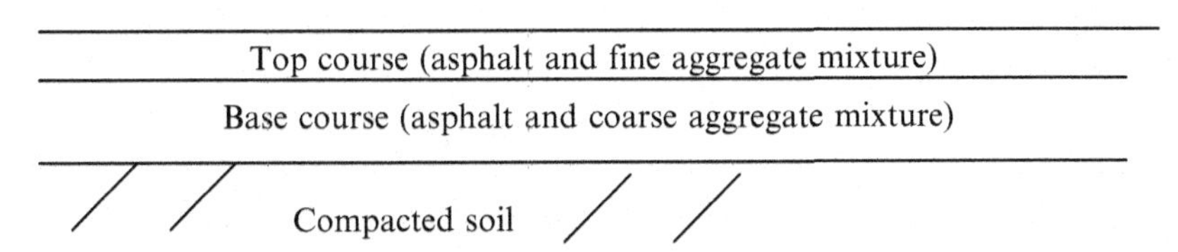

Fig. 20.16 Base course and top course.

Asphalt density in the field can be tested with a nuclear gauge. The most well known manufacturer of these nuclear gauges is Troxler Inc. The gauge indicates the density of asphalt. If the density of asphalt is less than the specified density, the asphalt needs to be rolled with a heavy roller.

20.5 QUALITY CONTROL PROCESS

It is important to control quality during a project. Typically, the project should have a quality control manual at the beginning of the project. The contractor should follow the quality control plan during construction of the project.

Items of a typical quality control program are:

- inspection checklist
- documentation and filing procedure
- testing program
- closeout procedure

Inspection program: The inspection checklist is developed to help the inspectors. The contractor is supposed to complete the project as per contract drawings and specifications. It is the responsibility of inspectors to make sure that the project is completed as per contract documents. The inspection checklist should contain items to be inspected and the inspection frequency. If the contractor is not following the contract documents, the inspector should generate a noncompliance report and distribute it to the engineer and the owner of the project.

Documentation and filing procedure:

Project documents include:

- noncompliance reports
- RFIs (request for information)
- change order requests
- safety violations
- daily field reports, weekly reports, and monthly reports
- test results
- permits
- closeout documents

A quality control plan should include a methodology to file project documents.

Testing program: A testing program is developed to facilitate the inspectors and project managers. The testing program should include what items are to be tested, the testing methodology, and the testing frequency.

CHAPTER 21

Temporary Structures

During construction, temporary structures are required. Let us assume a simple task of painting a building. How could workers go up to the upper levels to paint? Typically, workers stand on a temporary structure known as a scaffold (Fig. 21.1).

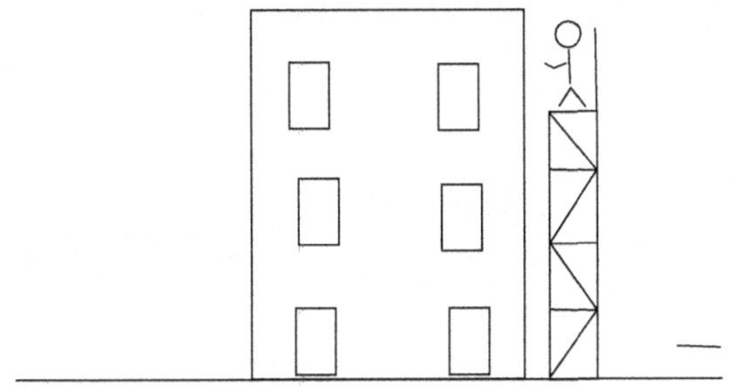

Fig. 21.1 The painting of a building. A scaffold is built for painters to stand on while painting the building.

21.1 SCAFFOLDS

Scaffolds are work platforms that enable workers to do their job at high elevations. The type of work can be brickwork, painting, steel work, concreting, or window installation.

Most scaffolds are made of steel pipes. In some countries bamboo is still used for scaffolds.

21.1.1 Pipe Scaffolds

Pipes are used to build scaffolds. Pipes are connected using special connectors. Platforms are provided for workers to stand on. Ladders are provided for workers to go from one platform to another one (Figs. 21.2 and 21.3).

Pipe scaffolds cannot be used for very tall buildings. Other methods such as outrigger scaffolds are used in such situations.

Construction Engineering Design Calculations and Rules of Thumb
http://dx.doi.org/10.1016/B978-0-12-809244-6.00021-4

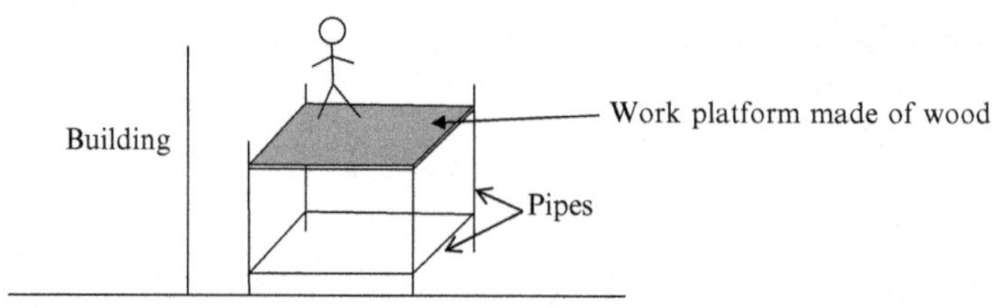

Fig. 21.2 Pipe scaffold with one platform.

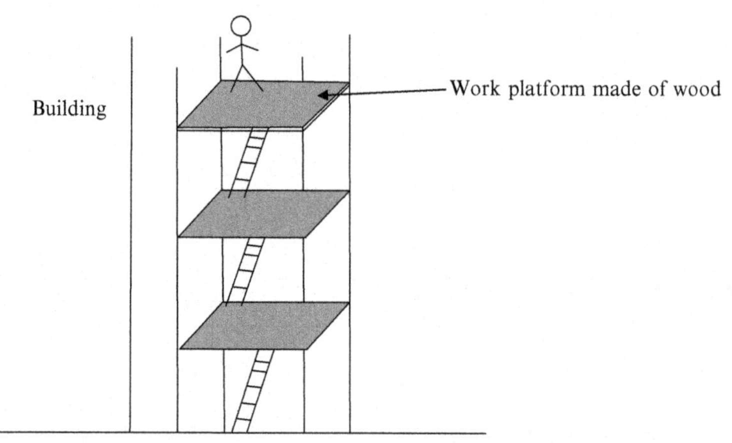

Fig. 21.3 Pipe scaffold with multiple platforms. Ladders are provided to climb to higher platforms.

21.1.2 Outrigger Scaffolds

Consider this scenario: Brickwork has to be done on the 70th story of a 70-story building. How could the workers get to the 70th story? Should they build a pipe scaffold from the ground all the way to the 70th story? That may not be very feasible. In this type of situation, outrigger scaffolds can be used.

Metal beams are attached to the building. These beams are used to build a work platform (Fig. 21.4).

In the case of outrigger scaffolds, metal beams or a metal structure are attached to the newly constructed building. This has to be done with the authorization of design engineers. This metal structure is known as outriggers. The protruding metal structure is used to build a work platform.

21.1.3 Modular Scaffolds

Pre-made modules are becoming common in many construction projects (Figs. 21.5 and 21.6).

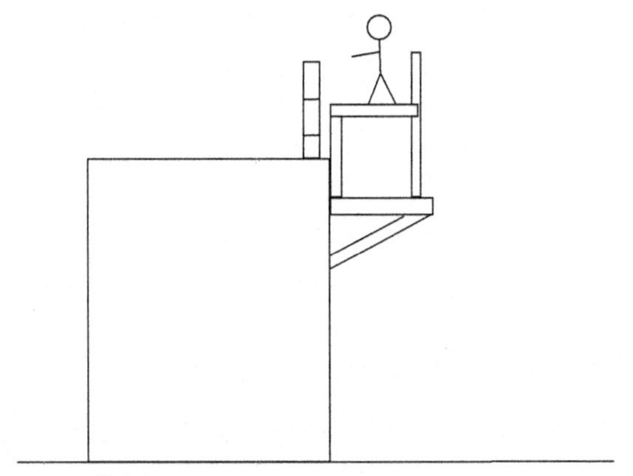

Fig. 21.4 Outrigger scaffold.

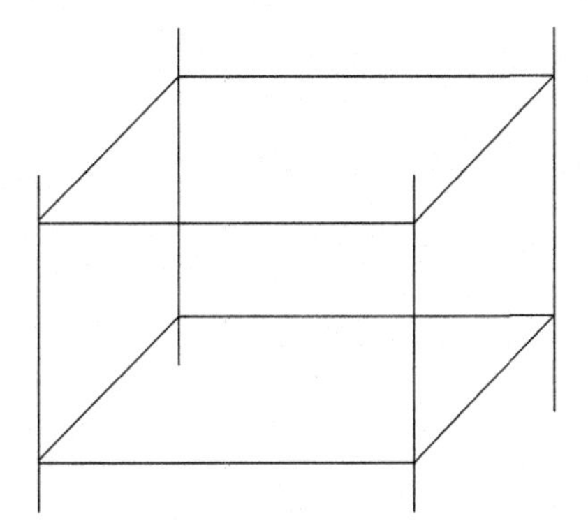

Fig. 21.5 Scaffolding modules.

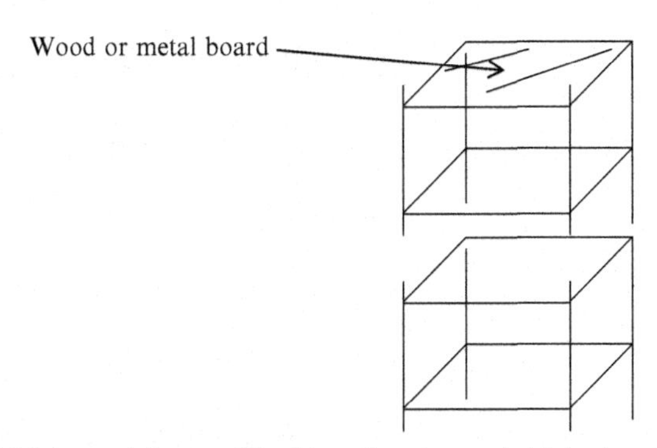

Fig. 21.6 Scaffolding modules are fitted together to reach high elevations.

Boards: Boards are made of metal or wooden planks attached to the scaffolding for people to stand and work.

Uprights also known as standards and poles are used to carry the load to base. False uprights are mainly used near entrances to the work platform. False uprights do not transfer any vertical loads to the ground. Though it may provide lateral support handrails, it does not provide any lateral supports to the scaffold system.

21.2 SHORING

Scaffolds are built for workers to work. Scaffolds act as work platforms for workers. On the other hand, shoring is done to support wet concrete. Once the concrete is hardened, the shoring is removed. Other than supporting wet concrete, shoring can be used to support weak columns.

Let us assume that an existing column in a building is deteriorated and has to be replaced. The procedure to remove an existing column and build a new column is shown in Fig. 21.7.

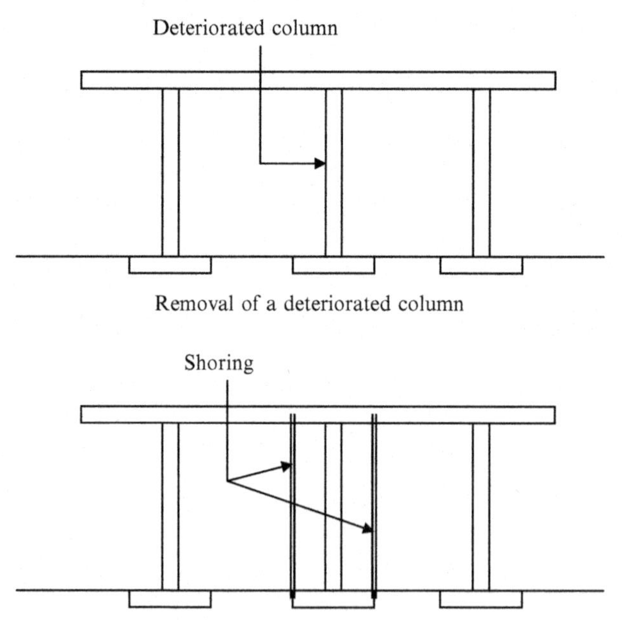

Fig. 21.7 Provide shoring prior to removal of the column.

Once proper shoring is provided and has been approved by the relevant authorities, the contractor can remove the existing load bearing column and construct a new one.

21.3 BRACING

Bracing is a support element provided to strengthen an existing structural element (Figs. 21.8 and 21.9).

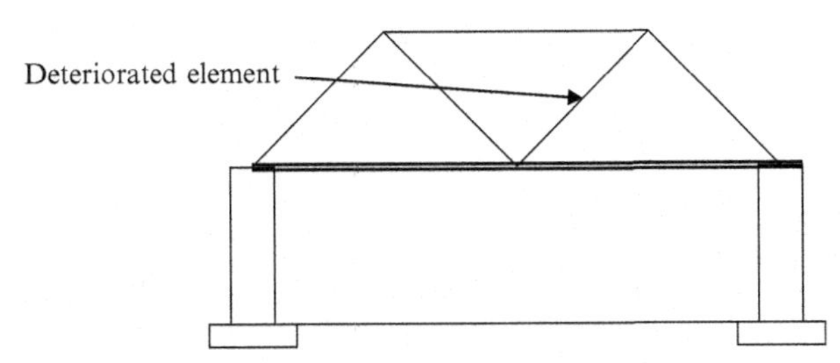

Fig. 21.8 Deteriorated element in a structure.

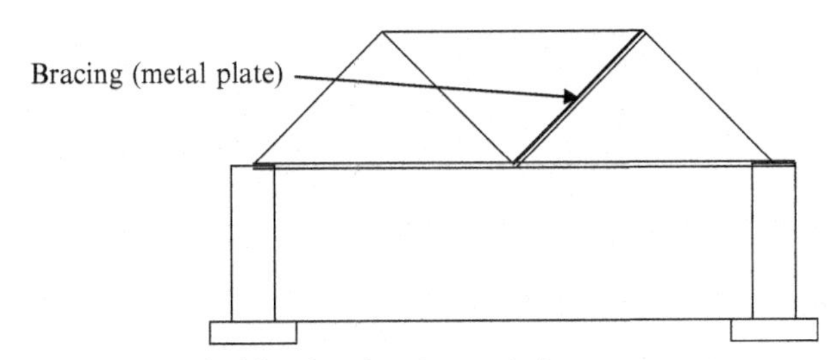

Fig. 21.9 Bracing is provided for the deteriorated element.

21.3.1 Bracing Masonry Walls

Masonry walls need to be braced during and after construction. All masonry walls have to be supported laterally. In a building, masonry walls are tied to beams, columns, and other walls. Until the masonry wall is laterally supported, it has to be braced.

In practice, bracing of masonry walls has to follow OSHA guidelines. For the exam purposes, NCEES recommends "Standard Practice for Bracing Masonry Walls Under Construction" by *Mason Contractors Association of America* (MCAA). Hence, you need to know both OSHA and MCAA guidelines (Fig. 21.10).

Masonry walls are made of bricks and mortar. Until a mortar brick joint is fully developed, masonry walls have little lateral stability. Even after the mortar is hardened, a standing masonry wall has little resistance against overturning. As per OSHA, any wall 8 ft or taller needs to be braced.

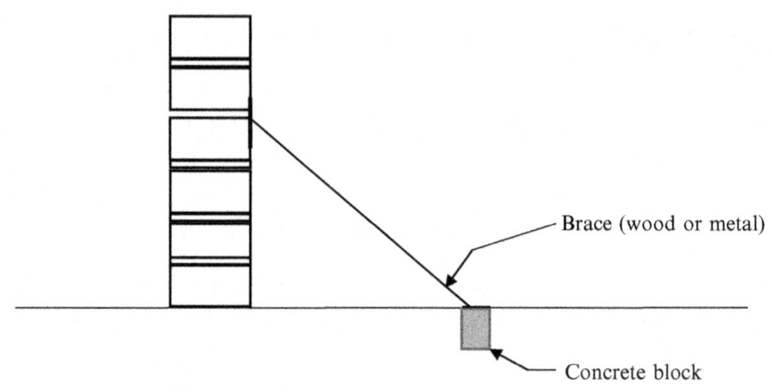

Fig. 21.10 Schematic diagram of a masonry wall bracing. Bracing is required to maintain the lateral stability of a masonry wall.

OSHA (1926.706 (b)) says the following:

> *All masonry walls over eight feet in height shall be adequately braced to prevent overturning and to prevent collapse unless the wall is adequately supported so that it will not overturn or collapse. The bracing shall remain in place until permanent supporting elements of the structure are in place.*

Masonry wall construction procedure: Typically, masonry walls are constructed using scaffolds and a worker platform. When the wall reaches 8 ft tall, the wall needs to be braced. The construction period is also known as the "*Initial Period.*" The initial period consists of 24 h, or the period that the wall is constructed, whichever is shorter. Bracing has to be installed after the initial period. In other words if wall construction is ongoing after 24 h, the bracing need to be installed after 24 h. If the construction of the wall is finished after 8 h, then bracing need to be installed after 8 h.

Intermediate period: After the initial period, the workers will be attaching the wall to other elements of the building such as columns, beams, and other walls. The bracing should be in place until the wall is properly attached to the columns, beams, and other walls. The bracing can be removed when lateral stability has been achieved.

Restricted zone: During the construction period (initial period) and the intermediate period, people who are not working on the wall should not go near the wall. A restricted zone is declared around the masonry wall. The restricted zone is equal to the height of the wall plus 4 ft on both sides of the wall (Fig. 21.11).

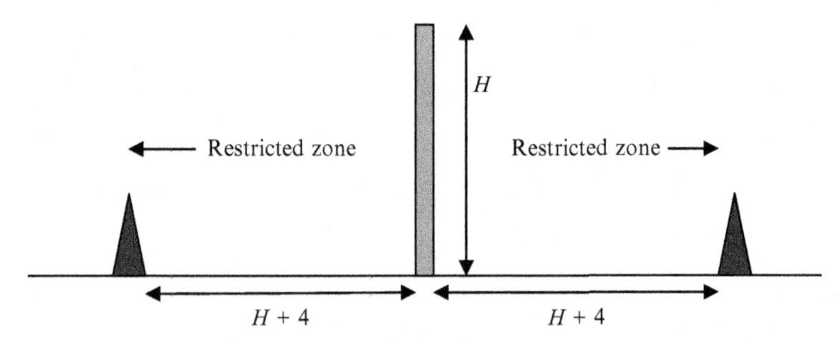

Fig. 21.11 Restricted zone.

Note: There is some confusion between OSHA and MCAA standard practice. MCAA standard practice says the restricted zone shall be established on both sides of the wall. OSHA says the following:

> *The limited access zone shall be established on the side of the wall which will be unscaffolded (1926.706(a)(3)).*

Limited access zone and restricted zone means the same thing. Only workers who are directly working on the wall can enter the restricted zone. For instance, a worker who is concreting a sidewalk near the masonry wall is not allowed inside the restricted zone.

As per MCAA, work on masonry walls should stop when the wind speed exceeds 20 mph during the initial period. On the other hand, work is allowed until the wind speed is 35 mph during the intermediate period.

As per OSHA, limited access zone should be established whenever masonry walls are built. OSHA does not specify a height limitation for the limited access zone.

Tables are given in "Standard Practice for Bracing Masonry Walls Under Construction" indicating the type of bracing required for a given wall height.

Factors that affect masonry bracing: Not all masonry walls are the same. Some masonry walls are fully grouted. Some are reinforced. Some are not reinforced. In addition, some have higher density.

Some factors that affect the lateral stability:

- the wall is reinforced or not,
- the wall is fully grouted or not,
- density of masonry,
- thickness of the wall,

- height of the wall, and
- wind speed.

Wind force calculation: MCAA gives the following equation to calculate the force acting on a wall:

$$w = 0.00256\,V^2$$

where w is the wind pressure in psf and V is the 5-s wind gust speed in mph.

If the wind speed is 20 mph, $w = 0.00256 \times 20^2 = 1.042\text{psf}$

Practice Problem 21.1

What is the wind force acting on a 12 ft tall wall when the wind is blowing at 20 mph? Assume the wind speed is the same on the wall from top to bottom. Length of the wall is 100 ft.

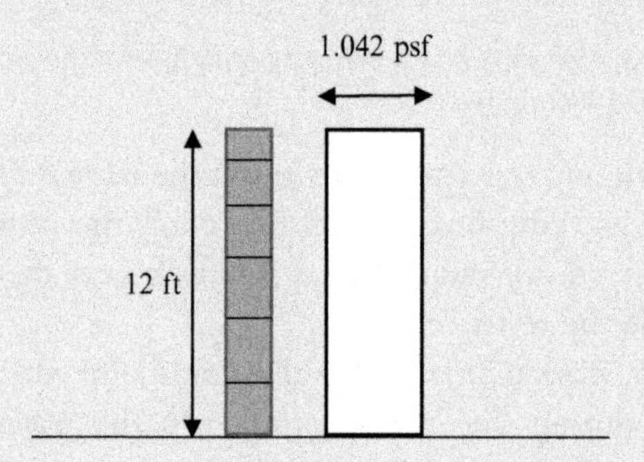

Solution

$$\text{Total force acting on the wall} = \text{Wind pressure} \times \text{Area of the wall}$$

$$\text{Total force acting on the wall} = 1.042 \times (12 \times 100)$$

$$\text{Total force acting on the wall} = 1250.4\text{lbs}$$

Three different bracing types are considered by MCAA standard practice.

- wood bracing,
- pipe bracing, and
- cable bracing.

21.4 COFFERDAMS

There are instances where construction has to take place near a river, lake, or ocean. Bridge piers, harbor structures, and flood control structures are some examples. In such situations water has to be kept away from the construction area. How can you concrete when water is pouring in? A structure has to be built to keep the water away. In such situations a temporary structure needs to be constructed to keep the water away. These temporary structures are known as cofferdams.

Cofferdams are temporary structures constructed to keep water out of the construction area. The majority of cofferdams are constructed in rivers mainly to build bridge piers. Thanks to improvement in caisson technology, in most cases cofferdams may not be necessary anymore.

21.4.1 Cofferdams in Bridge Pier Construction

Bridge piers are mostly constructed in rivers and water has to be kept out of the construction zone during construction. Typical bridge pier is shown in Fig. 21.12.

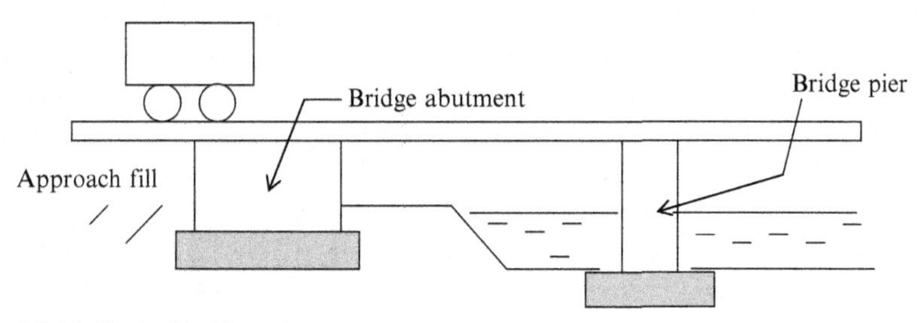

Fig. 21.12 Typical bridge pier.

Construction procedure of a typical cofferdam constructed using sheet piles for a bridge pier is shown in Fig. 21.13.

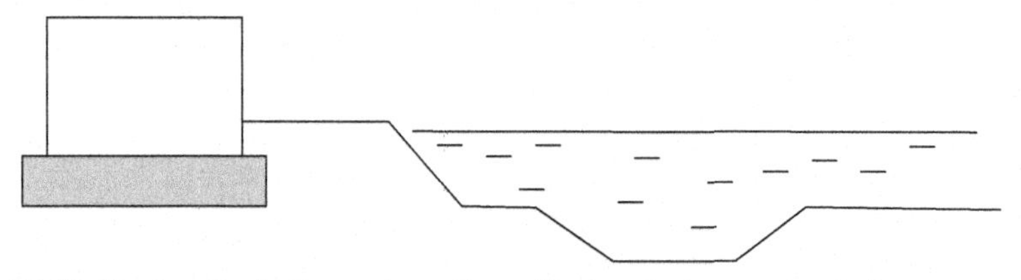

Fig. 21.13 Dredge the bottom where the cofferdam to be constructed.

STEP 1: The bottom is dredged to remove lose sediments and to obtain hard bottom surface (Fig. 21.14).

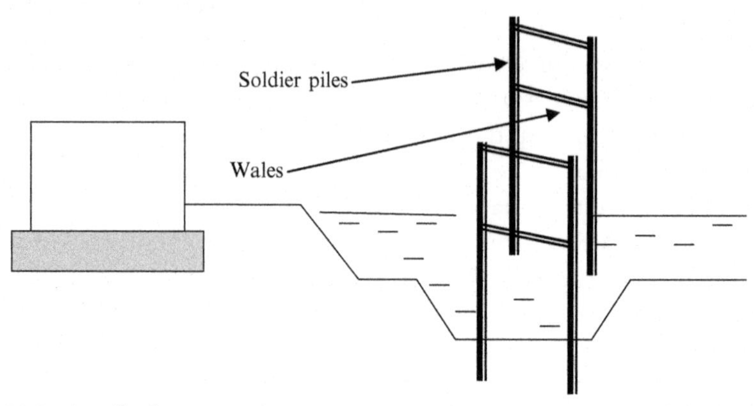

Fig. 21.14 Install piles.

STEP 2: Drive soldier piles and construct wales (horizontal beams).

Soldier piles are driven first. These piles should extend deep into the riverbed. Soldier piles will take most of the load. Wales or the horizontal beams are constructed to provide stability to the structure (Fig. 21.15).

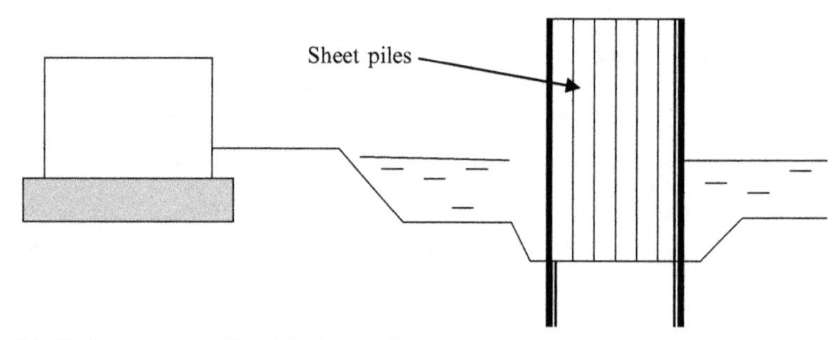

Fig. 21.15 Construct walls with sheet piles.

STEP 3: Drive watertight sheet piles between soldier piles and attach them to the soldier piles.

Watertight sheet piles are driven and attached to the structure. It is important to make sure that the cofferdam is stable during the construction process (Fig. 21.16).

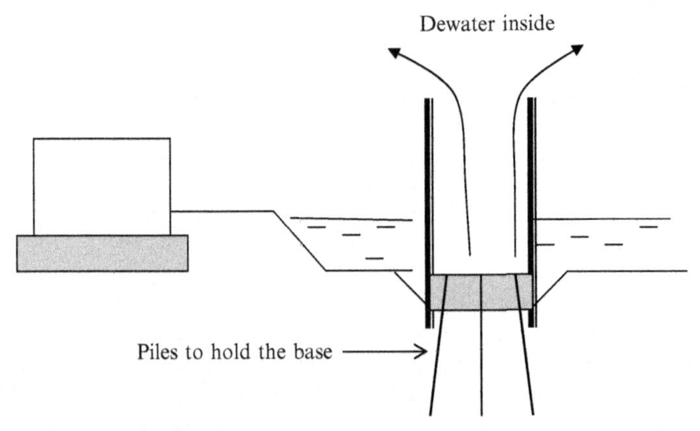

Fig. 21.16 Concrete the bottom.

STEP 4: Concrete the bottom. Then dewater inside the cofferdam (sheet piles are not shown). Piles are driven to hold the concrete base down.

Once the sheet piles are driven and attached to soldier piles, the base is concreted. The concrete base has to be designed to make sure that it will not fail due to the buoyant pressure of water. In some cases piles are driven prior to concreting the base to hold the base down. Piles can also be driven after constructing the base by coring holes through the concrete base. Driving of piles can be done from a barge from the top and extra length can be cut off.

Some engineers place gravel on top of the concrete to hold the base down instead of driving piles (Fig. 21.17).

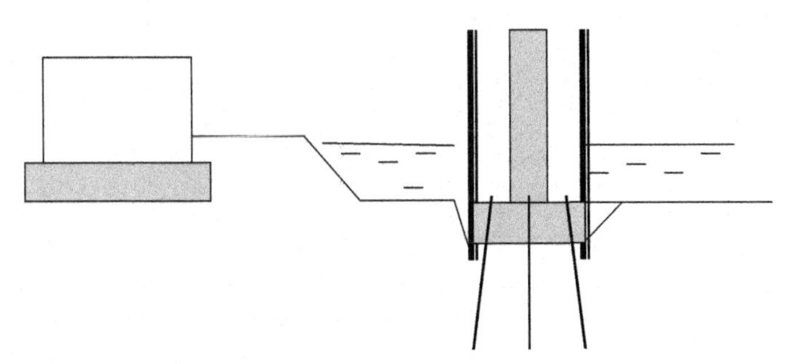

Fig. 21.17 Construct the pier.

Construction workers can go inside the cofferdam and build the pier:

It is needless to say that the construction methodology depends upon the site conditions. If rock is encountered, a concrete base can be placed on the rock. In this case piles may not be necessary to hold the base. Again, one has to be careful of cracks and fissures in the rock. If water can migrate through cracks and fissures in the rock, the concrete base will be subjected to uplift forces.

21.4.2 Forces Acting on Cofferdams

Force due to water flow: The main force on a cofferdam is the force due to flowing water. If the cofferdam is placed in the middle of a river, large forces could occur due to river flow. Force due to water flow is given by:

$$D = A \cdot C_d \cdot \gamma_w \cdot V^2/2g$$

where D is the force due to water flow; A is the projected area normal to the current; C_d is the drag coefficient (depends on the roughness of sheet pile material and overall shape of the structure); γ_w is the density of water; V is the velocity of flow; and g is 9.81 in SI units and 32 in fps units.

As can be seen from the above equation, the higher the velocity of flow, the higher the force on the cofferdam.

Static water pressure: Static water pressure occurs due to difference in hydrostatic heads.

Static water pressure is given by (Fig. 21.18):

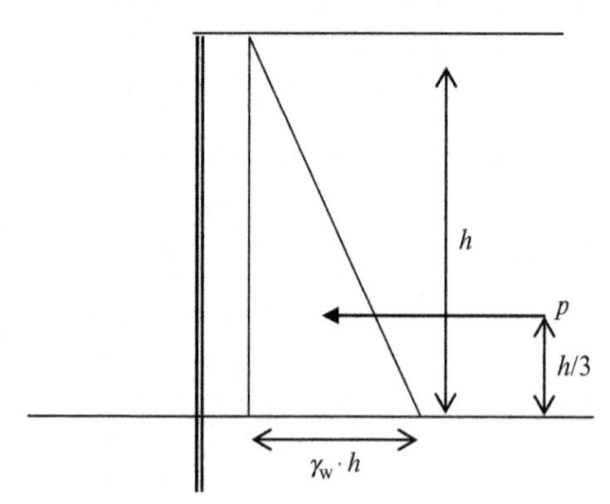

Fig. 21.18 Hydrostatic forces.

$$p = \gamma_w \cdot h$$

where p is the pressure due to water; γ_w is the density of water; and h is the hydrostatic head.

$$\text{Total force acting on the wall} = \text{Area of the triangle}$$
$$= \tfrac{1}{2} \times (\gamma_w \cdot h) \times h$$
$$= \tfrac{1}{2} \times \gamma_w \cdot h^2$$

The hydrostatic force acts $h/3$ distance from the bottom.

Force due to waves: Wave action can develop additional forces on cofferdams (Fig. 21.19).

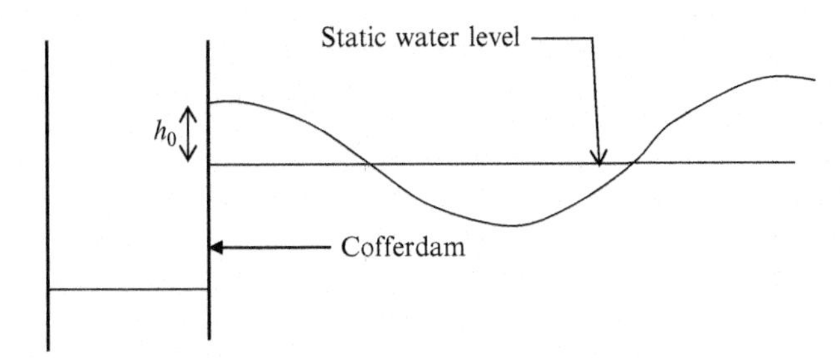

Fig. 21.19 Forces due to waves.

Additional hydrostatic head due to wave action also need to be computed.

21.4.3 Forces Due to Ice Action

When water becomes ice, volume increases. The increase in volume exerts additional pressure on sheet pile walls (Fig. 21.20).

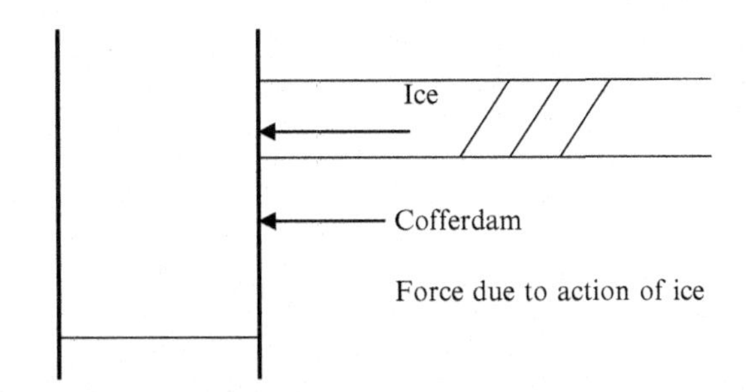

Fig. 21.20 Forces due to action of ice.

21.4.4 Cofferdam Types

Most common cofferdam types are:

(1) single wall cofferdams,
(2) double wall,
(3) cellular wall,

(4) diaphragm wall,
(5) earth type,
(6) timber crib, and
(7) rock fill.

Single wall cofferdams: As the name indicates, single wall cofferdams have only one wall. Typically, single wall cofferdams are built using sheet piles.

Double wall cofferdam: Double wall cofferdams are somewhat permanent in nature and are built to last for few years. When construction work can take many years, single wall cofferdams may not be suitable. Single wall cofferdams leaks and dewatering is required on a regular basis. This problem can be avoided with a double wall cofferdam.

- Single wall cofferdams can be built quickly with less cost.
- Dewatering and constant repairs are needed for single wall cofferdams.
- The cost is less for single wall cofferdams.
- Double wall cofferdams are costly and need less maintenance. In addition, dewatering inside the work site is negligible.

Cellular cofferdams: Cellular cofferdams are built when large areas need to be kept dry. Cellular structures can stand-alone and need not be braced.

CHAPTER 22

Loads During Construction

Loading during construction is different from permanent loading on a slab. During construction, workers will be utilizing their tools. In addition, concrete buggies, finishing equipment, hoses, and vibrators also will be on the slab.

ASCE Standard, SEI/ASCE 37, defines the following loads during construction. Note that load combinations and definitions are different during construction compared to a permanent situation (Fig. 22.1).

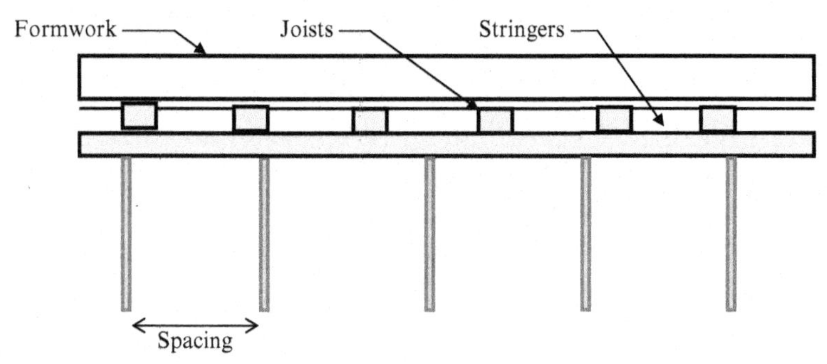

Fig. 22.1 Load due to wet concrete.

Dead load (D): Dead load due to permanent structure constructed at a given time. Shoring and other dead loads that are not part of the permanent structure are not considered for (*D*).

Construction dead load (C_D): Dead load due to shoring, scaffolds, and other construction related dead loads. These loads will be gone after the construction is completed.

Live load (L): Live load due to occupants during construction. In many cases this would be zero. Occasionally occupants may use the structure during construction. Construction workers and their equipment are not part of this load.

Construction personal and equipment load (C_P): Construction workers working on the structure, concrete buggies, concrete pumping hoses, equipment for formwork and shoring, generators, gang boxes, and compressors.

Construction Engineering Design Calculations and Rules of Thumb
http://dx.doi.org/10.1016/B978-0-12-809244-6.00022-6

Horizontal construction load (C_H): Construction activities can create horizontal loads. Moving vehicles, people, vibrating machines, and compressors can generate horizontal loads.

Wind load (W): Load due to wind acting on the structure. Wind load is mostly lateral and could uplift the structure as well. Lateral stability and resistance to uplift needs to be assessed. Wind load is a dynamic load. However, for computation ease, wind load is considered to be a static horizontal load.

Snow load (S) and ice load (I): Accumulation of snow on a structure during construction needs to be addressed. In some areas snow could turn into ice very quickly.

Earthquake load (E): Earthquake load is a dynamic load. However, for computation ease, earthquake load is considered to be a static load acting horizontally.

Material loads: Load due to material during construction is divided into two. Fixed material loads (C_{FML}) and variable material load (C_{VML}).

(C_{FML}) fixed material loads: During construction, materials have to be stockpiled in the structure that has been constructed. If the magnitude of the material load is fixed, then it is considered as C_{FML}. Load due to fuel and various other materials required during construction may be relatively fixed if they are replenished.

(C_{VML}) variable material load: Load due to some material may be variable. Steel, nuts, and bolts may be stored. Once they are constructed, it becomes part of the permanent dead load (*D*).

Horizontal construction loads (C_H): Moving wheelbarrows and moving personnel generate lateral loads on the structure. Wind load is not considered in this item since wind load is considered separately. It is assumed a moving person would exert 50 lbs lateral load on the structure.

Load combinations:

Load combination 1: $1.4D + 1.4C_D + 1.2C_{FML} + 1.4C_{VML}$

Load combination 2: $1.2D + 1.2C_D + 1.2C_{FML} + 1.4C_{VML} + 1.6C_P + 1.6C_H + 0.5L$

Load combination 3: $1.2D + 1.2C_D + 1.2C_{FML} + 1.4C_{VML} + 1.3W + 0.5C_P + 0.5L$

Load combination 4: $1.2D + 1.2C_D + 1.2C_{FML} + 1.4C_{VML} + 1.0E + 0.5C_P + 0.5L$

Load combination 5: $0.9D + 0.9C_D + 1.3W$

Load combination 6: $0.9D + 0.9C_D + 1.3E$

Apply all load combinations and find the highest load for design.

CHAPTER 23

Worker Health, Safety, and Environment

23.1 OSHA REGULATIONS (INTRODUCTION)

The president does not make the laws of the United States; rather, they are made by Congress. Once Congress agrees upon a piece of legislature (law) it has to be published. The Federal government uses the code of federal register (CFR) to publish the laws of the country. Health and safety of construction workers are addressed in CFR 1926.

CFR 1926 is written in legal jargon and may be difficult to understand. Fortunately, simplified versions of CFR 1926 are published by various organizations.

This chapter will provide an overview of CFR1926.

CFR 1926 is divided into different subparts.

23.2 SAFETY TRAINING

As per OSHA, It is the responsibility of the employer to provide safety training for workers. Special types of construction work, such as confined space entry, working in high-rise buildings, and working in hazardous environments require special training. For instance, an OSHA 40 hr hazardous training certificate is required for workers who work in hazardous environments.

Exposure to noise: 1926.52 deals with exposure to noise. OSHA provides a table where noise is considered to be a hazard.

For an example, as per Table 23.1, if a worker is subjected to a noise of 102 decibels for more than 1.5 h, the employer need to provide hearing protection.

Construction Engineering Design Calculations and Rules of Thumb
http://dx.doi.org/10.1016/B978-0-12-809244-6.00023-8

Table 23.1 Permissible noise exposure (OSHA 1926.52)

Duration per day, hours	Sound level dBA slow response
8	90
6	92
4	95
3	97
2	100
1 1/2	102
1	105
1/2	110
1/4 or less	115

23.3 WORKING NEAR WATER: OSHA 1926.106

Employees working over or near water shall be provided with US Coast Guard-approved life jacket or buoyant work vests.

23.4 SIGNS SIGNALS AND BARRICADE: 1926 SUBPART G

Signs are a very effective way to communicate hazards to workers in a construction site.

23.5 DANGER SIGNS

As per OSHA, danger signs should be used when there is an immediate hazard. Danger signs have a white and black background with red letters.

Caution signs: Caution signs are used when there is a potential hazard due to unsafe practices such as smoking or not wearing PPE, etc. Caution signs have a yellow and black background with yellow letters.

OSHA provides the color and designs for danger and caution signs.

23.6 TOOLS (HAND TOOLS AND POWER TOOLS)—SUBPART I

OSHA indicates the all hand and power tools and similar equipment, whether furnished by the employer or the employee, shall be maintained in a safe condition. Power tools need to have guards. All tools should be used in conjunction with proper personal protective equipment.

OSHA has named machines that need mandatory guards. They are Guillotine cutters, shears, alligator shears, powered presses, milling machines, power saws, jointers, and portable power tools.

23.7 ELECTRICAL SAFETY

Electricity (introduction): Electricity is the flow of electrons through a conductor. Copper wires are used as conductors.

Electric Circuit: Electric circuit shown below has an electric source, light bulb, and a switch (Fig. 23.1).

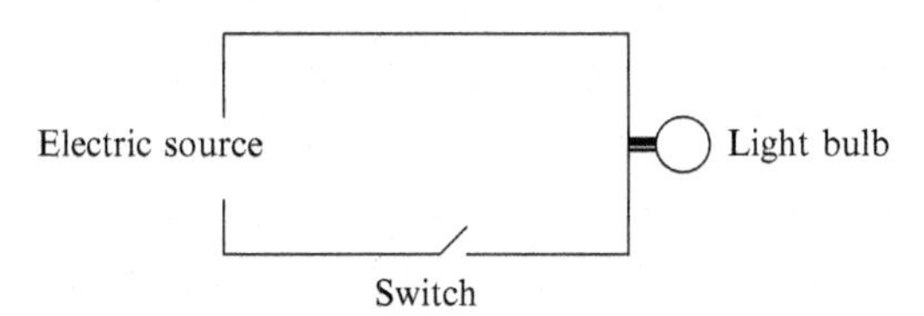

Fig. 23.1 Electric circuit.

An electric source could be the supply from an outside source or generator.

Switch boards: electric panels: Buildings can contain a large number of circuits. In such situations there will be large number of switches. These switches are included in metal boxes. Such boxes are known as switchboards (or electric panels).

Typically, these boxes are waterproof. Only qualified electricians are allowed to make repairs or changes to switch boards (Fig. 23.2).

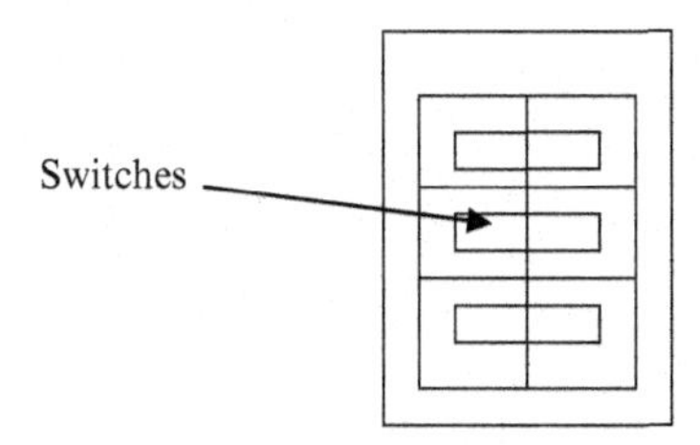

Fig. 23.2 Electric panel.

Panel directory: A panel directory indicates the circuits allocated to a given switch. For instance if repairs need to be done to the boiler, one can switch off the power to the boiler. If there is no directory indicating what switch is for what purpose then one may not be able to figure out how to switch off the electricity for the boiler. In such situations electrician has to conduct various tests to figure out the proper switches (Table 23.2).

Table 23.2 Color code for electric wires

Ground wire	Green
Phase 1	Black
Phase 2	Red
Phase 3	Blue
Neutral	White

Circuit protective devices: circuit breakers: Circuit breakers shut off a circuit when the current is above the normal limit. Circuit breakers are essential for the safe operation of an electric circuit.

Ground wires: Circuits are grounded in case there is a short circuit.

Hazards of electricity: Electricity is an invisible killer. An innocent looking cable could kill you in an instant. When electricity travels through the human body, the organs in the body would be damaged. The damage is dependent upon the current (Fig. 23.3).

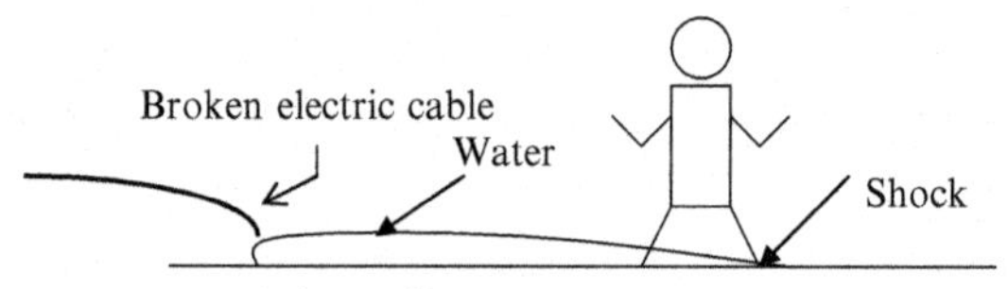

Fig. 23.3 Electricity can travel through water.

One could be electrocuted by standing on water where there is contact with an electric circuit.

Rubber mats should be placed in areas where there is a possibility of a stray current. Electricity rarely travels through rubber mats.

Electricity and burns: Another hazard of electricity is arcing. If you are closer to a high voltage circuit, it can arc (Fig. 23.4).

Severe burns can occur due to electric arcing.

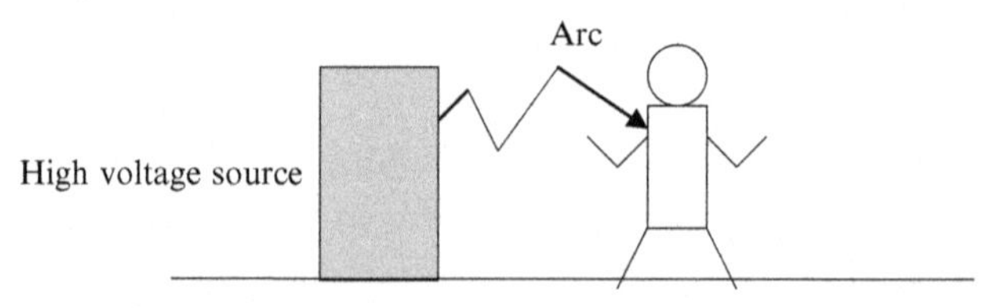

Fig. 23.4 Danger of arcing.

23.8 SCAFFOLDS

When workers have to work at elevated location scaffolds are used. Scaffolds are temporary structures to facilitate workers to work at elevated heights. Let us imagine painting a seven story building. In this case, scaffolds are erected outside the building so that workers can paint the building.

Scaffolds erected using pipes are known as pipe scaffolds.

OSHA indicates that all scaffolds should have a capacity of carrying its own weight plus four times the working weight. Ropes and connecting hardware should be able to withstand six times the working load. OSHA has a higher safety factor for ropes since ropes are prone to damage over time.

Compliance with manufacturer's specifications: The contractor shall comply with the manufacturer's specifications and limitations. Rated load capacities, recommended operating speeds, and special hazard warnings or instructions shall be posted.

Inspection: All hoists shall be inspected and tested at not more than 3-month intervals. The employer shall prepare inspection reports.

Most hoists use wire ropes. Wire rope shall be taken out of service if there are six randomly distributed broken wires in one lay or three broken wires in one strand in one lay. Alternatively, wear of one-third the original diameter of outside individual wires. In addition, wire ropes shall be taken out of service if there is any evidence of heat damage.

23.9 OVERHEAD HOISTS

Manufacturer's specifications and safe loads shall be followed by the operator. The hoist shall be located in such a manner that the operator can stand clear of the load. Inspection of the hoist shall be conducted as per manufacturer's guidelines.

23.10 MOTOR VEHICLES, MECHANIZED EQUIPMENT, AND MARINE OPERATIONS: OSHA 1926 SUBPART O

This section may not be too important for the exam. But I recommend you read it at least once.

23.11 EXCAVATIONS: OSHA 1926 SUBPART P

Excavation safety problems are commonly found in PE construction exams. Hence, it is important to pay attention to excavation safety.

Protecting utilities: Prior to excavation work, the employer shall contact the utility companies and inform of the excavation work. In most cases, they would come to the site and mark out utility locations. If a certain utility company is unable to provide the utility mark up, the contractor shall use detecting devices to locate utilities.

Means of egress: Exit locations need to be provided every 25 ft, in trenches deeper than 4 ft. Stairways, ladders, ramps, or other safe means of egress shall be provided.

Oxygen level: If the trench or excavation may possibly contain low level of oxygen (less than 19.5 oxygen) or high level of hazardous gases, these situations shall be tested before employees enter excavations greater than 4 ft in depth.

Inspections for cave-ins: Excavations need to be inspected daily prior to work for possible cave-ins.

When excavation protection is not needed:

As per OSHA, excavations protection is not needed when following conditions are met:

- Excavation is conducted in stable rock
- Excavation is less than 5 ft and a competent person had determined that there is no risk of cave-in.

It should be mentioned here that in many cases contractors assume that all excavations less than 5 ft may not need excavation support. This is a false assumption. Competent person needs to certify that excavation does not require support, even if the excavation is less than 5 ft.

Sloping and benching: Sloping and benching can be done instead of protection of slopes for some situations.

OSHA provides Table 23.3 for sloping and benching (Fig. 23.5).

Table 23.3 Maximum allowable slopes

Soil or rock type	Maximum allowable slopes (H:V) (1) for excavations less than 20 ft deep (3)
Stable rock	VERTICAL (90 degrees)
Type A	3/4H:1V (53 degrees)
Type B	1:1 (45 degrees)
Type C	1 ½H:1V (34 degrees)

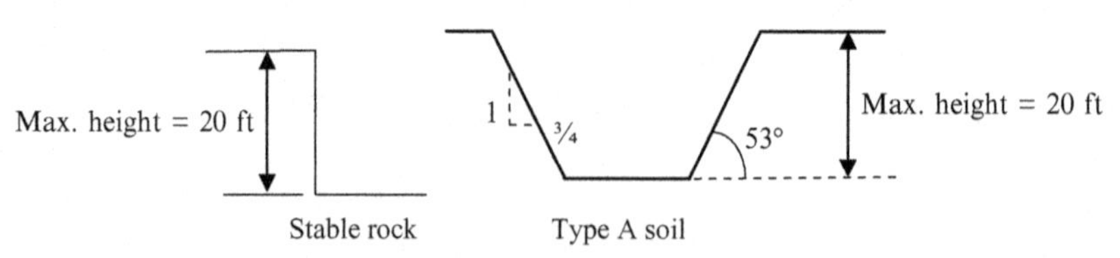

Fig. 23.5 Stable rock and type A soil.

Stable rock: Excavations in stable rock do not need any support and can have a vertical surface as shown in Fig 23.5 as long as the height is less than 20 ft.

Excavations in type A soil: Type A soils are clayey soils with an unconfined, compressive strength of 1.5 tsf or greater. Unconfined compressive strength test in Fig. 23.6.

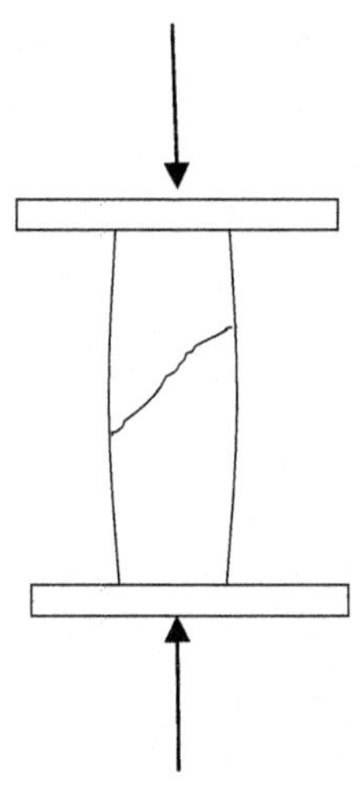

Fig. 23.6 Unconfined compressive strength test.

An unconfined compressive strength test is conducted by placing a soil sample between two plates and applying a force. This test can be done only for clay soils. Clay soils with high cohesion would fail at a higher load. The higher the unconfined compressive strength test value, the higher the resistance to slope failure.

In type A soils, the maximum slope angle allowed is 53 degrees up to a height of 20 ft (53 degrees is same as 3/4H:1V).

In addition to simple slope as shown above, OSHA also provides benching methods.

Excavations in type B soil: Type B soils shall have a unconfined compressive strength less than 1.5 tsf and greater than 0.5 tsf.

In type B soils, the maximum slope angle allowed is 45 degrees up to a height of 20 ft (45 degrees is same as 1H:1V). Since type B soils have a lesser-unconfined compressive strength value, the slope shall be less steep than type A soils (Fig. 23.7).

Excavations in type C soil: Type C soils shall have an unconfined compressive strength less than 0.5 tsf.

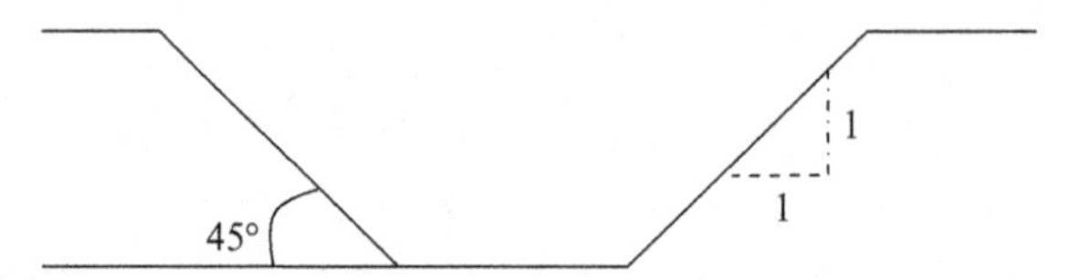

Fig. 23.7 Excavation with 45 degree slope.

In type C soils, the maximum slope angle allowed is 33.7 degrees up to a height of 20 ft (33.7 degrees is same as 1.5H:1V) (Fig. 23.8).

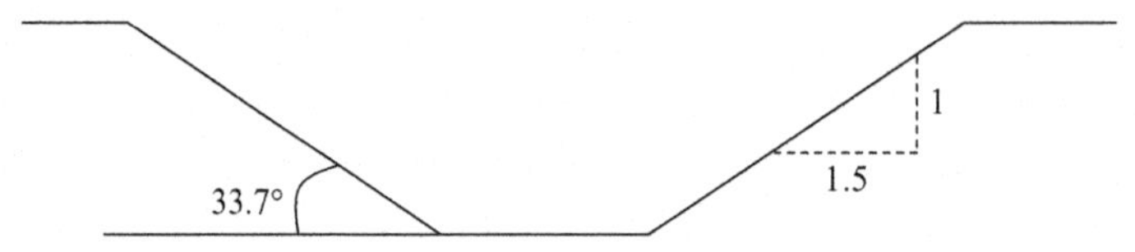

Fig. 23.8 Type C soil excavation.

Timber shoring: In some cases sloping may not be feasible due to nearby buildings, roads, and various other obstructions. In such situations shoring is done. Timber shoring is still very popular and probably the cheapest.

Timber shoring picture is shown in Fig. 23.9.

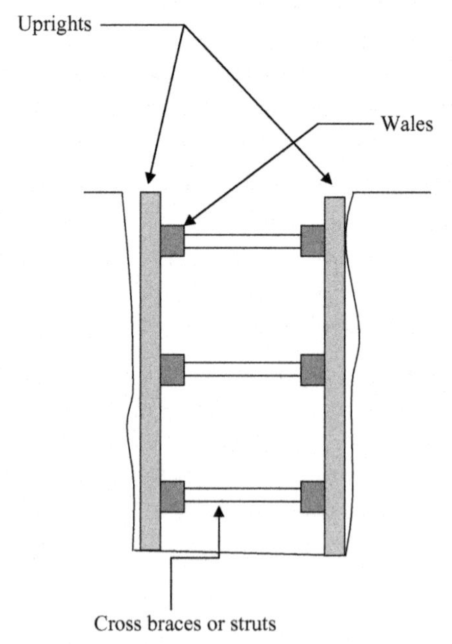

Fig. 23.9 Elements of a timber shoring system.

Elements of a timber shoring system:

- Cross braces (also known as struts)
- Uprights
- Wales (sometime not needed)
- Sheathing (sometimes not needed)

Cross braces, uprights, and wales are timber elements. Sheathing is timber boards or plywood.

OSHA provides tables to design timber shoring.

Design of cross braces: Use Table C.1.1 in OSHA.

Portion of the table is shown in Table 23.4.

Table 23.4 Design of cross braces

		Width of trench					
Depth of trench	**Hor. spacing**	**Up to 4 ft**	**Up to 6 ft**	**Up to 9 ft**	**Up to 12 ft**	**Up to 15 ft**	**Vert. spacing (ft)**
5–10 ft	Up to 6 ft	4×4	4×4	4×6	6×6	6×6	4
	Up to 8 ft	4×4	4×4	4×6	6×6	6×6	4
	Up to 10 ft	4×6	4×6	4×6	6×6	6×6	4
10–15 ft	Up to 6 ft	4×4	4×4	4×6	6×6	6×6	4
	Up to 8 ft	4×4	4×4	6×6	6×6	6×6	4
	Up to 10 ft	6×6	6×6	6×6	6×8	6×8	4
15–20 ft	Up to 6 ft	6×6	6×6	6×6	6×8	6×8	4
	Up to 8 ft	6×6	6×6	6×6	6×8	6×8	4
	Up to 10 ft	8×8	8×8	8×8	8×8	8×10	4

Practice Problem 23.1

A trench with a depth of 18 ft need to be constructed. The width of the trench is 10 ft. What is the size of the cross bracing required? The contractor would like to place cross bracing every 10 ft.

Continued

Solution

The depth is 18 ft. Hence, locate the row with depth 15–20 ft. The width is 10 ft. Hence, locate the column "up to 12 ft." Now the contractor has three choices. The contractor wishes to place cross bracing every 10 ft.

Hence the size of the cross bracing required = 8 × 8.

Nominal size 8 × 8 (8 in. × 8 in.) is not the actual size. Actual size is 7.5 in × 7.5 in.

Design of uprights: OSHA provides tables to design wales, cross bracings and uprights for all types of soil (Table 23.5).

Table 23.5 Design of uprights and wales

Depth of trench	Wale size	Wale vertical spacing (ft)	Maximum allowable horizontal spacing of uprights and size				
			Close	4 ft	5 ft	6 ft	8 ft
5–10 ft	Not required	–				2 × 6	
	Not required	–					2 × 8
	8 × 8	4			2 × 6		
	8 × 8	4				2 × 6	
10–15 ft	Not required	–					3 × 8
	8 × 8	4			2 × 6		
	8 × 10	4				2 × 6	
	10 × 10	4					3 × 8

Aluminum hydraulic shoring: Instead of timber, there is another alternative. That is aluminum hydraulic shoring. These shoring can be reused many times. Though the initial cost is high, cost per given year may be less with this type of shoring.

The length of hydraulically operated cross bracings can be adjusted to tighten the cross bracings. OSHA Table D1.1 is used to design aluminum struts and uprights.

Trench shields: In urban environments, trench shields are much easier and faster.

23.12 CONCRETE AND MASONRY CONSTRUCTION: OSHA 1926 SUBPART Q

Many workers perform concrete and masonry construction tasks. Concrete is considered to be the most popular construction material in the world.

General safety requirements:

Construction loads: Construction loads include tools, machines, and construction workers. OSHA states that no construction loads shall be placed on a concrete structures unless the employer determines, based on information received from a person who is qualified in structural design, that the structure or portion of the structure is capable of supporting the loads.

Reinforcing steel: All protruding reinforcing steel need to be capped to avoid impalement.

Post-tensioning operations: Post tensioning is the process of applying a compressing concrete beams and slabs using tendons. Note that post tensioning is different than pre-stressing. Post tensioning tendons do not have a bond with concrete. On the other hand, pre-stressed tendons are bonded to concrete.

Post tensioning method: A concrete beam is cast with tendons inside. Then after the concrete is hardened, tendons are pulled and attached to the beam (Fig. 23.10).

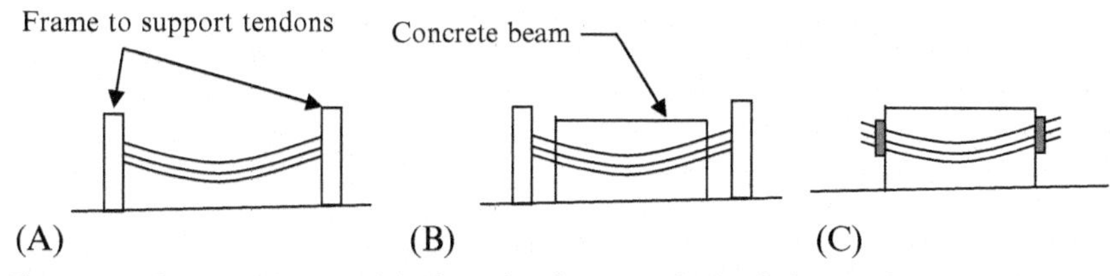

Fig. 23.10 Post tensioning. (A) Place the frame and install the tendons. Tendons are inside a duct. (B) Cast the concrete beam and wait till the beam is hardened. (C) After the beam is hardened, pull the tendons and tighten them at the end. Once the tendons are released, the concrete beam will be compressed.

The left photo shows post tensioning tendons and ducts prior to casting the beam. The right photo shows the tendons, duct, and locking mechanism.

No employee (except those essential to the post-tensioning operations) shall be permitted to be behind the jack during tensioning operations. Signs and barriers shall be installed to maintain the post tensioning area free of visitors and other construction workers who are not involved in post tensioning activities.

Concrete buckets: Concrete buckets are widely used to pour concrete in high-rise buildings. As per OSHA, no employee shall be permitted to ride concrete buckets.

No employee shall be permitted to work under concrete buckets while buckets are being elevated or lowered into position.

Power concrete trowels: Powered and rotating type concrete toweling machines shall be equipped with an automatic shutoff control switch that will automatically shut off the power whenever the hands of the operator are removed from the equipment handles.

Formwork: Formwork drawings shall be available in the job site.

Shoring and reshoring: Shoring equipment shall be inspected immediately prior to, during, and immediately after concrete placement. Reshoring shall be erected, as the original forms and shores are removed whenever the concrete is required to support loads in excess of its capacity.

Removal of formwork: Many contractors like to remove formwork as early as possible to accelerate the project. However, as we all know, wet concrete has less strength than hardened concrete. Seven-day-old concrete has approximately 65% of the 28-day strength. As per OSHA, removal of forms should be done as per plans and specifications. In case plans and specifications do not address this issue, the concrete should be tested as per ASTM method for strength. Forms shall be removed after required strength is achieved.

Precast concrete: Concrete structures can be built two ways.

- cast in situ
- precast

A large majority of concrete structures are built using cast in situ method. This is the standard method where formwork and rebars are installed and concreted. In the case of precast concrete, concrete elements are manufactured in a plant and brought to the site and installed. It is fair to say that precast concrete is increasingly becoming popular.

It is important to adequately support the precast units during construction. Precast units are held in place by fasteners and inserts. As per OSHA these fasteners that are used to stop precast elements from overturning shall be capable of supporting at least two times the maximum intended load.

On the other hand, lifting inserts shall be capable of supporting at least four times the maximum intended load.

No employee shall be permitted under precast concrete elements during construction except the workers who are involved with precast work.

Masonry construction: Masonry structures are very unstable before the mortar is hardened. Hence, OSHA recommends a limited access zone to be established. Only the workers who are actively engaged in masonry construction will be allowed in this zone. A limited access zone shall be equal to the height of the wall to be constructed plus four feet, and shall run the entire

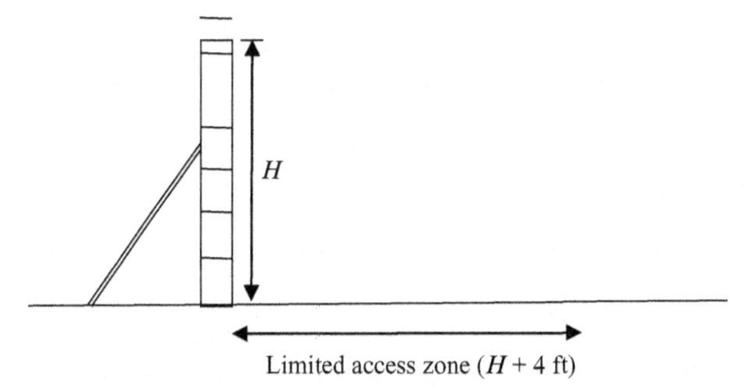

Fig. 23.11 Masonry construction.

length of the wall. The limited access zone shall be established on the side of the wall, which will be unscaffolded (Fig. 23.11).

23.13 STEEL ERECTION SAFETY: OSHA 1926 SUBPART R

Steel erection is considered to be one of the most dangerous activities.

Start of erection: A steel erection contractor shall not erect steel unless he has received written notification that the concrete in the footings has attained at least 75% of the design strength or sufficient strength to support the loads imposed during steel erection.

Here it is important to note that one may not need to wait till 75% of the design strength is achieved. If the structural engineer decides that the footings have enough strength to hold the steel structure, then the erector can start work.

Site access: Steel erection requires bring in cranes, steel beams, columns, and various other deliveries. Adequate access roads, flagmen, fences, and pedestrian control mechanisms should be in place prior to start of steel erection.

Site-specific erection plan: A site-specific erection plan shall be developed by a qualified person and be available at the work site.

Guidelines for establishing a site-specific erection plan are:

Preconstruction meeting: A preconstruction meeting shall be held between the erector, fabricator, and other parties involved before the start of steel erection.

Components of a site-specific erection plan:

Sequence: The sequence of erection activity shall be developed. This sequence will contain the areas that are to be started first and what beams and columns are to be erected. Typically, the structure will be divided into portions and each portion and floors will be numbered. Then a sequence will be developed.

When developing the sequence of erection, one may need to consider following aspects:

- material deliveries
- material staging and storage
- coordination with other trades
- other construction activities such as excavation, concrete, fencing, retaining wall construction, etc.

Crane selection: An erection plan should address the process of crane selection. The site should be prepared for the crane and there needs to be a path for overhead loads.

Fall protection: Fall protection procedures should be established.

Certifications: All steel erecting workers shall be certified for their trade.

Emergency action plan: A procedure policy that is implemented during an emergency shall be developed. Emergency action plan shall address issues such as access to first aid equipment, exit locations, and address of the nearest hospital.

Rigging safety: Rigging is the process of lifting and moving construction equipment and material. Cranes are widely used to lift and move in construction sites. As per OSHA, cranes should be visually inspected prior to every shift.

Items need to be observed in a crane prior to the start of a shift:

- all control mechanisms for maladjustments
- safety devices, boom angle indicators, boom stops, boom kick out devices, anti-two block devices, and load moment indicators
- air, hydraulic, and other pressurized lines for deterioration or leakage
- hooks and latches for deformation, cracks, or wear
- wire ropes for damage
- electrical apparatus for malfunctioning
- tires for proper inflation and condition
- ground conditions around the hoisting equipment for proper support, including ground settling under and around outriggers, ground water accumulation, or similar conditions
- the hoisting equipment for level position

OSHA states that the headache ball (also known as the overhaul ball), hook, or load shall not be used to transport personnel.

A headache ball is device that has a swivel inside. If the wire rope twists, the swivel mechanism will make sure that the load will NOT twist. In the past headache ball and loads were used to transport people.

As per OSHA, cranes may be used to hoist employees only if there is a personnel platform.

Working under loads: Sometimes it may be necessary to be under the load. As per OSHA, only the workers who are hooking, unhooking, or connectors are allowed to be under the load.

Structural stability: During erection, the erection crew will assemble steel beams and columns with few bolts. The bolting crew will then follow and complete all the bolts. Next comes the welding crew followed by the detailing crew. Hence, it is possible there are many floors that are not fully completed.

OSHA states the following:

> *There shall be not more than eight stories between the erection floor and the upper-most permanent floor, except where the structural integrity is maintained as a result of the design.*

Let us see what this means. During the erection, a temporary floor is erected for the workers to use. OSHA states that at any given time, when the erection is taking place at a certain floor, a permanent floor shall be installed eight stories below.

As an example, let us say that erectors are erecting at floor 20 and they have installed temporary floors up to floor 16. On the other hand, they have only erected the permanent floors up to floor 10. In this case, the erector does not comply with the OSHA rule. To be in compliance, the erector should stop erecting and complete the permanent floors up to floor 12.

OSHA also says;

> *At no time shall there be more than four floors or 48 feet (14.6 m), whichever is less, of unfinished bolting or welding above the foundation or uppermost permanently secured floor, except where the structural integrity is maintained as a result of the design.*

What does this OSHA rule means?

Let us say that the bolting crew is bolting an unfinished floor at floor 15. They have fully completed all the floors up to floor 10. In this case, the contractor is not in compliance. As per OSHA When a crew is working on the 15th floor, they should have finished the 11th floor completely.

Tripping hazards: Shear connectors (such as headed steel studs, steel bars or steel lugs), reinforcing bars, deformed anchors, or threaded studs shall not be attached to the top flanges of beams, joists, or beam attachments so that they project vertically from or horizontally across the top flange of the member until after the metal decking, or other walking/working surface, has been installed.

Covering roof and floor openings: Openings in decks are needed for ducts, elevator shafts, and pipes. As per OSHA, openings shall have covers that are capable of supporting twice the weight of the employees, equipment, and materials that may be imposed on the cover at any one time.

Also all covers shall be secured when installed to prevent accidental displacement by the wind.

All covers shall be painted with high-visibility paint or shall be marked with the word "HOLE" or "COVER" to provide warning of the hazard.

Column stability: As per OSHA, all columns shall be anchored by a minimum of four anchor bolts.

Each column anchor rod (anchor bolt) assembly, including the column-to-base plate weld and the column foundation, shall be designed to resist a minimum eccentric gravity load of 300 pounds located 18 in. from the extreme outer face of the column in each direction at the top of the column shaft.

Columns shall be set on level finished floors, pre-grouted leveling plates, leveling nuts, or shim packs, which are adequate to transfer the construction loads.

All columns shall be evaluated by a competent person to determine whether guying or bracing is needed; if guying or bracing is needed, it shall be installed.

Anchor bolts shall not be repaired, replaced, or field-modified without the approval of the project structural engineer of record.

Prior to the erection of a column, the controlling contractor shall provide written notification to the steel erector if there has been any repair, replacement, or modification of the anchor rods (anchor bolts) of that column.

Other subparts of OSHA 1926 are:

Subpart S: Underground construction, caissons, cofferdams and compressed air

Subpart T: Demolition

Subpart U: Blasting

Subpart V: Power transmission

Subpart W: Overhead protection

Subpart X: Ladders

Guard rails: Guardrails should be 39–45 in. in height.

Guardrails must be able to withstand a force of 200 lbs applied within 2 in. of the top of the guardrail.

23.14 CFR 1926 SUBPART N: CRANES, DERRICKS, HOISTS, ELEVATORS AND CONVEYORS

Cranes and derricks are needed to lift steel beams, equipment, and material. Hoists are used to lift material and personnel. Conveyor belts are used for transporting material.

23.15 SAFETY MANAGEMENT

Safety management is an important aspect of any project. Some projects such as high-rise buildings and bridges may have more stringent safety management plans. Safety management plans should address following issues.

(1) safe operating procedures
(2) safety training of workers
(3) safety inspections and protocols

Some organizations, such as oil refineries, conduct exams for workers. OSHA-approved courses are available for construction and hazardous waste management workers.

23.16 SAFETY STATISTICS

Incidence rate means number of injuries or lost workdays per 100 full time workers per year. One worker works 40 h a week. Assuming he works 50 weeks, one worker would work 2000 h per year.

Hundred workers would work 200,000 h per year (2000 × 100).

$$\boxed{\text{Incidence rate} = N \times 200{,}000/\text{EH}}$$

N = number of injuries,
EH = number of work hours per year.

Practice Problem 23.2

A company has 300 workers working 45 h per week. The work crew works 50 weeks. There were nine injuries. What is the incidence rate?

Solution

$$\text{Total number of work hours (EH)} = 300 \times 45 \times 50 = 675{,}000$$

$$\text{Incidence rate} = N \times 200{,}000/\text{EH} = 9 \times 200{,}000/675{,}000 = 2.67$$

CHAPTER 24

Temporary Traffic Control

Temporary traffic control or TTC is required during construction activities. Prior to establishing TTC devices, a TTC plan needs to be prepared. The plan should consider passenger traffic, commercial traffic, and pedestrians. Commercial traffic may not be able to go over certain bridges due to weight restrictions. Hence, TTC planners need to be aware of special considerations for commercial traffic when diverting traffic.

When developing a TTC plan, one has to assume that drivers would reduce the speed only if they clearly perceive a need to do so.

Seven fundamental principles to follow when developing a TTC:

(1) General plans or guidelines should be developed to ensure safety of motorists, bicyclists, and pedestrians.

(2) Road user movement should be inhibited as little as practical.

(3) Motorists, bicyclists, and pedestrians should be guided in a clear and positive manner while approaching and traversing TTC zones and incident sites.

Which means signs should be visible and clear. Cones and other traffic guidance equipment should be properly used.

(4) Routine day and night inspections of TTC elements should be performed:

Cones and signs could be misplaced due to wind, rain, and kids playing around. Workers should routinely inspect cones and signs.

(5) Attention should be given to maintenance of roadside safety during the life of the TTC zone.

(6) All personnel involved in developing and maintaining a TTC program should be adequately trained.

(7) Good public relations should be maintained. This can be done by providing advance notices of road closing and diversions so that motorists can plan alternate routes.

24.1 TRAFFIC CONTROL DEVICES

Traffic signs: Traffic signs are a major part of any TTC plan. Traffic signs will let the motorists know what to expect.

Construction Engineering Design Calculations and Rules of Thumb
http://dx.doi.org/10.1016/B978-0-12-809244-6.00024-X

Crash attenuators: When workers are working in the side of the road, a crash attenuator will protect the workers (Fig. 24.1).

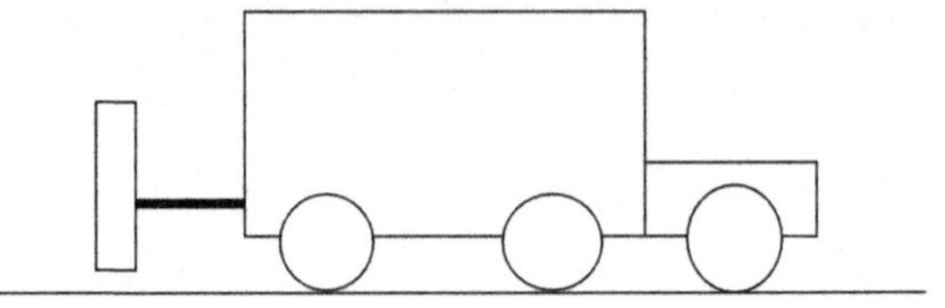

Fig. 24.1 Crash attenuator.

24.2 COMPONENTS OF TEMPORARY TRAFFIC CONTROL ZONES

TTC (temporary traffic control) zones are divided into four areas:

(A) the advance warning area
(B) the transition area
(C) the activity area
(D) the termination area

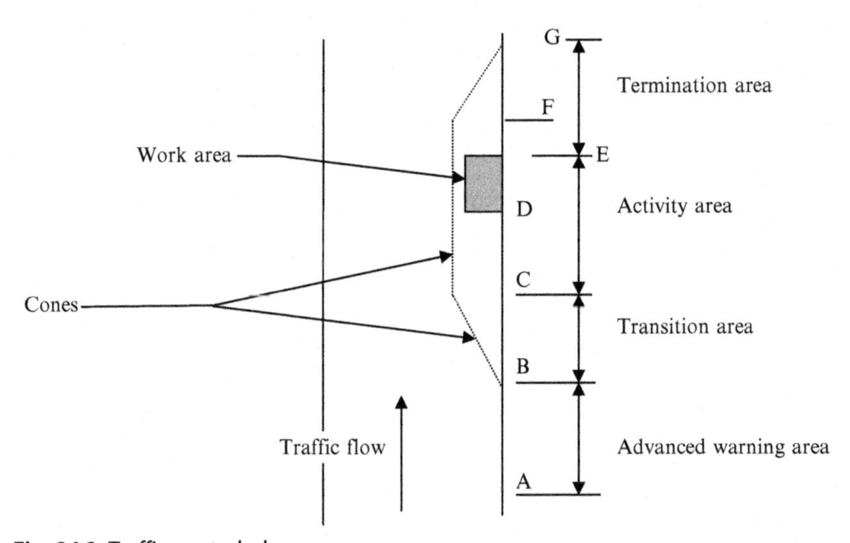

Fig. 24.2 Traffic control plan.

Fig. 24.2 shows the four temporary traffic control zones. These four are further divided as shown below.

- *A to B: Advance warning area:* An advance warning area consists of signs such as "Construction Ahead," "Your Tax Dollar at Work," or various signs indicating that construction zone is ahead.

- *B to C: Transition area:* In this zone, cones will be placed to guide the traffic.
- *C to D: Buffer zone:* A buffer zone is created to provide a safety area for workers. In case a runaway driver to come through the cones, the workers will be able to see the vehicle that is coming towards them. In some cases, speed attenuators are placed in this zone.
- *D to E: Work zone:* Workers will be working in the work zone.
- *E to F: Buffer space:* Another safe space for workers to move around.
- *F to G: Downstream taper:* Traffic is guided back.

24.3 DEVELOPMENT OF A TTC PLAN

A TTC plan should be developed by qualified personnel. Depending upon the complexity of projects, TTC plans can be simple to very complicated. Generally, TTC plans are costly. Placing and maintaining cones, barriers, and pillow trucks on a daily basis costs a significant amount of money for contractors.

24.3.1 Flaggers

One lane two-way traffic (conditions for no flagmen): As per MUTCD, if the workspace is on a low-volume street or the road is short and road users from both directions are able to see the traffic approaching from the opposite direction through and beyond the worksite, the movement of traffic through a one lane, two-way constriction may be self-regulating. In other words, a flagman is not necessary (Fig. 24.3).

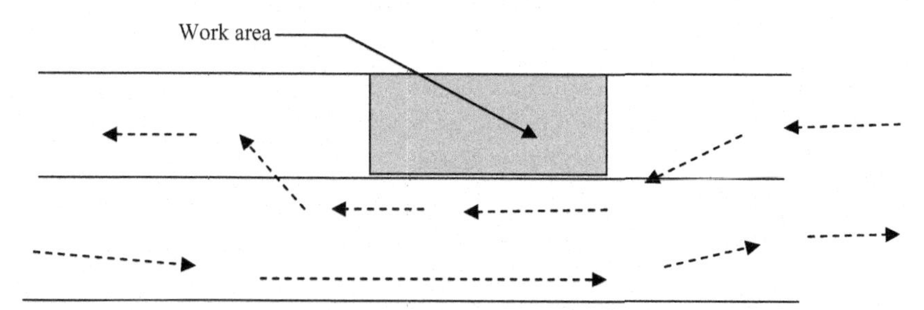

Fig. 24.3 Traffic around work area.

If the traffic is high or drivers cannot see each other due to a curve then flaggers are necessary.

One lane two-way traffic (conditions for one flagman): One flagman may be used only if the flagman can see one end to the other. When the road construction area is too long, the flagman may not be able to see the traffic coming from the other end.

One lane two-way traffic (conditions for two flagmen): When one flagman cannot see the full length of the road, then two flagmen should be used (Fig. 24.4).

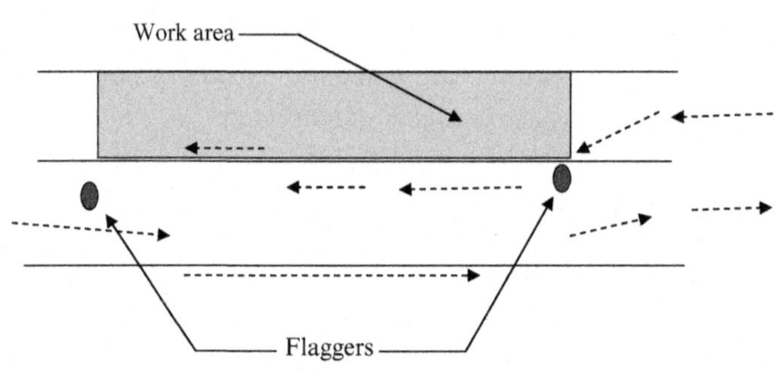

Fig. 24.4 Flaggers near a construction site.

6D.01: Pedestrian safety around construction sites:

MUTCD 6D.01 deals with safety of pedestrians near construction work.

MUTCD provides following three guidelines to follow when dealing with pedestrians.

Guideline A: Pedestrians should not be led into conflicts with vehicles, equipment, and operations.

Guideline B: Pedestrians should not be led into conflicts with vehicles moving through or around the worksite.

Guideline C: Pedestrians should be provided with a convenient and accessible path that replicates as nearly as practical the most desirable characteristics of the existing sidewalk(s) or footpath(s). In addition, the pedestrian route should not be severed and/or moved for non-construction activities such as parking.

MUTCD Chapter 6E: Flagger control:

MUTCD Chapter 6E mainly deals with flagger control. This is a long chapter with subject matter ranging from flagger qualifications, flagging signs, flagger assistance devices, and flagger stations.

Flagger qualifications:

(A) Ability to receive and communicate specific instructions clearly, firmly, and courteously;

(B) Ability to move and maneuver quickly in order to avoid danger from errant vehicles;

(C) Ability to control signaling devices (such as paddles and flags) in order to provide clear and positive guidance to drivers approaching a TTC zone in frequently changing situations;

(D) Ability to understand and apply safe traffic control practices, sometimes in stressful or emergency situations; and

(E) Ability to recognize dangerous traffic situations and warn workers in sufficient time to avoid injury.

24.4 HAND SIGNALING DEVICES

MUTCD recommends STOP/SLOW sign as the primary device of the flagger.

CHAPTER 25

Construction Management

25.1 GENERAL INTRODUCTION TO CONSTRUCTION MANAGEMENT

Construction projects originate with a need. Just imagine a fast growing company like Facebook. Due to their growth, they need more computer programmers, engineers, managers, secretaries, and hardware specialists. These people need a place to work. The upper management of Facebook decides that they need a new building. They consult with an architect, who draws a building. Then come the engineers. Structural engineers design the beams, columns, and slabs. Mechanical engineers design the boilers, hot water piping, heating systems, air handlers, and air conditioning units. Electrical engineers design the lighting, power outlets, switches, and transformers. Network engineers design the computer networks, TV cables, and telephone lines (Fig. 25.1).

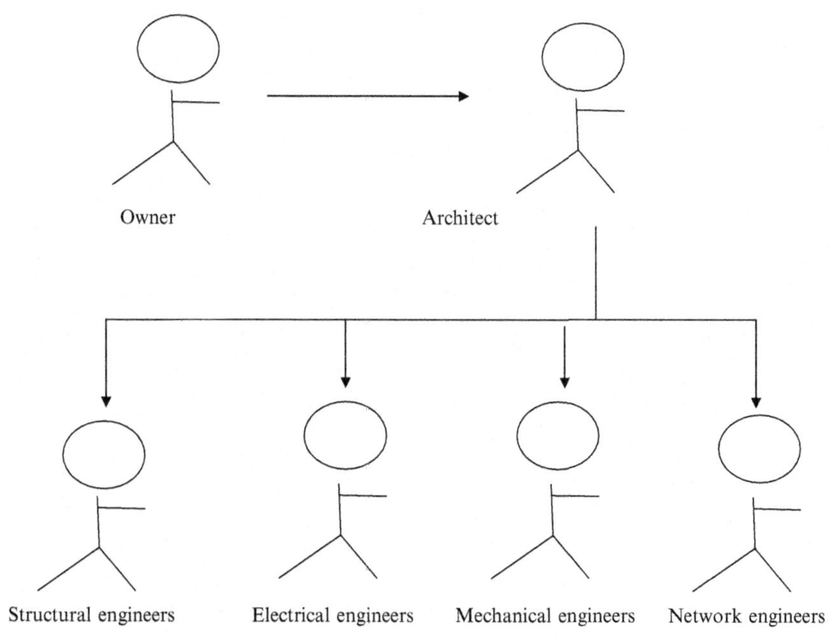

Fig. 25.1 Project hierarchy.

Construction Engineering Design Calculations and Rules of Thumb
http://dx.doi.org/10.1016/B978-0-12-809244-6.00025-1

Design completion: After all the engineers get together and completed the design, a set of drawings and specifications would be produced. Typically, after the drawings are completed, builders would be asked to bid for the project. In government jobs the lowest bidder gets the job.

What is the difference between design drawings and specifications?

Design drawings indicate dimensions and material to be used. For example let us look at a steel building. The drawings show steel beams and their dimensions. There are hundreds of different steel types available in the market based on the quality. The specifications would provide the steel type, steel manufacturer, ASTM number, and a plethora of other information.

25.2 DESIGN DRAWINGS

All construction managers should be able to read and thoroughly understand construction drawings. We will later study construction drawings in detail. For now let us look at a simple example (Fig. 25.2).

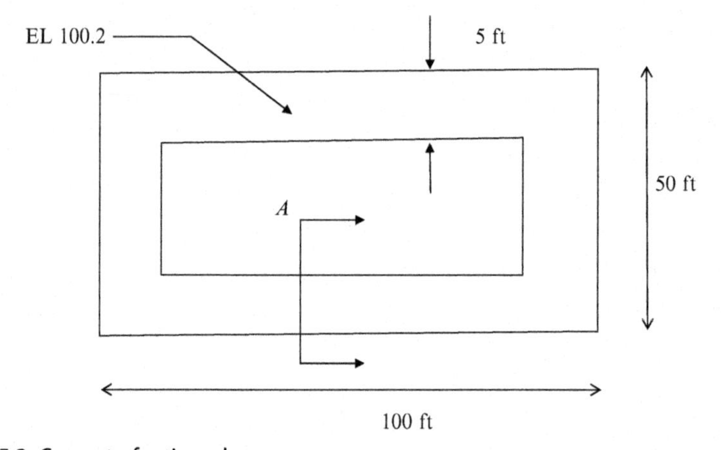

Fig. 25.2 Concrete footing plan.

Let us look at Fig. 25.2. It shows a plan view of a concrete footing. A footing is a foundation where walls are resting. Later you will learn plans, elevations, and sections. Now you know what an actual footing look like.

25.3 SPECIFICATIONS

Fig. 1.1 shows a drawing of a footing. Some information is difficult to provide in a drawing. For example, material types, material strength, and accepted manufacturers are provided separately in a book known as a specification book. Specifications are also part of the contract.

Example of specifications: The following is a typical example that you probably might see in specifications.

> *Structural steel as per ASTM A992, 50% recycled content, welding as per AWS (American welding society guidelines), bolts as per ASTM A325.*

As you could see this type of information will not be provided in drawings. Instead material types, strength requirements, and various other information is generally provided separately.

Specifications generally do not provide dimensions. Dimensions are provided in drawings.

What information is provided in specifications?

- material properties required
- material quality standards such as ASTM
- tests that need to be done
- sample requirements

25.4 BIDDING PROCESS

Once the design drawings and specifications are completed, the project would be advertised. Builders would be asked to bid for the project (Fig. 25.3).

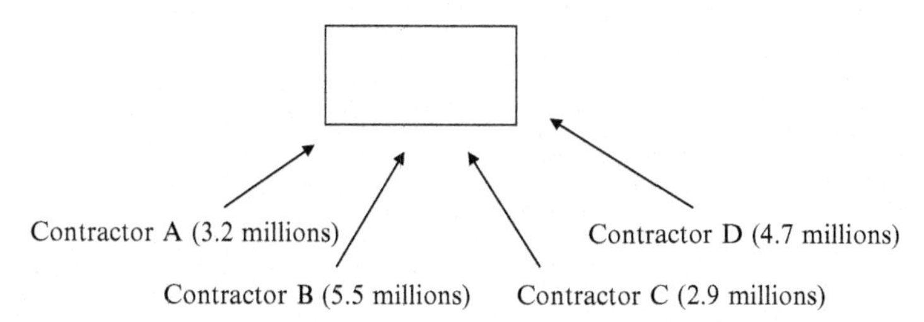

Fig. 25.3 Bidding process.

Typically, the lowest bidder gets the job. In Fig. 25.3 Contractor C will get the job. The lowest bidder does not get the job all the time. In some situations, the lowest bidder may not have enough experience to perform the project. For example, a builder who has been building highways bids for a building project. Though the builder may be financially sound, lack of experience may be a concern for the owners. Also in some cases builders may have a bad safety record in the past or have large debts.

Some of the reasons that may disqualify a low bidder are given as follows:

- The builder has financial troubles.
- The builder has no prior experience in type of work to be performed.
- The builder has no experienced personnel to run the project.
- The builder has a bad safety record.
- The builder has a bad reputation with neighbors.
- The builder has too many other projects and may not give enough care for the project.
- The builder's main office is far away. (This may not be an issue for some projects.)
- The design firm who designed the project is an affiliate of the builder. (If the design firm and the builder are the same company, it might create a conflict of interest. But in some cases, this may not be an issue.)

25.5 CONSTRUCTION PROJECT MANAGER WORKING FOR A GENERAL CONTRACTOR

Assume that contractor C (the low bidder), was given the job. Next, contractor C will hire a project manager who knows about construction. He or she would read and understand the drawings and specifications, and start building.

25.6 DUTIES OF A CONSTRUCTION PROJECT MANAGER

- develop a project plan
- develop a schedule with the help of schedulers
- procure subcontractors
- order material
- rent necessary equipment
- hire workers
- develop submittals and shop drawings
- start construction
- responsible for safety
- negotiate change orders

All the previously mentioned activities will be discussed in great detail later.

Don't worry. A project manager will have experienced people to help him. Schedulers to develop schedules, CAD operators to generate shop drawings and field foremen to develop work plans. There are also safety officers to help with safety issues.

25.7 CONSTRUCTION MANAGEMENT FIRMS

General contractors also known as builders. They build a project. Construction managers are hired by clients to manage the project. Construction managers ensure builders build the project as per contract drawings and specifications. In addition, the required city permits are obtained. Construction managers would ensure builders follow proper safety procedures.

Some of the responsibilities of the construction manager given as follows.

- ensure that the builder follow the drawings and specifications
- ensure that the builder is using approved material
- ensure that the builder perform the work in a safe manner
- ensure that the builder do not conduct any illegal activities such as paying workers off the books
- ensure that the builder provide required safety training to workers
- ensure that the builder obtain required construction permits, asbestos abatement permits
- ensure that the builder provide enough manpower to complete the job on time

25.8 START OF A PROJECT

25.8.1 Client Coordination

There are many housekeeping tasks that need to be done prior to start of a project. Before we go any further, let us look at a kitchen remodeling job in a house. Let us say you need to remodel your kitchen. You hire a contractor who does this type of work. As the client, you would like to know the following.

Example: kitchen remodeling project:

(1) How many days or weeks would it take to complete the job. In construction jargon, this is called the "schedule."

(2) How much will it cost? This is called the "cost estimate."

(3) What time does the contractor begin working each day? This is important since you have a job and you have to go to work. You arrange a security guard to be home while the contractor is working. The workers should come at the given time. Not earlier.

(4) What time should they finish work? You do not want them working later than 5:00 pm since you are home after work and you would like your privacy.

(5) Workers should not walk around in the house. For example, they should not be going in to bedrooms and living rooms. The workers will be given an entrance and pathways to the kitchen and they should come from that entrance and work and leave from the same way. Any worker who does not follow these rules will be banned from working in the project.

(6) Workers should not be using your bathrooms. They should have their own portable toilet. You would provide a location in the yard to place the portable toilet.

(7) You would provide a location in the yard to keep the tools, wood, steel, and other equipment.

(8) Workers should not damage the yard or the house.

(9) The contractor should provide a contact number in the case of an emergency. Emergency such as a pipe leak or electrical outage due to work of the contractor.

(10) If, in any case, the final cost of the project would be higher than the agreed amount, the contractor should inform you. Contractor should not do any additional work without your prior approval.

Problems that could happen during a project:

Practice Problem 25.1

Cost is higher than the initially agreed amount.

This is a common problem. It is very rare for a project to be completed within the initial budget. In this case, the reason for the higher cost is investigated.

Let us say that the high cost is due to the fact that there was a hidden wall that had to be repaired. The contractor could not have seen this damaged wall. Hence, the additional cost is legitimate.

Or the high cost is due to building cabinets that do not fit. In this case, the contractor has to eat the additional cost.

There could be some gray areas. Let us say contractor came to work with five people and the security guard was not home and the doors were locked. The contractor did not have access to the kitchen. Five workers had to go home that day. In that case, it is fair for the contractor to ask the cost for the lost time. You may claim the loss from the security guard or the company he works.

Practice Problem 25.2

The job takes longer than initially agreed. Now this is a problem. First of all you need to pay the security guard more money than previously planned. In addition, it is an inconvenience for you.

A job could get delayed due to following reasons.

(1) *Contractor's fault:* The contractor did not have enough men on the job. Or contractor did not have correct material.

(2) *No fault of the contractor:* It could be a natural event such as flood. Or it could be a labor strike for which the contractor has no control.

(3) *Owner's fault:* A project can get delayed due to the fault of the owner. Maybe owner wanted a different style of cabinets than what was agreed upon. Sometimes, an owner may impose more restrictions on work after the contract was written.

In large projects it is not easy to pin down the reasons for a delay. It could be a mix bag of everything.

25.9 EMERGENCY CONTACT NUMBERS

Emergencies are not planned but they do happen. Construction is basically working with heavy equipment and machinery. In such situations, personnel on-site should know who to notify. Hence, emergency contact list should be prepared and circulated among interested parties.

Sample example of emergency contact number list is given in Table 25.1.

Table 25.1 Emergency contact list
Project: San Francisco Triangle Company, Basement Renovation Project

Project number: 129-67				
Name	**Company**	**Position**	**Cell phone**	**E-mail**
Allen Dupree	LBC contracting	Safety officer	311-454-0999	adupree@lbc.com
Joc Franks	LBC contracting	Super	311-456-1290	ajfrank@lbc.com
Harry Lam	SCD const. mgt	Proj. mgr	512-810-8103	hlam@scdc.com
Call 911 for emergency				

In the past, the nearest hospitals, police stations and fire departments are included. Now, in most situations 911 can be called for all emergencies.

25.10 PROGRESS MEETINGS

Progress meetings should be held weekly, bi-weekly, or monthly depending upon the nature of the project. During progress meetings following items are discussed.

- *Progress of the project:*
  All parties are anxious to complete the job. Project progress will be hotly discussed. Any delays, causes for the delay, and how to avoid them in the future will be part of the discussion.
- *Safety incidents:*
  If any safety incidents happen, this would be discussed. Also how to avoid such incidents in the future will be part of the discussion.
- *Cost overruns:*
  Clients like to know any cost overruns. Additional costs could occur due to no fault of the contractor. For example, contaminated soil or groundwater not known at the start of the project may create additional costs to the contractor. These type of issues need to be discussed and an understanding needs to be achieved among all parties.
- *Communication protocol:*
  Communication breakdowns will create major headaches. There may be issues that were not properly communicated to the construction manager, builder, or the owner. In addition, some information needs to be communicated in writing.

25.11 BASELINE SCHEDULE

It is important to know what the builder is planning to do. A schedule with activities and durations needs to be provided by the builder. The schedule would look something like before.

Baseline schedule (example):

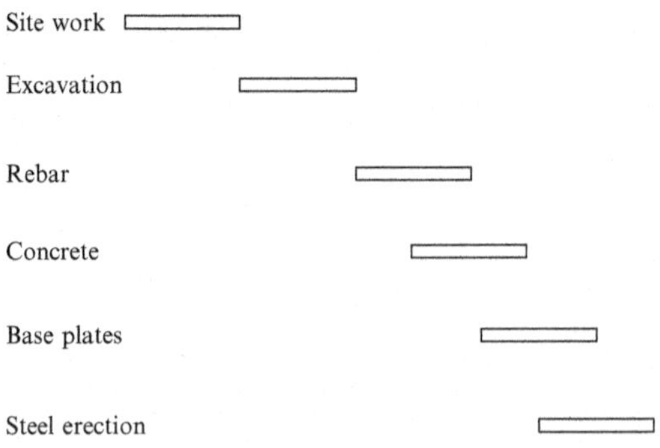

Site work

Excavation

Rebar

Concrete

Base plates

Steel erection

The schedule discussed earlier shows various activities and durations. Some activities are done in a parallel manner. Rebar installation and concreting overlap with each other. The builder can install rebars in one part of the building while concreting another part of the building. Similarly base plates can be installed in a portion of the building while concreting is done in another part of the building.

25.12 MONTHLY LOOK AHEAD SCHEDULES

Monthly look ahead schedules are produced every month indicating what activities will be conducted in the coming month. In some projects these schedules can be developed bi-weekly or weekly depending upon the project needs.

Need for monthly look ahead schedules (example): Let us imagine a road construction project. The builder would decide to lay asphalt in a certain road. Assume the road will be closed for a period of 2 days. Many city agencies require ample notice prior to close down of a road. Police also need to be informed. So as nearby schools and hospitals need to be informed. Also the residents living in the area need to be informed of the road closing. For this reason the builder need to provide a monthly look ahead schedule so that affected parties can be informed in a timely manner.

25.13 SAFETY AT WORKPLACE

"Safety First" is the motto of many builders. Separate chapter will deal with workplace safety. Some important points will be pointed out in this chapter. OSHA (Occupational Safety and Health Agency) is responsible for providing safety rules and regulations in the United States. States and cities have their own safety regulations. Generally they are more stringent than OSHA regulations. Typically, hard hats, boots, and safety vests are required on construction sites. When drilling or welding is conducted, eye protection is also required. If workers are working at high elevations, fall protection is needed. Builders need to follow the safety regulations.

25.14 SECURITY OF WORKPLACE

Workers should be provided with a secure environment to work. Gates and security guards should be provided depending upon the site situation.

25.15 PHOTOS OF EXISTING CONDITIONS

In many situations existing conditions need to be photographed. If there is a damage to a certain existing structure, this could be verified.

Example: A bridge builder is building a bridge near a building. During the construction of the bridge, the building owners claimed that the building was damaged due to construction activities. The builder had taken photographs of the building prior to start of construction. Fig. 25.4 shows that the cracks were present before the construction of the bridge. Hence, the builder was vindicated.

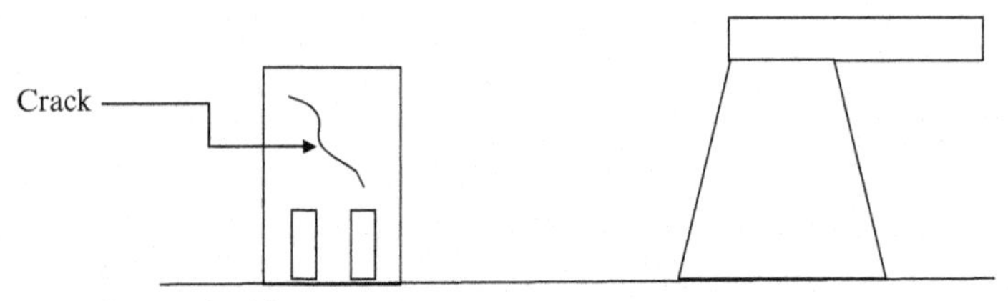

Fig. 25.4 Cracks in a building.

Progress photos: Progress photos are taken to show the progress of work. These photos can be sent to interested parties. In addition, they will be useful for planning and also to resolve any disputes.

25.16 QUALIFICATION OF STAFF

Most clients would like to know who will be involved in the project. Resumes of construction managers, superintendents, safety officers, construction directors, and other key personnel need to be provided to the owner. Many government agencies conduct background checks in addition to their qualifications. These background checks include checking of criminal history and fraud activities of the worker.

25.17 PERMITS

In the United States, approval must be obtained from a government agency for all construction projects. Typically, every city has a building department to regulate construction activities. Plans have to be submitted to the building department of the city or town and obtain approval. Approval comes in the form of a permit. Government approval is needed for many construction activities. In addition, in some cases more than one government agency has to approve a given construction. Hence, you may need bunch of permits.

25.18 CLEANLINESS

A construction site has to be kept clean. Material should be stockpiled in an orderly manner. Tripping hazards need to be avoided. Mud, water, dirt, and other construction debris should be removed from the site on a regular basis.

25.19 REQUEST FOR INFORMATION

There are situations where drawings and specifications do not provide adequate information to build. In such situations additional information is requested from the design team. Some of the reasons for request for information (RFI) are;

- *Missing dimensional information:* Drawings do not provide adequate dimensions
- *Conflicting dimensional information:* Dimensions in plain view are different than dimensions in sections
- *Missing material type information:* Specifications do not provide adequate information regarding type of material to be used

Let us first look at missing dimensional information.

Practice Problem 25.3

Plan view and section view are given for a water tank (Fig. 25.5).

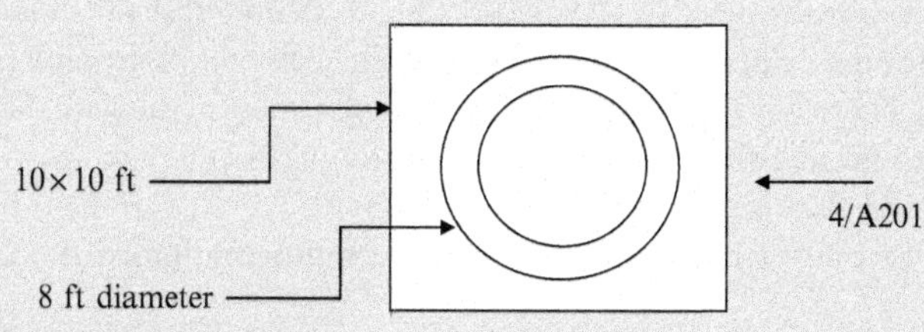

Fig. 25.5 Plan view (3/A-201).

Note: A-201 is the drawing number and 3 is the detail number in the drawing. The arrow shown on right indicates the elevation view. This elevation view is given in detail 4 in drawing A-201 (Figs. 25.6 and 25.7).

Continued

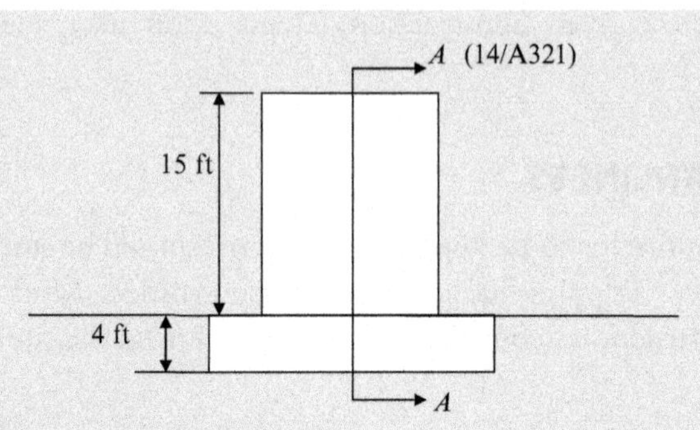

Fig. 25.6 Elevation view (4/A-201).

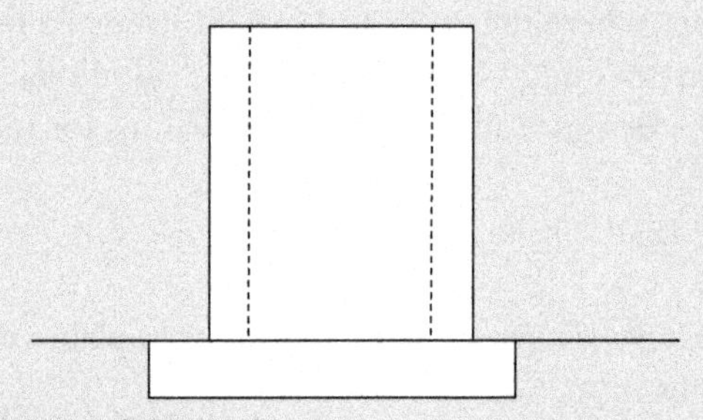

Fig. 25.7 Section A-A (14/A-321)

A section is cut in the elevation view. This section is shown in detail 14 in drawing A321.

Is there enough information to build the water tank?

If not, write a RFI to the design engineer.

Solution

The base of the water tank is a square of 10 × 10 ft. The height of the base is 4 ft. If not shown otherwise, it is reasonable to assume that the tank would be placed at the center of the base. The outside diameter of the tank is given to be 8 ft. There is no way to deduce the wall thickness of the tank. Hence, a RFI has to be sent to the design engineer.

A typical RFI is shown below.

In most cases it is important to provide following information in the RFI:

- identify the project name
- project address
- project number
- date sent
- relevant drawing numbers

- relevant detail numbers
- location
- reason for the RFI

Project Name:	Maplewood Park Water Tank
Address:	326 Prince Avenue, Maplewood, OR
Project Number:	MP005
Date Sent:	Sep. 21, 2015
RFI Number:	0022

Location: Water tank No.2

Relevant drawing numbers: A-201, A-321
Relevant detail numbers: 3/A201, 4/A201, 14/A-321
Reason for the request of information: Missing dimension

After reviewing drawings A-201 and A-321, we were unable to find the wall thickness of the water tank. Please provide the wall thickness of the water tank or direct us to the drawing to obtain this information. Please provide an answer within 3 days, since we have mobilized labor and material to build this tank.

Let us look at another example.

Practice Problem 25.4

Plan view and two elevation views of a building are shown in Figs. 25.8–25.10.

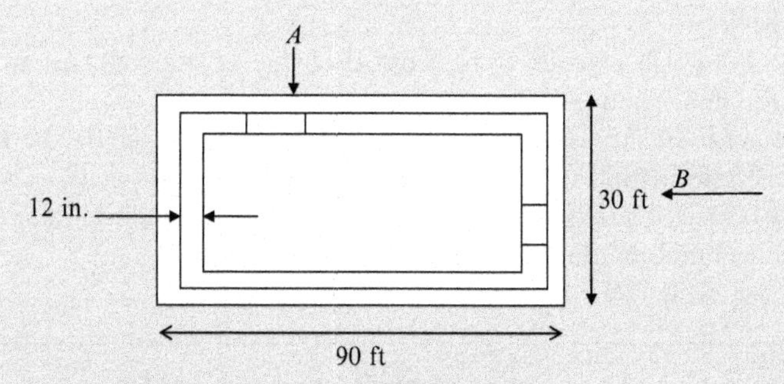

Fig. 25.8 Gymnasium—plan view (3/A-200).

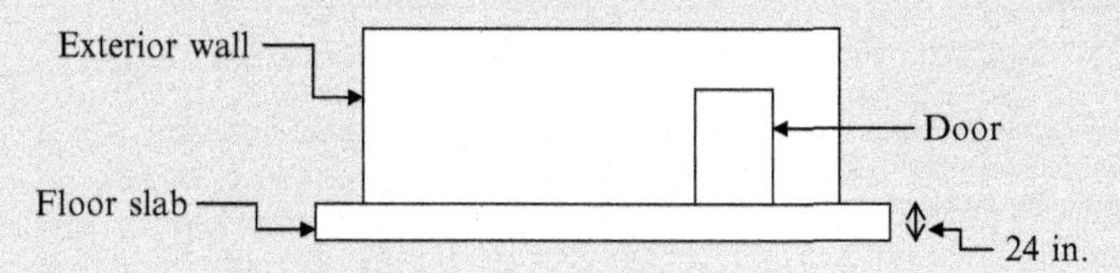

Fig. 25.9 Gymnasium—elevation A (4/A-200).

Continued

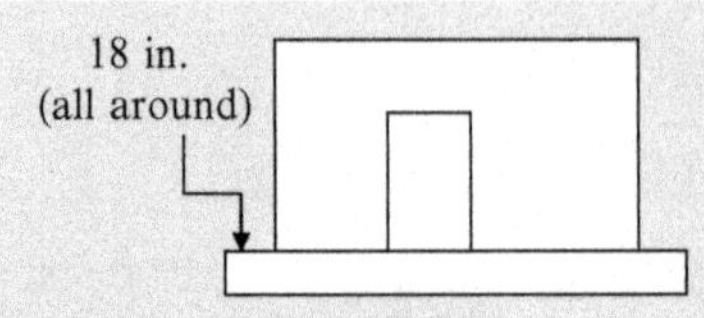

Fig. 25.10 Gymnasium—elevation B (5/A-200).

The floor slab, exterior walls, and doors are shown. Write an RFI for missing information. The following information is available.

Project Name:	Highland Building Project
Address:	226 King Avenue, Maplewood, NH
Project Number:	JP005
Date Sent:	Sep 21, 2015
RFI Number:	0012

Solution

Let us look at the elements shown in these drawings.

- floor slab
- exterior walls
- doors

Floor slab: The length of the floor slab is 90 ft and width is 30 ft. The thickness of the slab is 24 in. All dimensions to construct the floor slab are available.

The floor slab extends 18 in. from the edge of the wall. All around means the dimension given is valid for all sides.

Exterior walls: Thickness of the exterior wall location is given to be 12 in. Hence all information to build the wall is available.

Doors: Height and width of doors are not given. In addition, the exact location to build the doors are also not shown.

Hence, an RFI has to be generated.

Project Name:	Highland Building Project
Address:	226 King Avenue, Maplewood, NH
Project Number:	JP005
Date Sent:	Sep. 21,2015
RFI Number:	0012

Location: Gymnasium

Relevant drawing numbers: A-200
Relevant detail numbers: 3/A200, 4/A200, 5/A-200
Reason for the request of information: Missing dimension

After reviewing drawing A-200, we were unable tofind the width and height of doors. Also, we are unable to find information related to the location of doors. Please provide distance to doors measured from the edge of the wall.

25.20 SHOP DRAWINGS

What are shop drawings?

Who submit them?

Design drawings and specifications are given to the builder. The builder has the freedom to make changes to design drawings. These changes has to be approved by the designer. There are number of reasons why this freedom is given to the builder.

(1) Design drawings may not be accurate. There may be some changes that need to be done prior to construction.

(2) Slight changes have to be made based on field conditions.

(3) There are many ways to construct the same thing. For instance let us assume a beam has to be connected to a column. The builder has the freedom to use five 1.0 in. bolts or ten ½ in. bolts. But the builder need to submit shop drawings indicating what he is planning to do.

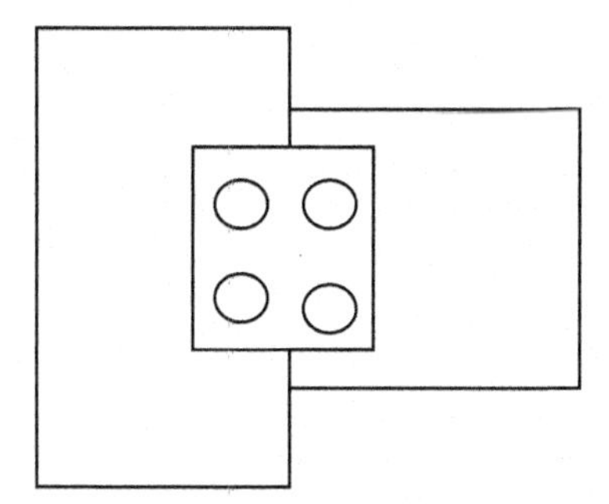

Fig. 25.11 Design drawing (what the designer proposed).

Fig. 25.11 shows the design drawing that came from the design engineer. It shows two steel plates connected using a third plate and four large bolts.

The builder can suggest to connect the two plates with six smaller bolts. This may be acceptable to the design engineer (Fig. 25.12).

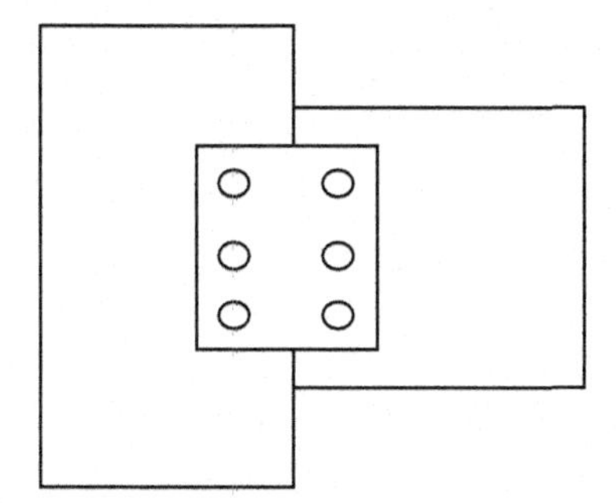

Fig. 25.12 Shop drawing (proposed by the contractor).

There are many ways to skin the cat. Both methods may be acceptable.

The builder should submit drawings to the design engineer for review. These drawings are called shop drawings.

Let us look at another example.

Fig. 25.13 shows two water tanks.

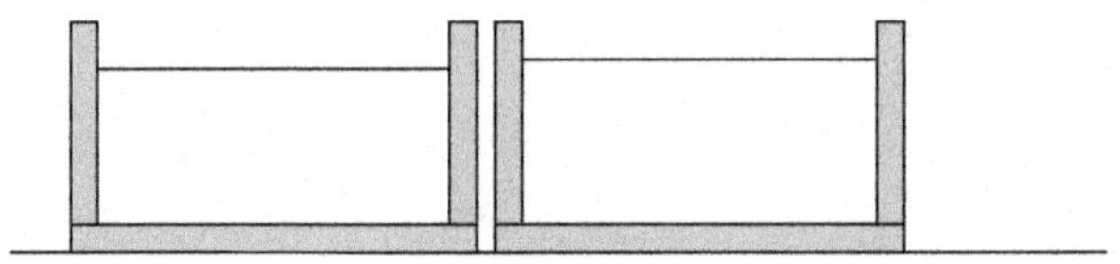

Fig. 25.13 Design engineer proposes two water tanks as shown.

The builder looks at the drawing and finds an easier way to construct (Fig. 25.14).

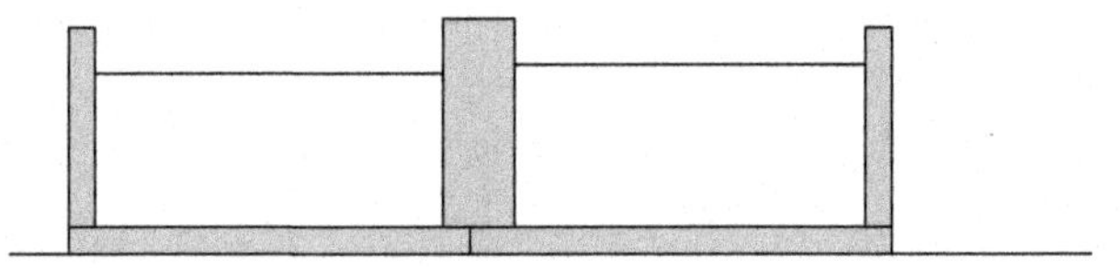

Fig. 25.14 Builder's proposal.

The builder proposes to build one wall at the center. This is easier for the builder. In reality there is no problem with this. The builder's proposal will be submitted in the form of a shop drawing to be reviewed by the design engineer.

In some cases a design engineer may not agree with the builder. Let us look at an example where design engineer may not agree.

Fig. 25.15 shows a ramp for wheelchair users.

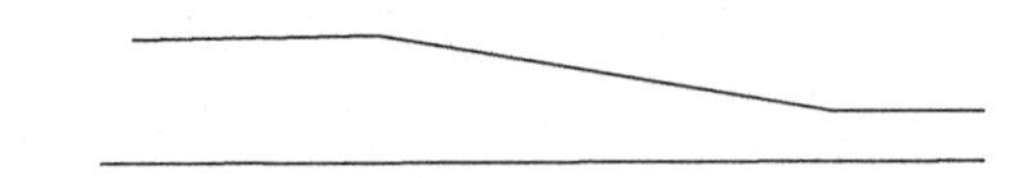

Fig. 25.15 Ramp shown in design drawings.

The builder proposes a much steeper ramp (Fig. 25.16).

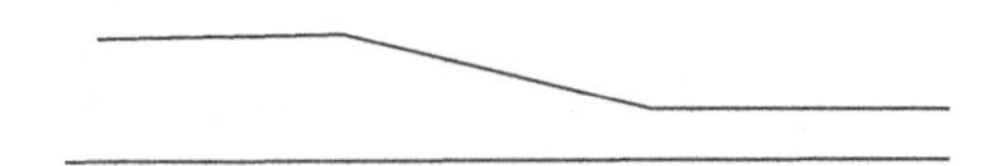

Fig. 25.16 Steeper ramp proposed by the builder in shop drawings.

The design engineer may reject this proposal stating that the steepness of the ramp would be uncomfortable for wheelchair users.

25.21 CHANGE ORDERS

What is a change order? A change order is requested by builders when there is an additional cost that was not expected at the start of the project. Some legitimate reasons for change orders are:

(1) The owner changed the original drawings or specifications (scope change).
(2) There are unexpected field conditions.
(3) There is an omission in design drawings.
(4) There is an error in design drawings.

Now let us look at these items in detail.

25.21.1 The Owner Changed the Original Drawings or Specifications (Scope Change)

The original contract drawings and specifications are known as bid set. The builders agree to build as per the bid set. Let us say after the contract is signed, the owner would like to add an extra floor. This is an owner initiated scope change. In this case the builder is entitled to additional money.

Example: The following floor plan was agreed upon by the builder and the owner. After the price is negotiated, the owner would like to add an additional wall as shown in Fig. 25.17.

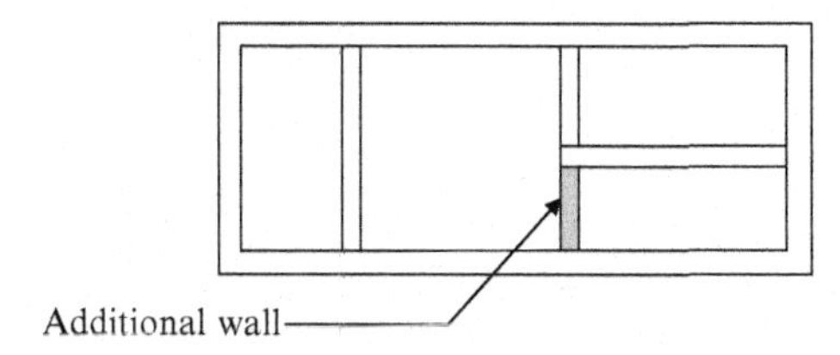

Fig. 25.17 Plan view.

The contractor is entitled to a change order to construct the additional wall.

25.21.2 Unexpected Field Conditions

Sometimes unexpected field conditions could occur. Let us assume that during excavations, chemical contaminated hazardous soil was found.

The owner would like the builder to excavate and remove the contaminated soil. The original bid documents had no mention of removing contaminated soil. The builder would be entitled to a change order. Let us look at another example. Let us assume original bid set indicated that the contractor has to blast and remove 20 tons of rock for a basement excavation. Once the excavation was completed, the contractor has removed 30 tons of rock. The contractor is entitled to a change order to blast and remove additional 10 tons of rock.

25.21.3 Omission in Design Drawings

In some situations, important information may not be in design documents. For example, an asphalt driveway in front of a garage is not shown in the design drawings (Fig. 25.18).

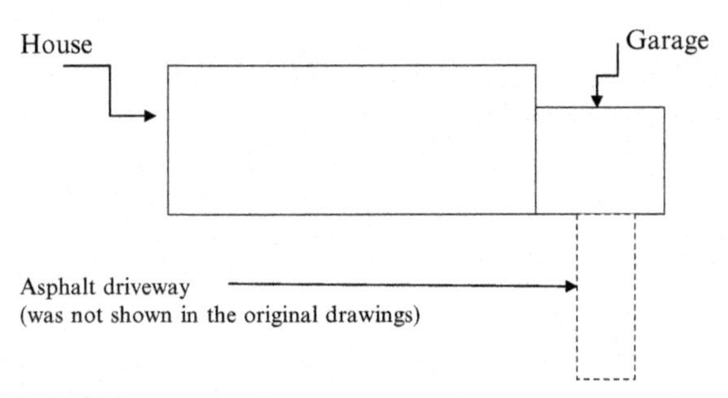

Fig. 25.18 Asphalt driveway.

This can be considered as a design omission. Hence, the contractor will be able to request a change order construct the asphalt driveway.

25.21.4 Error in Design Drawings

Sometimes design drawings contain errors. For example, the thickness of a concrete slab may be shown as 12 in. But the real thickness should be 18 in. Hence, the contractor is entitled to a change order for the difference of cost (Fig. 25.19).

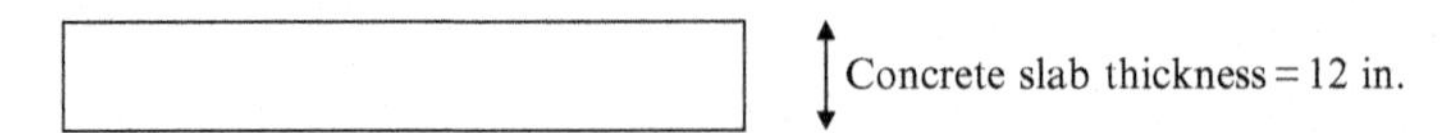

Fig. 25.19 Concrete slab thickness. The design engineer admit that the thickness of the slab is erroneous and need an 18 in. thick slab. The cost difference between construction of an 18 in. slab and a 12 in. slab shall be considered as a change order.

CHAPTER 26

Case Studies

In this chapter different projects will be discussed. The discussion will be based on construction methods, problems faced, solutions, equipment used, and cost issues. It is fair to say there are many ways to build a project. There are many types of equipment from which to choose, and many different methods to pursue. For instance, let us consider building of a concrete wall. One foreman can order concrete from a nearby concrete yard. He also can mix concrete on-site. The placement of concrete can be done using buggies or with a pump. A wall can be done in portions or in one concrete pour. The wall can be completed in 1 week or 1 month (Fig. 26.1).

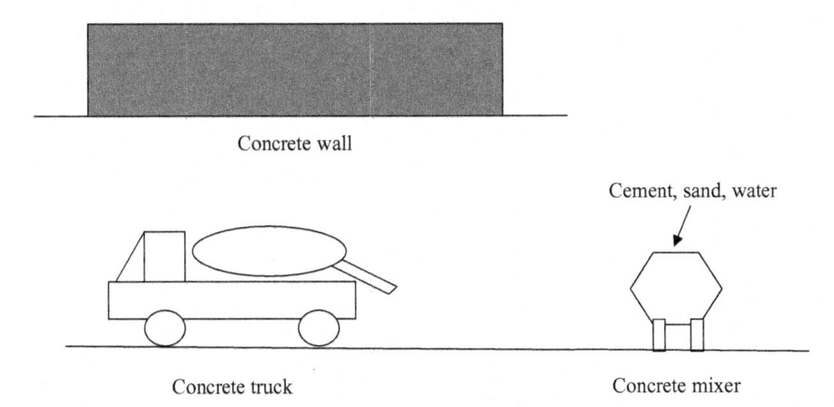

Fig. 26.1 Concrete can be brought in using concrete trucks or mixed on site.

Also the wall can be built in 1 week using 20 workers or can be built in 2 weeks using 10 workers. In some situations it would be cheaper to use less labor and complete the job in a longer duration. Accidents can happen when there are many people in a construction site. In addition, if the work was cancelled due to rain or other event, many workers had to be paid for half day. On the other hand if the work is behind schedule, then it is necessary to catch up by employing large number of workers.

Sequencing: Work can be sequenced in many different ways. For instance, let us look at the concrete wall again. Fig. 26.2 shows a cross section of the wall.

Construction Engineering Design Calculations and Rules of Thumb
http://dx.doi.org/10.1016/B978-0-12-809244-6.00026-3

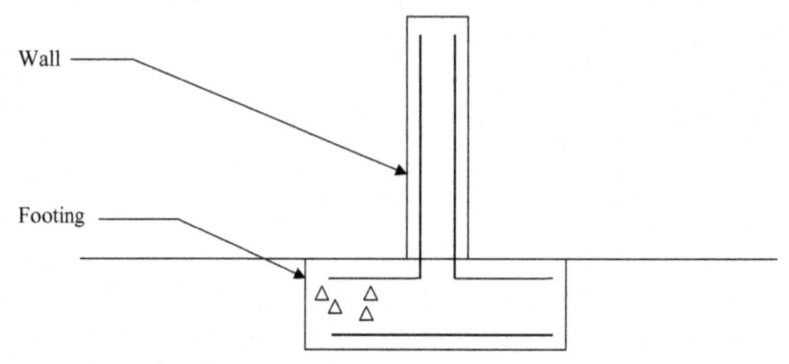

Fig. 26.2 Concrete footing.

Let us assume the wall is 500 ft long. The wall can be sequenced in many different ways. Let us look at some of the possibilities.

Method 1:

(a) excavate a 500 ft long trench for the footing,
(b) build a 500 ft long footing, and
(c) build the 500 ft long wall.

The schedule for this method is shown below.

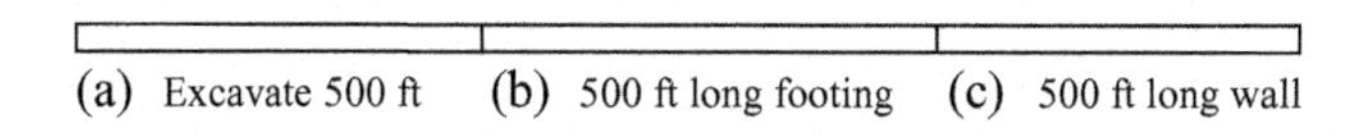

Method 2:

(a) excavate a 250 ft long trench for the footing,
(b) build a 250 long footing,
(c) excavate the remaining 250 ft while building the footing,
(d) build the next 250 ft of the footing, and
(e) build the 500 ft long wall.

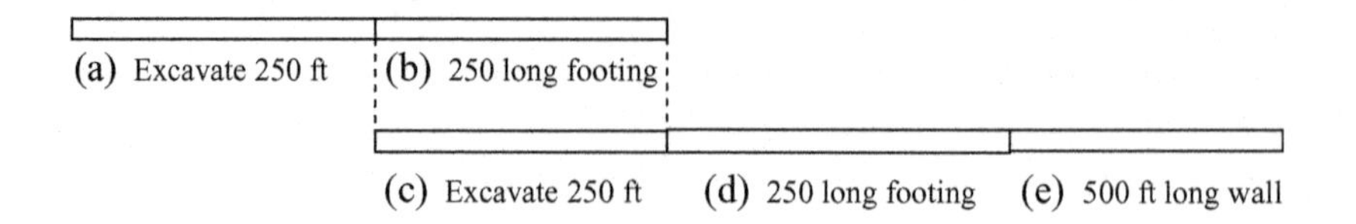

Can you think of two more methods to construct the wall?

Equipment usage: Many different types of equipment are available in the market. If we look back at our wall example, one foreman can use two small backhoes for excavation. Another foreman may use one large backhoe.

As you can see in a larger project, with hundreds of different activities, there are thousands of different ways to complete the project.

26.1 CASE STUDY 1

Five story steel frame building construction in Delaware: The building was a moment frame structure with a footprint of 200 ft × 100 ft. In other words, the connections are rigid. Let us discuss moment frame structures.

Steel structures need to resist vertical loads and lateral loads (Fig. 26.3).

Vertical loads in a structure are:

- dead load (load of slabs, beams, and columns) and
- live load (people, furniture, partitions).

Lateral loads in a structure are:

- wind loads,
- seismic loads, and
- soil loads in basement walls.

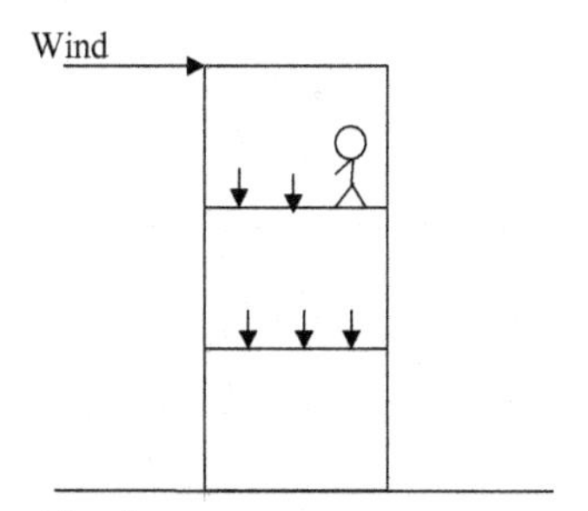

Fig. 26.3 Vertical and lateral loads on a structure.

Vertical loads typically transferred to slabs and then to beams and columns. Braces are required to resist lateral loads. The following figure shows a structure that would move laterally due to lateral forces (Fig. 26.4).

To avoid lateral movement, braces are installed (Fig. 26.5).

Steel frame buildings with braces are known as "braced frame buildings." There is another way to resist lateral forces. Rigid connections can be constructed so that connections would not yield to lateral forces (Fig. 26.6).

Rigid connections are created by welding steel elements together. Let us look at a rigid connection and a pin connection.

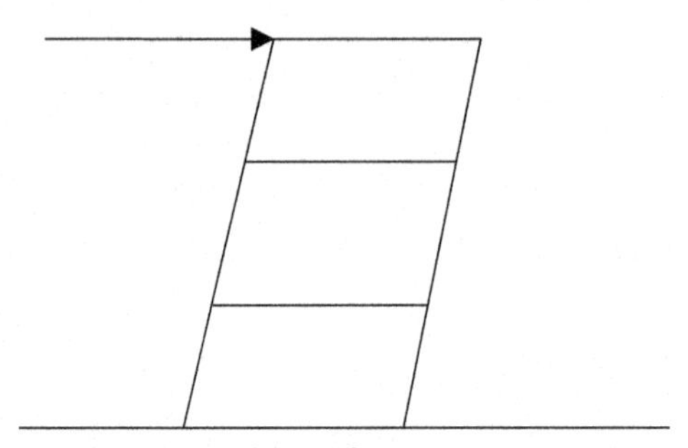

Fig. 26.4 Lateral movement due to lateral forces.

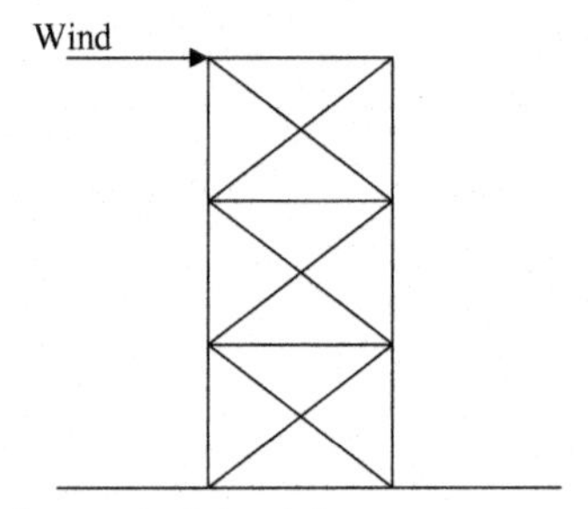

Fig. 26.5 Braces are added to resist lateral forces.

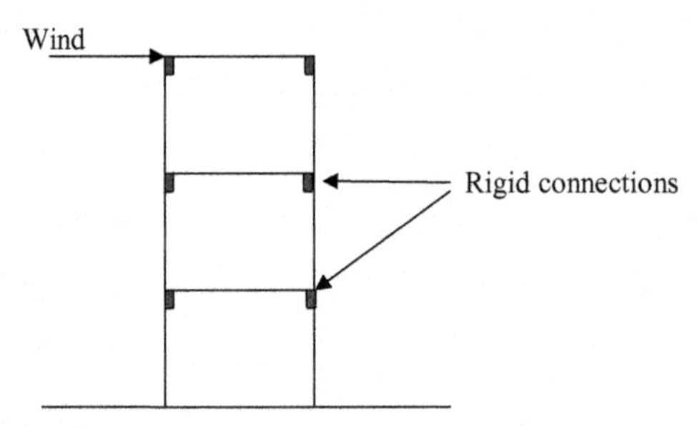

Fig. 26.6 Rigid connections.

Fig. 26.7 shows a pinned connection. They are also known as shear connections. These are not rigid connections. Now let us look at a rigid connection (Fig. 26.8).

Welds creates additional rigidity to the connections. Structural engineers would design the size of the weld based on seismic and wind forces.

The building had a footprint of 200 ft × 150 ft. The building was steel frame structure. Steel columns were placed on column footings. Columns footings were 5 ft × 5 ft with a depth of 4.5 ft (Fig. 26.9).

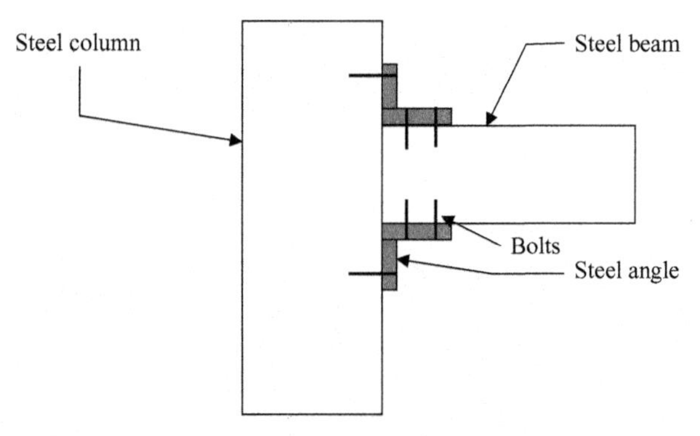

Fig. 26.7 Pinned connection.

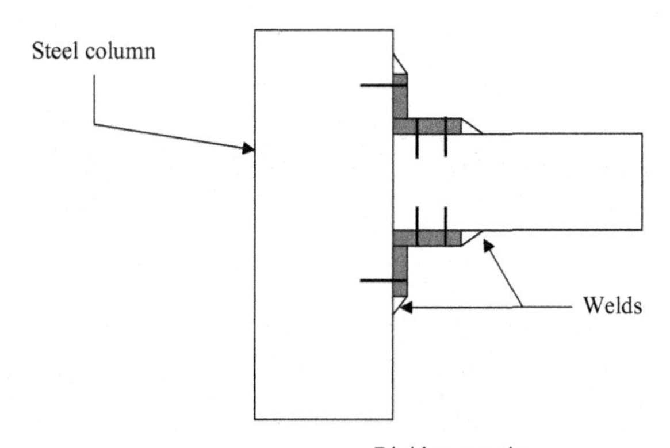

Fig. 26.8 Rigid connection.

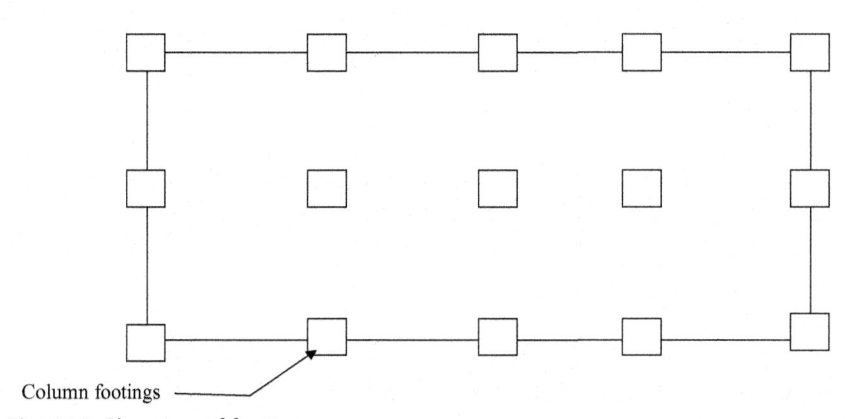

Fig. 26.9 Plan view of footings.

Frost depth: The frost depth in the region was found to be 3.5 ft. The footings were placed 4.5 ft below grade. During the winter, the top soil layer would freeze. During the summer months, the top soil layer would thaw. Hence, the footings have to be placed below the frost depth (Fig. 26.10).

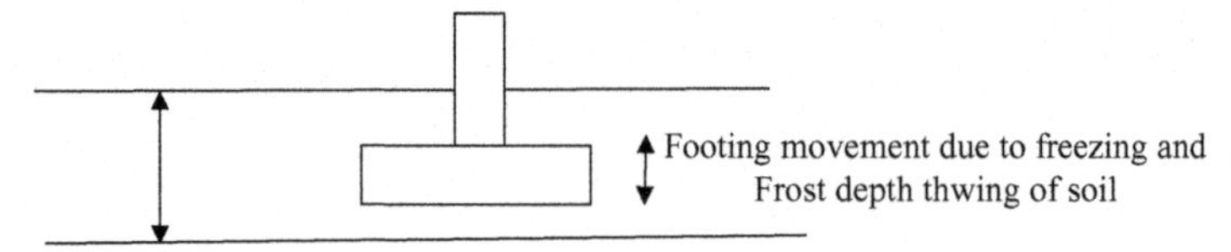

Fig. 26.10 Wrong way to place a footing.

As you could see from the above figure, the footings will move due to freezing and thawing of soil. Hence, the footings have to be placed below the frost depth (Fig. 26.11).

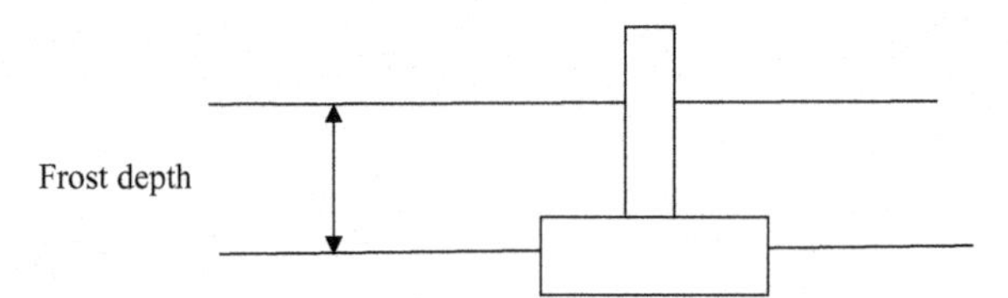

Fig. 26.11 Correct way to place a footing (below the frost depth).

Grade beams were constructed between footings to place walls (Fig. 26.12).

Grade beams are tied to the footings through rebars. The load on the grade beams due to the walls will be transferred to the footings. Heavy wall loads should not be transferred to slab on grade. This can be done if the soil is strong enough. Typically, large loads are transferred to a grade beam. Then the load will be transferred to footings (Fig. 26.13).

As shown on the above figure, the wall load will be transferred to the grade beam. The grade beam will be transferred to the wall load to footings. Grade beams and footings are connected through rebars (Fig. 26.14).

Construction of steel columns: Anchor bolts were installed by drilling holes. Base plates are placed on metal shims. Below the base plate is grouted. Columns are welded to the base plates (Fig. 26.15).

Issues with column erection: The crane was not able to drive along the north edge of the building due to unstable soil conditions. Hence, the slab had to be constructed in pieces (Figs. 26.16–26.19).

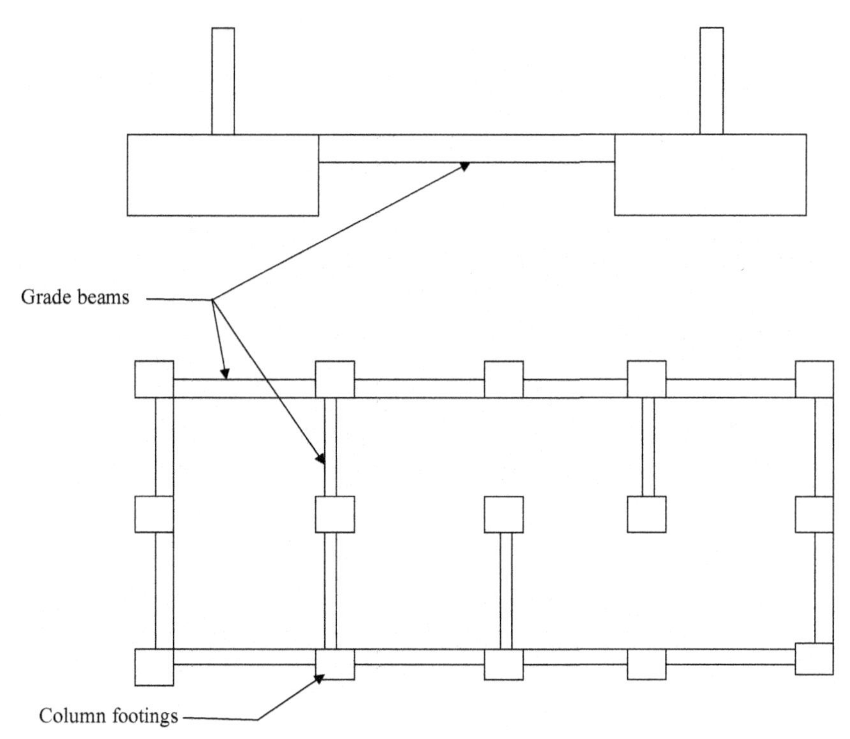

Fig. 26.12 Grade beams.

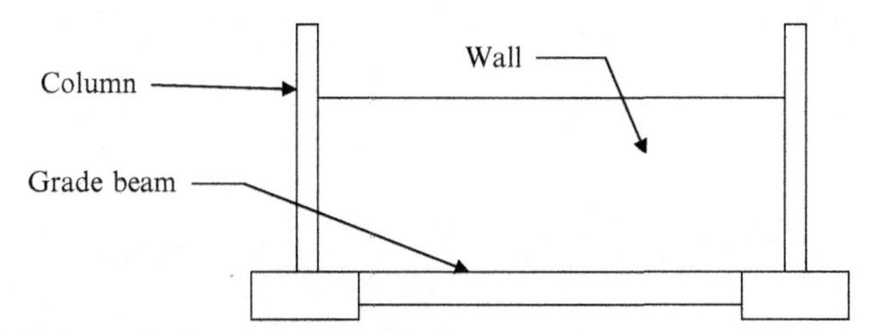

Fig. 26.13 Walls are built on top of grade beams.

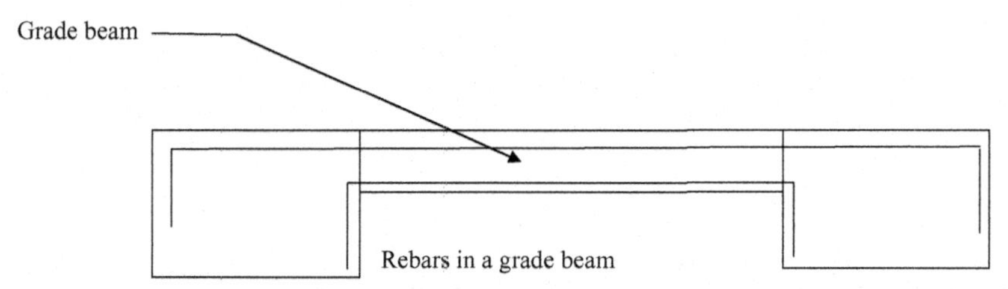

Fig. 26.14 Rebars or reinforcement bars connects grade beams to footings as shown above.

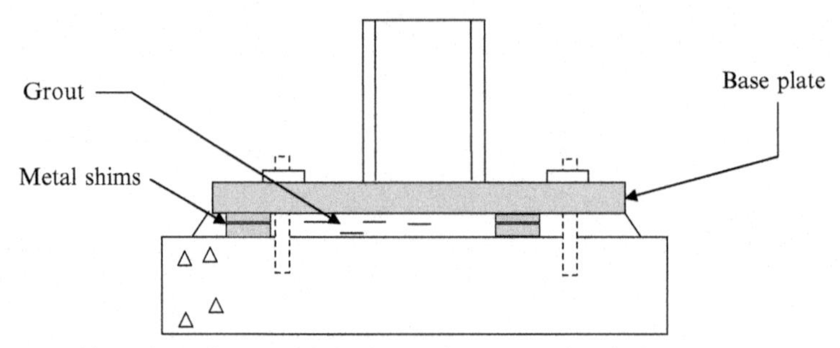

Fig. 26.15 Place the column and the base plate on anchor bolts.

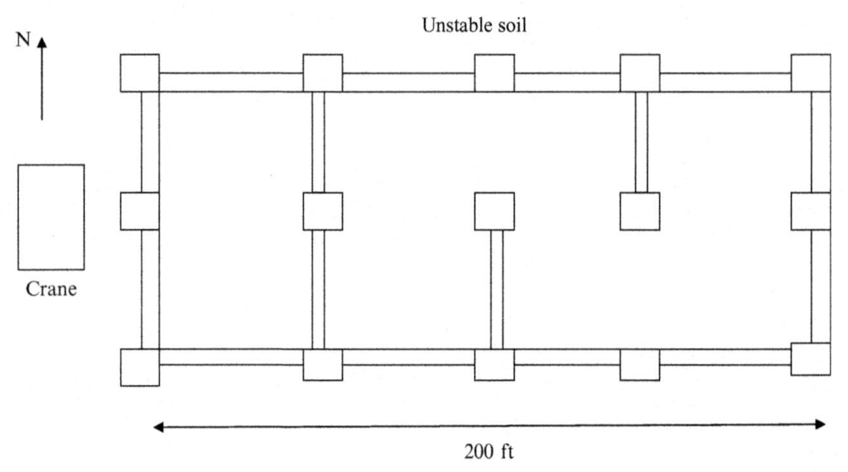

Fig. 26.16 The crane cannot reach the columns at the middle. Hence the slab had to be constructed in pieces.

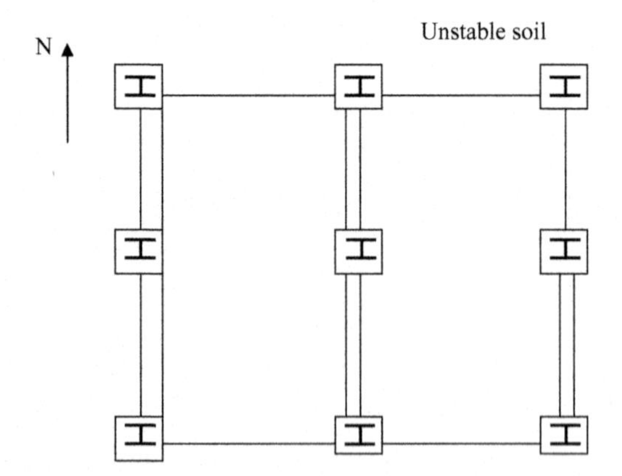

Fig. 26.17 Construct portion of the slab. Install all columns.

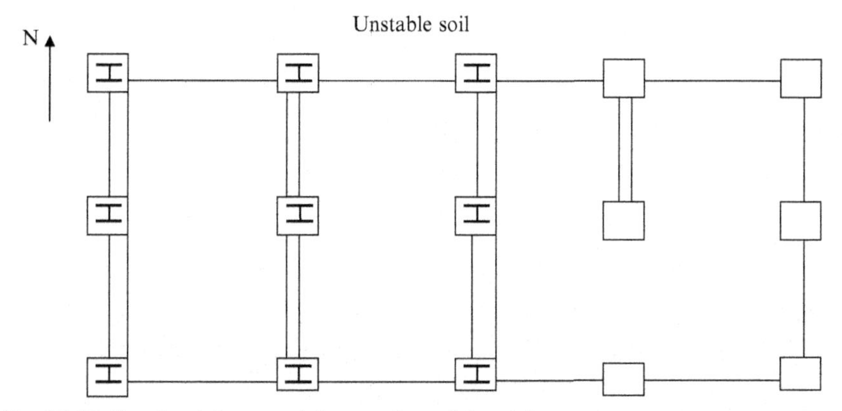

Fig. 26.18 Construct the remaining portion of the slab.

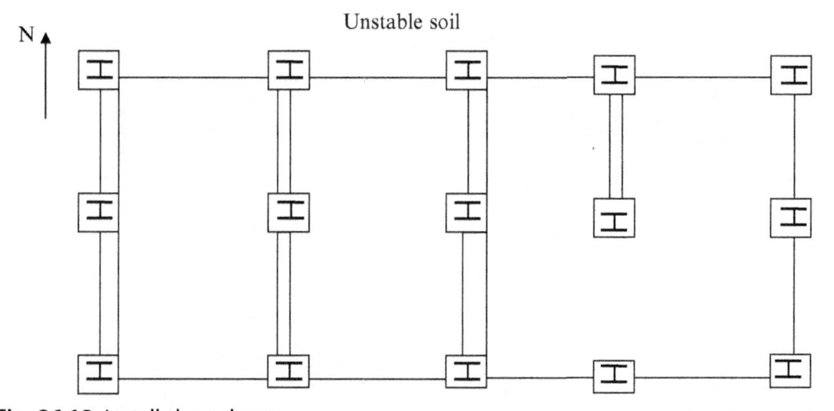

Fig. 26.19 Install the columns.

It should be mentioned here that constructing the slab in two parts delayed the schedule. But this was a much better option since larger cranes would have cost much more.

Column—beam connections: Most column—beam connections were rigid connections. Most connections were welded. Typical column to beam connection is shown below (Figs. 26.20 and 26.21).

A welding crew used a man lift or a bucket lift to reach the connections.

Field welding versus shop welding: Field welding is more expensive that shop welding. Hence, the contractor welded flange plates to the beam in the shop. X-ray tests on these welds were done on the shop. A flange plate was welded to the column in the field. The cost of welding is estimated using pounds of weld material used.

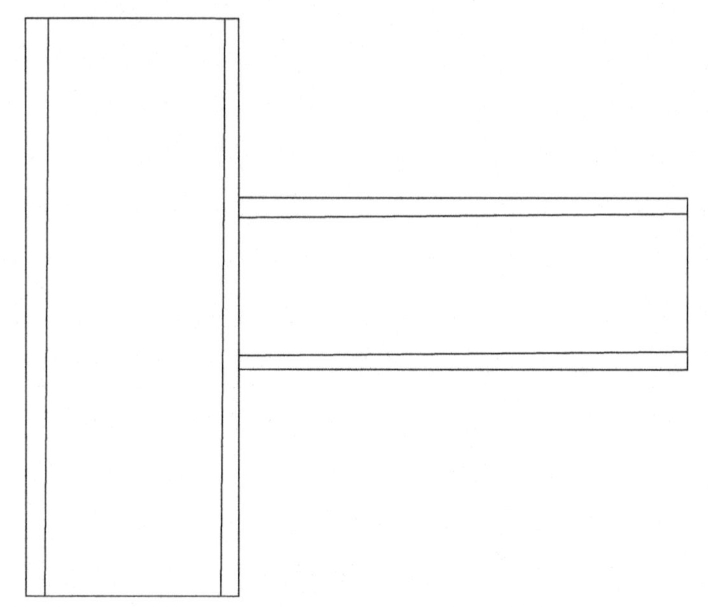

Fig. 26.20 Column and beam shown without connection details.

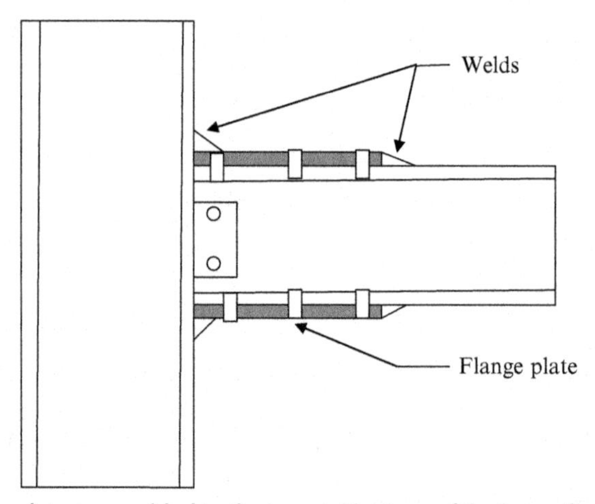

Fig. 26.21 Two plates are welded to the top and bottom of the beam flanges. They also get bolted as shown in the figure.

Erection drawings: Steel contractor developed erection drawings prior to installation of steel. Erection drawings showed what column and beam to be used at each location. The beams and columns were marked. The erectors find the correct steel element and connect them together.

Accuracy of surveys: Measurements should be extremely accurate. Once the steel is fabricated, they need to fit. Let us assume that the fabricated beams are 30 ft in length. But, column to column distance is 30 ft 2 in. In this case the beams will not fit. Additional connection detail may be used to rectify the situation.

Item 1 (metal deck and shear wall configuration): Shear walls were provided to resist seismic forces. A metal deck had to be tied to a concrete shear wall providing protection against seismic activity. This was a very challenging task since rebars had to go through the pour stop of the metal deck. (see figure below). Instead of concreting the metal deck and the shear wall separately, monolithic construction was completed. The monolithic construction was a better solution since the gap between the concrete wall and the concrete filled metal deck was eliminated (Fig. 26.22).

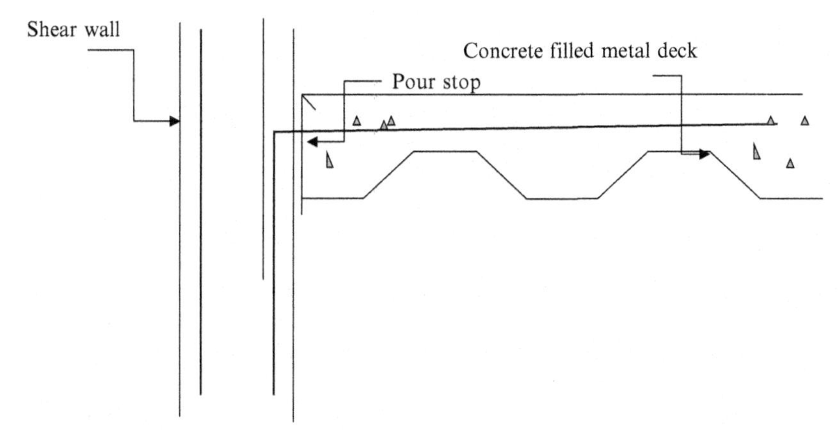

Fig. 26.22 Design drawings show a gap between slab and shear wall.

It was estimated that at least 2 weeks would be needed to make holes in the pour stop and insert the rebars through holes. This was a challenge. Hence, it was decided to build the slab and the shear wall as a monolithic unit (Fig. 26.23).

Pour stop was removed and the deck was supported by formwork for the shear wall. Both the shear wall and the slab was concreted monolithically.

After concreting the first section of the shear wall, second section was completed (Fig. 26.24).

The revised procedure was much easier to construct. Also this method saved time as well.

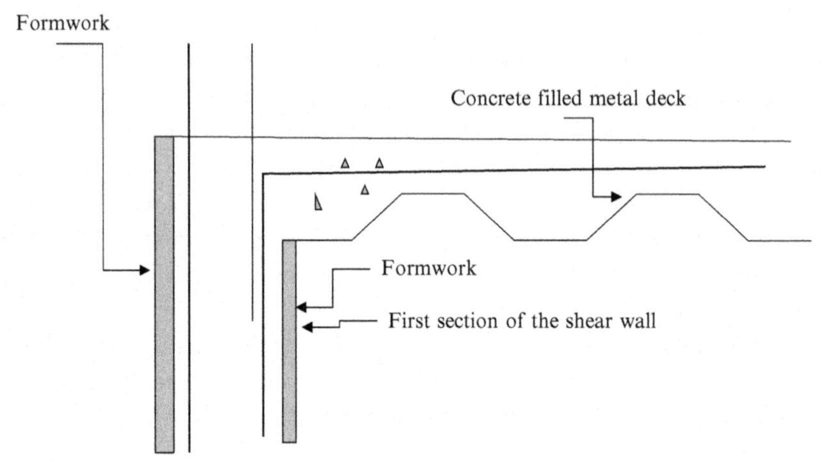

Fig. 26.23 Formwork for the shear wall.

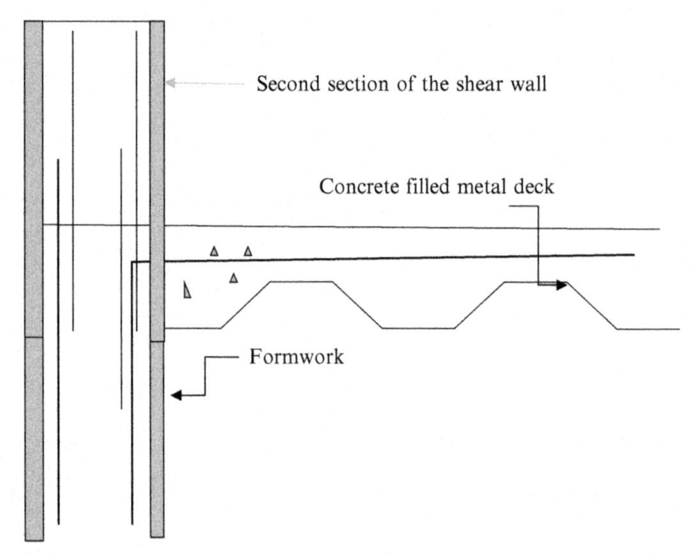

Fig. 26.24 Concrete the slab and the shear wall together.

Concrete slab connection using knee walls for seismic diaphragm effect: Fig. 26.25 shows the elevator machine room and the surrounding slabs.

Fig. 26.25 shows the top slab of the elevator shaft. The dotted line shows the bottom slab in the elevator shaft. Two slabs were connected through a knee wall. A knee wall is essential to transfer lateral loads from one slab to the other since the lower slab was not able to resist the chord forces due to lateral loads. In the general sequence of the operation, the lower slab had to be poured and then the knee wall had to be built and finally the top slab had to be poured. This sequence of operations would have taken weeks since curing of the lower slab and knee wall is needed prior to constructing the top slab. Hence, two slabs were poured at the same time and the knee wall was built later. This way two slabs were available for the mechanical and electrical trades and also to the masonry contractor (Figs. 26.26 and 26.27).

Fig. 26.25 Top slab and a bottom slab.

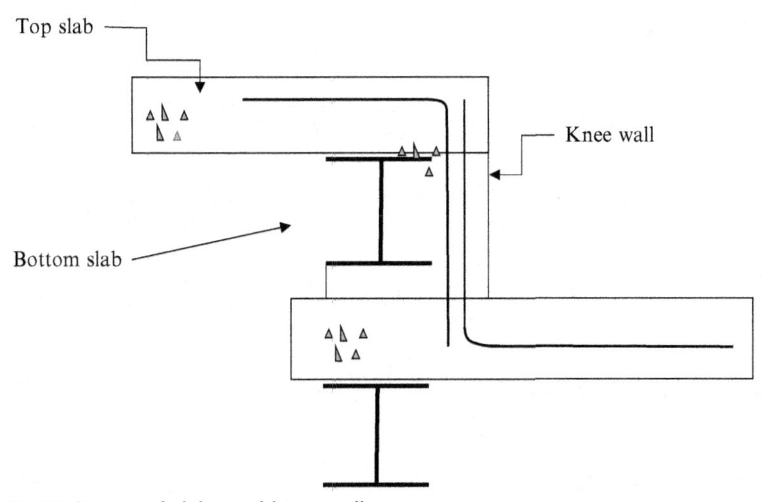

Fig. 26.26 Proposed slabs and knee wall.

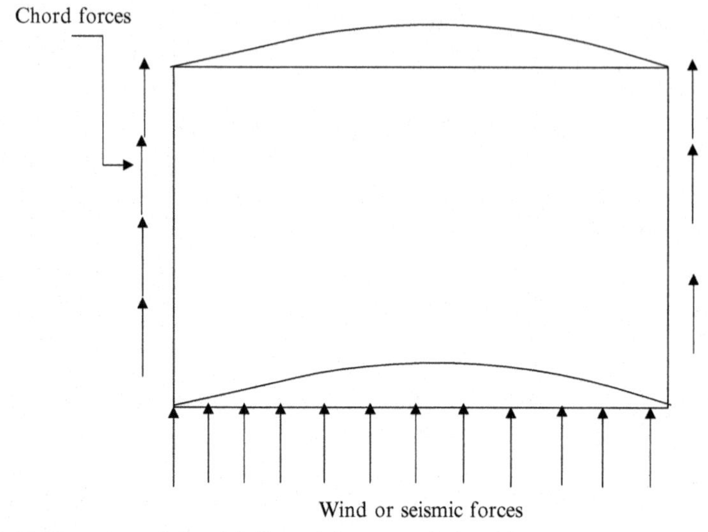

Fig. 26.27 Plan view of the slab, lateral forces, and chord forces.

Construction procedure to save time: See Figs. 26.28–26.31.

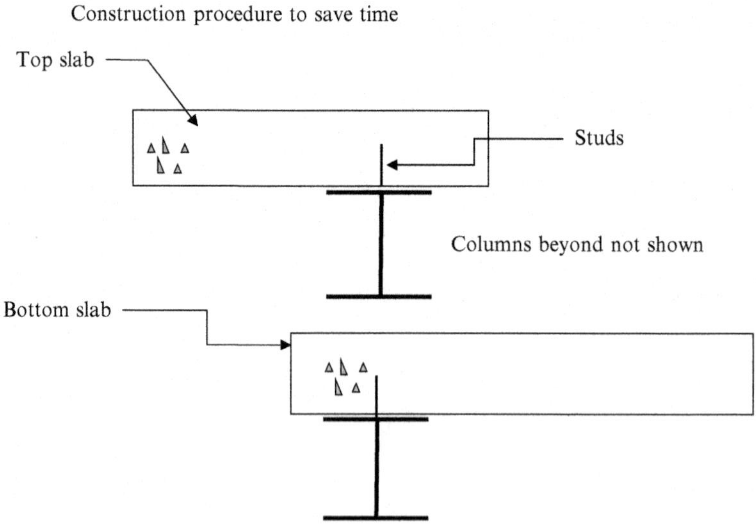

Construction procedure to save time

Top slab

Studs

Columns beyond not shown

Bottom slab

Fig. 26.28 Pour the two slabs first.

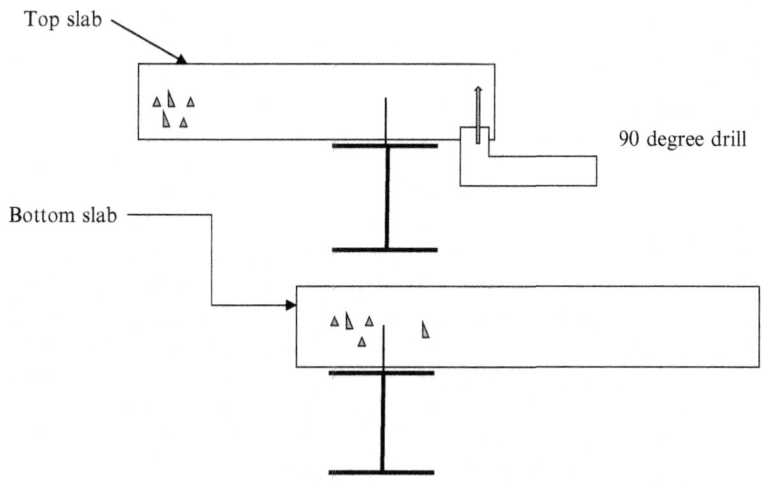

Fig. 26.29 Drill holes in top and bottom slab using 90 degree drills.

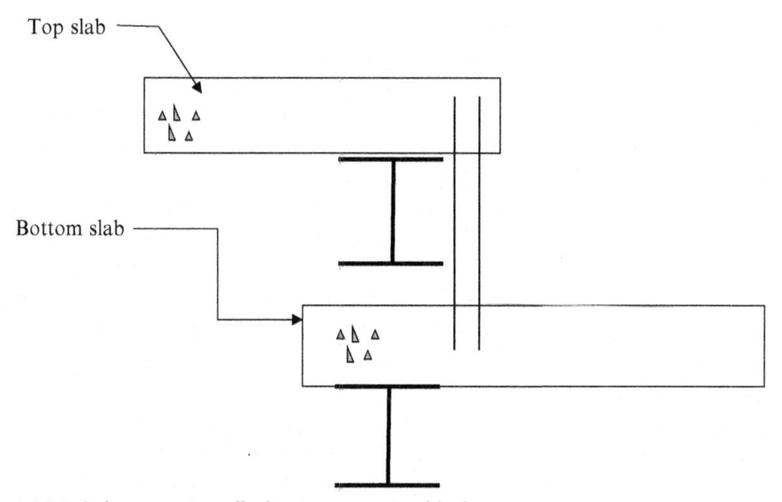

Fig. 26.30 Rebars are installed using oversized holes.

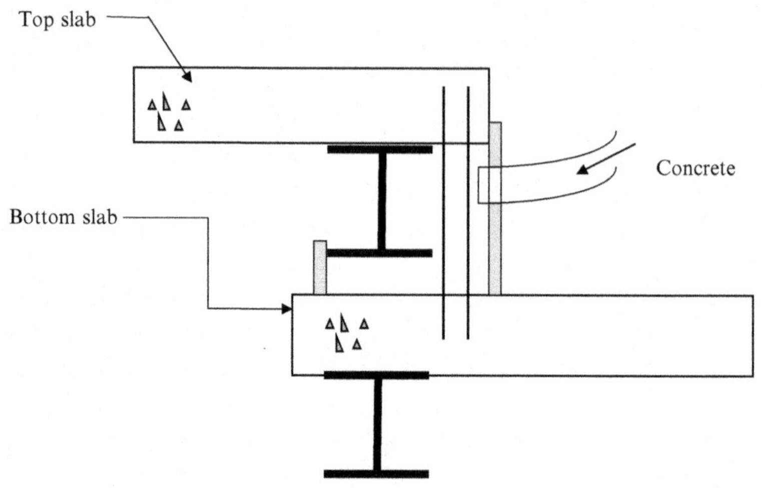

Fig. 26.31 Build formwork and concrete the knee wall. A hole was created in the formwork for the concreting purpose.

26.2 CASE STUDY 2

The building was an office structure located in Teaneck, New Jersey. The structure was three stories tall and designed to be built using concrete. Columns, beams, and slabs would be concrete.

STEP 1: The site was cleared of all plants and brushes. A backhoe was used to remove boulders and uproot small trees. A grapple was used to remove roots. Then dozers were used to level the ground (Fig. 26.32).

Fig. 26.32 Remove trees and brusshes.

STEP 2: The building pad was made level by cutting and filling activities. Most areas were cut with dozers. A backhoe was used for few days to remove hard soil. 50% of the soil was reused. Geotechnical inspectors had to approve the soil that can be reused. Mostly sandy and gravelly soil was reused. Clay and silt was taken out of the site (Fig. 26.33).

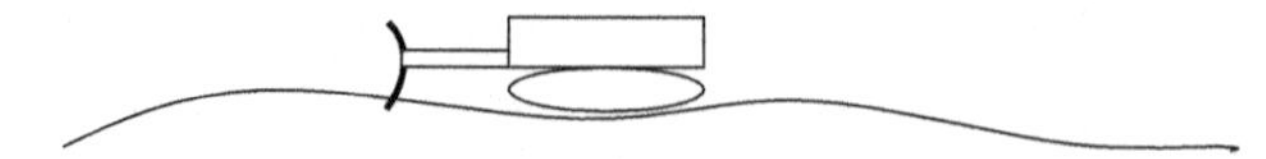

Fig. 26.33 Cut and fill the site.

The proper grade was obtained using laser beams. A laser was installed at the edge of the site. A receiver is located in the grading machine. The driver of the grading machine can see a monitor indicating how much cut of fill required in a given location (Fig. 26.34).

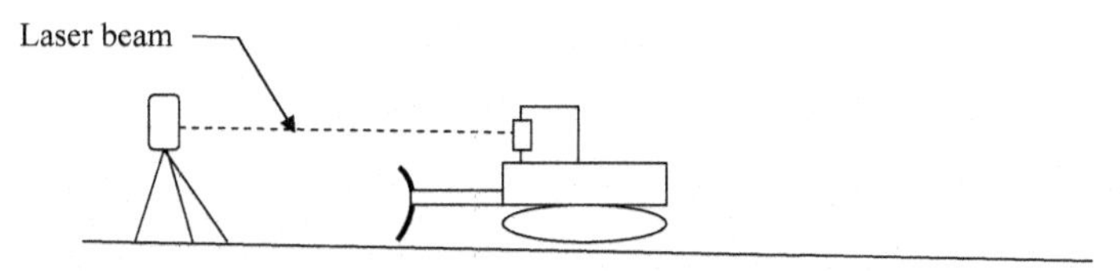

Fig. 26.34 Laser beam is used to get the exact elevation.

LCD monitor in the cabin of the dozer: The LCD monitor in the cabin of the dozer shows the blade in a dashed line. The required grade is shown in a solid line. The operator would know which areas to cut and which areas to fill (Fig. 26.35).

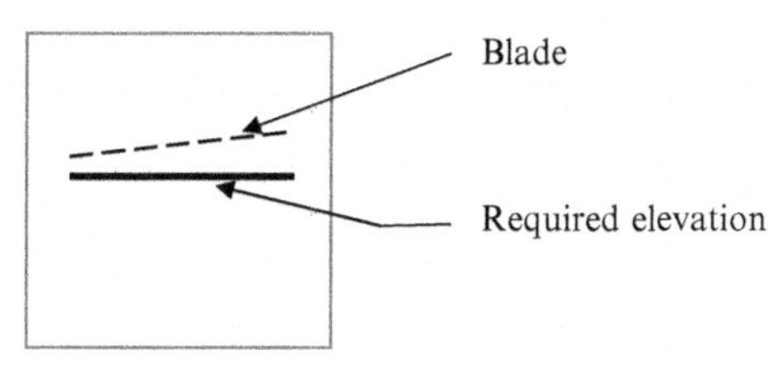

Fig. 26.35 Typical LCD monitor for the laser controlled grading.

Note that the monitor display is different from equipment to equipment. STEP 3: After grading the site, the whole site was compacted with rollers. A large 15-ton roller was used to compact the building pad (Fig. 26.36).

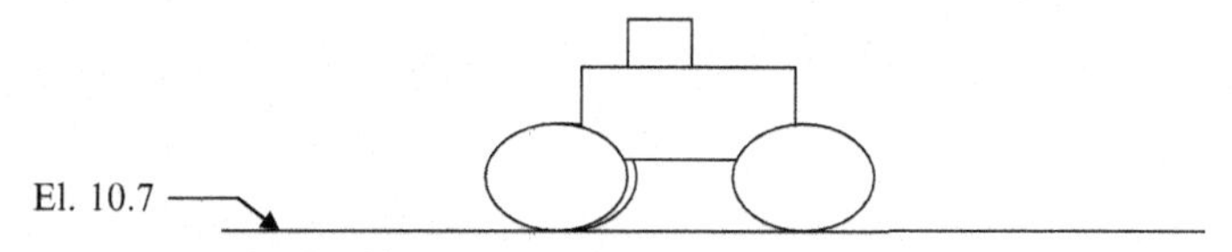

Fig. 26.36 Compact the building pad using rollers.

STEP 4: *Excavation for footings:* After the building pad is compacted, footings were excavated (Figs. 26.37 and 26.38).

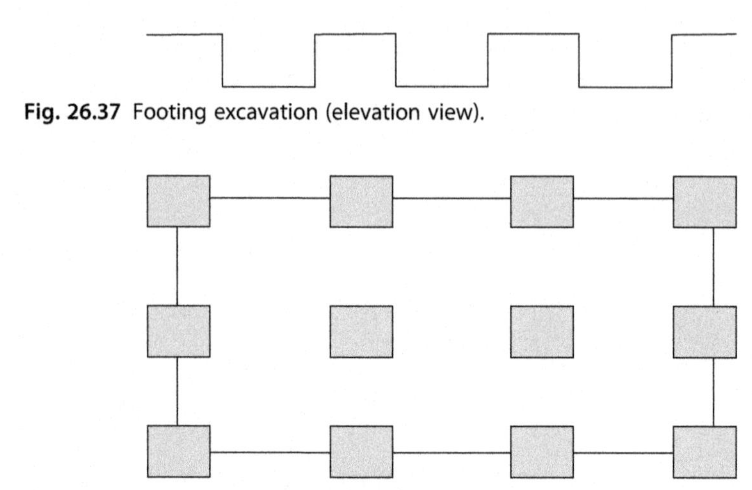

Fig. 26.37 Footing excavation (elevation view).

Fig. 26.38 Footing excavation (plan view).

Excavation work was done with a backhoe. After the footing was excavated, inside the excavation was compacted with a vibratory plate.
STEP 5: *Placement of a gravel layer:* A four in thick, ¾-in stone layer was placed at the bottom of the excavation. Gravel can spread the load evenly to the ground below (Fig. 26.39).

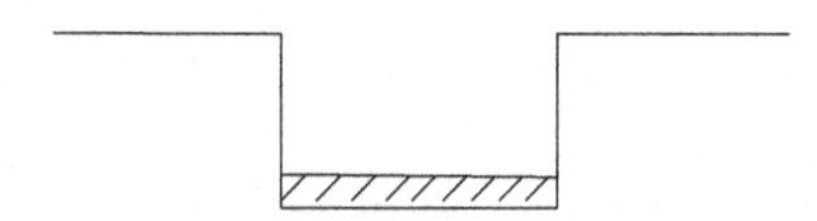

Fig. 26.39 Gravel was placed with a backhoe and workers spread them evenly with shovels.

Water seepage: Water was seeping to the excavation from sides. Water was removed with a submersible pump (Fig. 26.40).

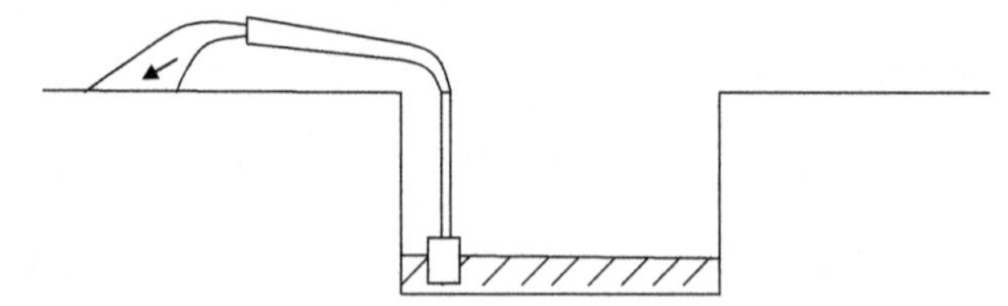

Fig. 26.40 Pumping of water.

Forming of the footings: Footings were formed for concreting. Metal formwork was used since they are faster to install. Wooden formwork was cheaper but can be used only few times. On the other hand, metal formwork can be used numerous times. Also much smooth surface can be achieved (Fig. 26.41).

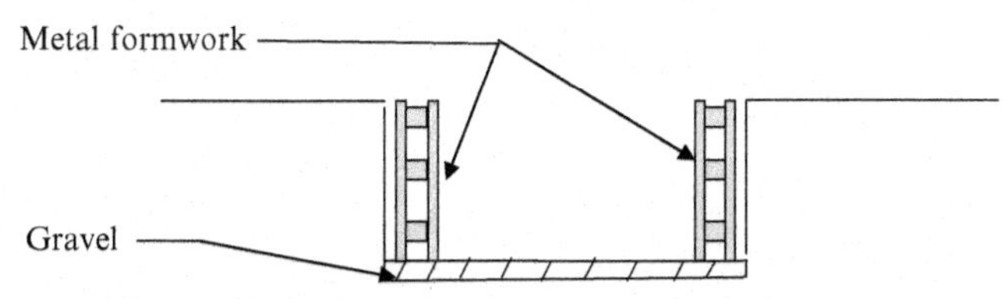

Fig. 26.41 Metal formwork for footings.

STEP 6: After construction of the footing formwork, rebars were installed. Rebars were placed on chairs. Rebar chairs are needed to obtain proper concrete cover requirements (Fig. 26.42).

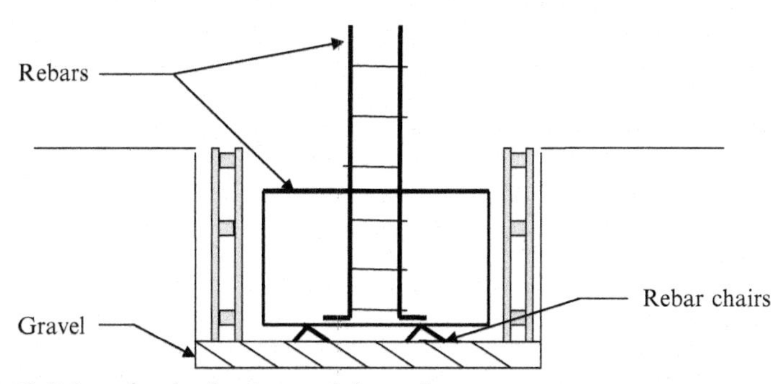

Fig. 26.42 Rebars for the footing and the wall.

Why do we need concrete cover? Adequate concrete cover is needed to protect rebars from water and other chemicals. Typically 1–2 in of concrete cover is provided (Plate 26.1).

STEP 7: *Concrete the footing:* After completion of the rebars, the footing was concreted. In this site a concrete pump was utilized. Concrete cylinders were taken and tested on 7 days, 28 days, and 52 days, respectively. The required concrete strength was 4000 psi. Concrete strength at couple of locations was less than 4000 psi. The design engineer required Windsor probes to be conducted. Windsor probe tests came up higher than 4000 psi.

Plate 26.1 Rebars installed for a column footing is shown above. Formwork and a ladder to go to the pit is also seen.

A Windsor probe is a gun that sends out bolts into the concrete. If the bolt embed more into the concrete, it would mean the concrete is not hard enough (Fig. 26.43).

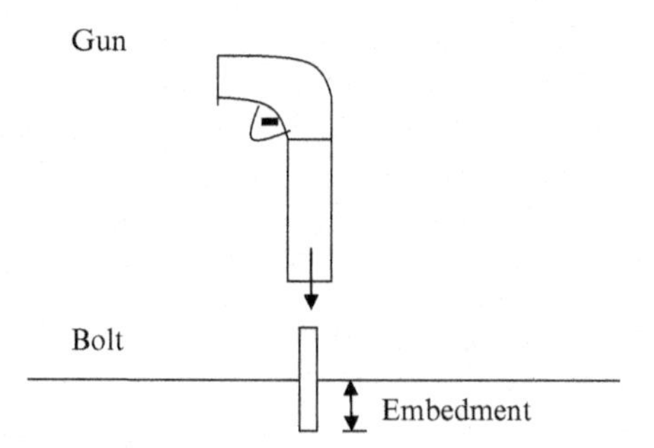

Fig. 26.43 Windsor probe to check the strength of concrete.

Find the embedment of the bolt into concrete. The harder the concrete, the lesser the embedment. From the embedment, the concrete strength can be computed (Fig. 26.44).

The next step is to remove the formwork and backfill the sides with gravel or soil (Fig. 26.45).

As you can see rebars are left high to be connected with the column. In case enough splice length is not provided rebars have to be drilled in with epoxy.

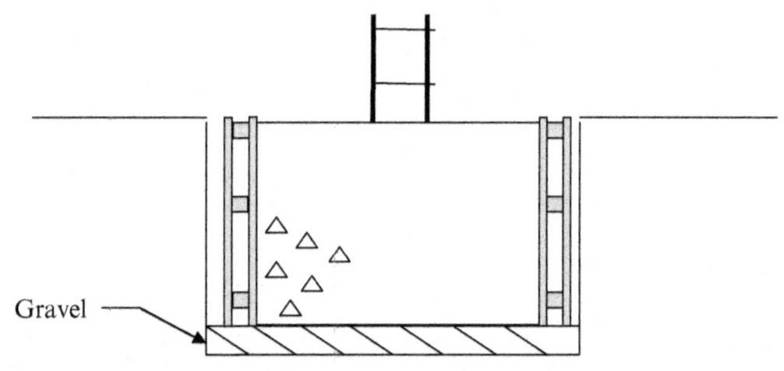

Fig. 26.44 Concreted footing.

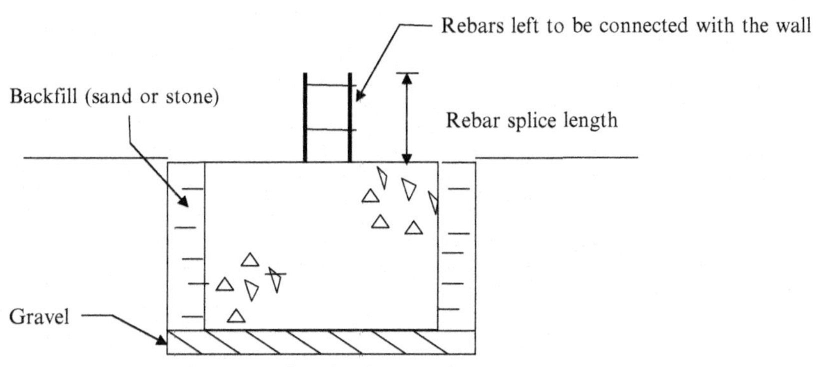

Fig. 26.45 Formwork removed and sides backfilled.

STEP 8: *Install rebars for columns:* Typically, a rebar cage for columns are completed at the ground level. Then lifted with a crane and placed inplace (Figs. 26.46 and 26.47).

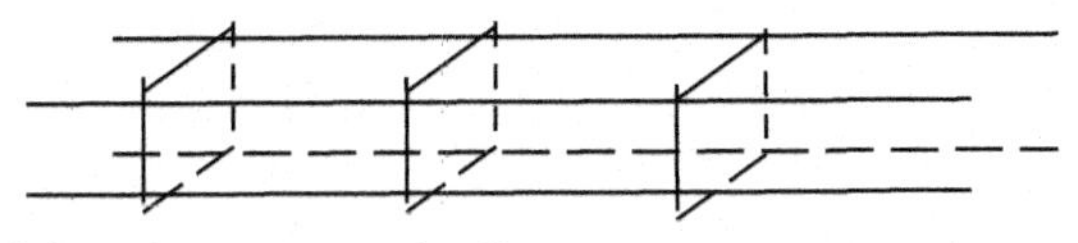

Fig. 26.46 Build the rebar cage on the floor.

Once the rebar cage is placed and secured, the formwork is placed. In many cases column formwork is built on ground and placed (Fig. 26.48 and Plate 26.2).

Column formwork can be built around the rebar cage or can be built on the ground and place with a crane. If a crane is available it is economical to build the column formwork on the ground and place it.

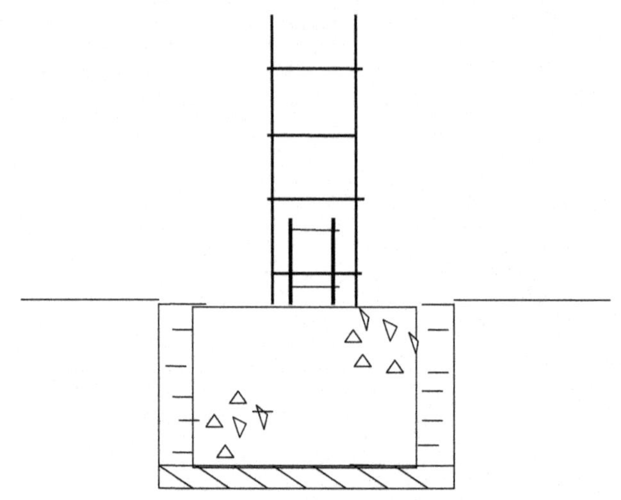

Fig. 26.47 Place the rebar cage on the column using a crane.

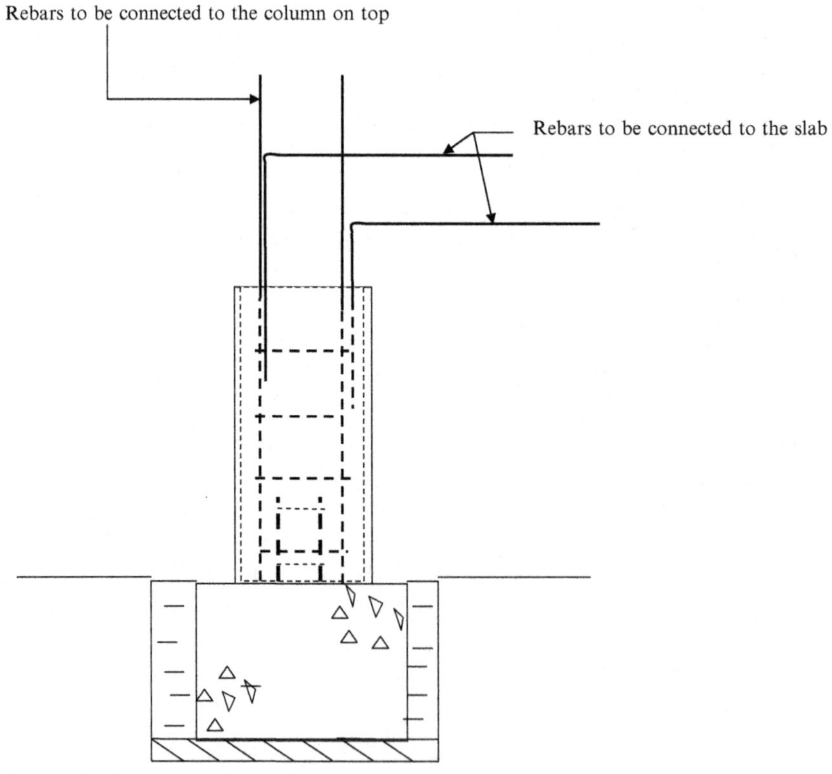

Fig. 26.48 Place the column formwork.

Plate 26.2 Rebar cage built on ground and placed on the column.

It is important to provide set of rebars to the upper column and also the slab above (Plate 26.3).

Plate 26.3 In this photograph, extended rebars for the column above is seen. Also bent rebars are for the slab. Plywood deck for the slab is already constructed.

STEP 9: *Build the deck for the slab:* A plywood deck is supported by two sets of beams and shoring. Shoring are metal posts. The height of these metal posts can be adjusted. Immediately below the plywood deck lies a set of beams. Another set of beams lies below the top beams. Shorings are metal posts. These metal posts can be adjusted (Fig. 26.49 and Plates 26.4 and 26.5).

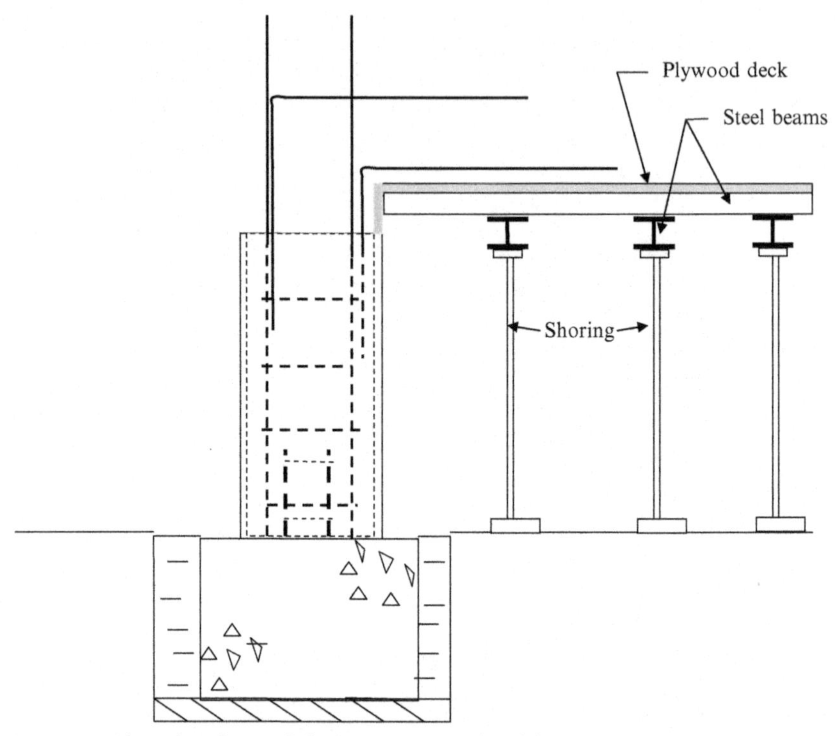

Fig. 26.49 Place the plywood deck to concrete the slab.

Plate 26.4 In this photograph, plywood deck is shown. Bent rebars coming from walls also can be seen.

Plate 26.5 Another photograph of the plywood deck to be concreted can be seen. Column rebars continue upwards. Wall rebars are bent to be attached with slab rebars.

STEP 10: *Place rebars for the slab:* The next step is to place rebars for the deck. Rebars are placed on metal chairs (Fig. 26.50).

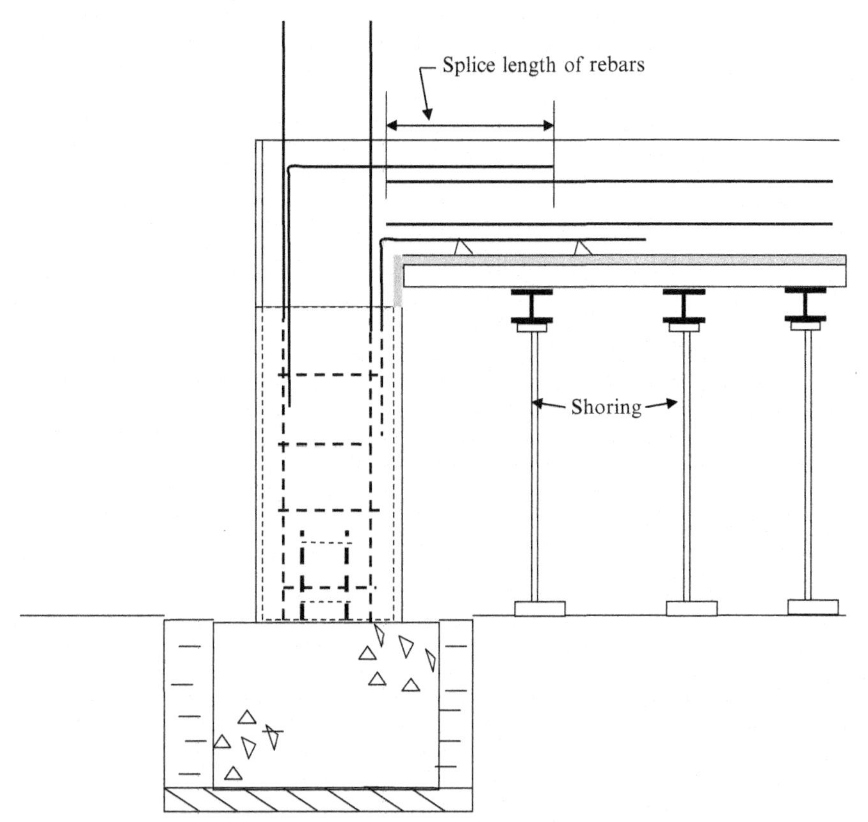

Fig. 26.50 Install rebars for the slab.

Slab rebars are needed to transfer the load to rebars in the walls and columns. A sufficient amount of splice or overlap is required to transfer the load. Splice length depends on the size of rebars. The design engineer would specify the splice length required.

STEP 11: *Insufficient splice length:* During construction it was found that the splice length was not sufficient. There are two ways to address this issue.

(1) welding of rebars and

(2) mechanical connections.

Welding of rebars is an expensive process. Hence, mechanical connections were attached to the rebars (Fig. 26.51).

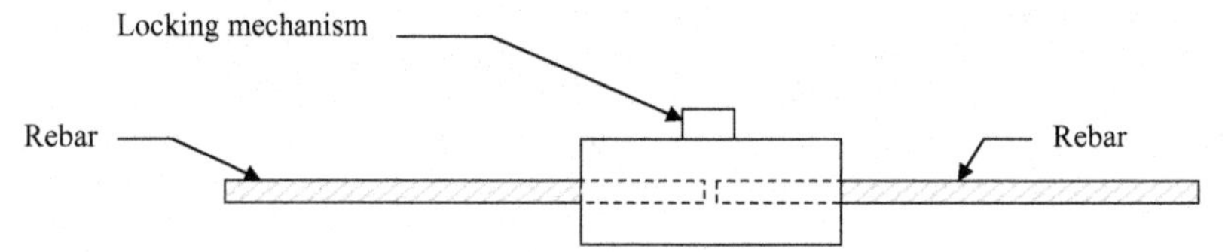

Fig. 26.51 Mechanical rebar connection.

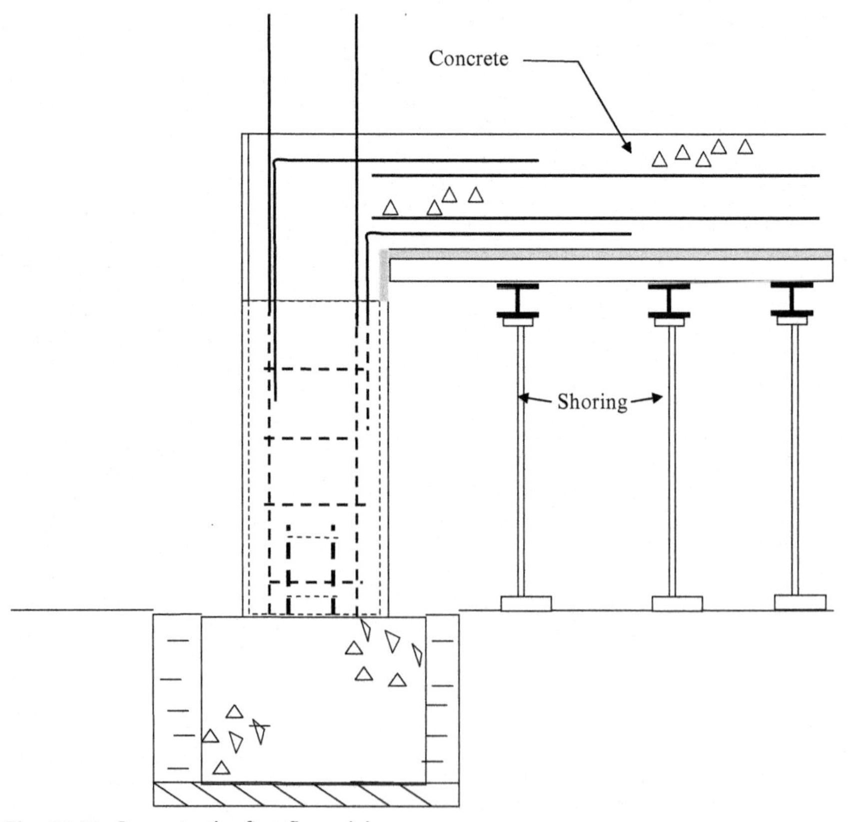

Fig. 26.52 Concrete the first floor slab.

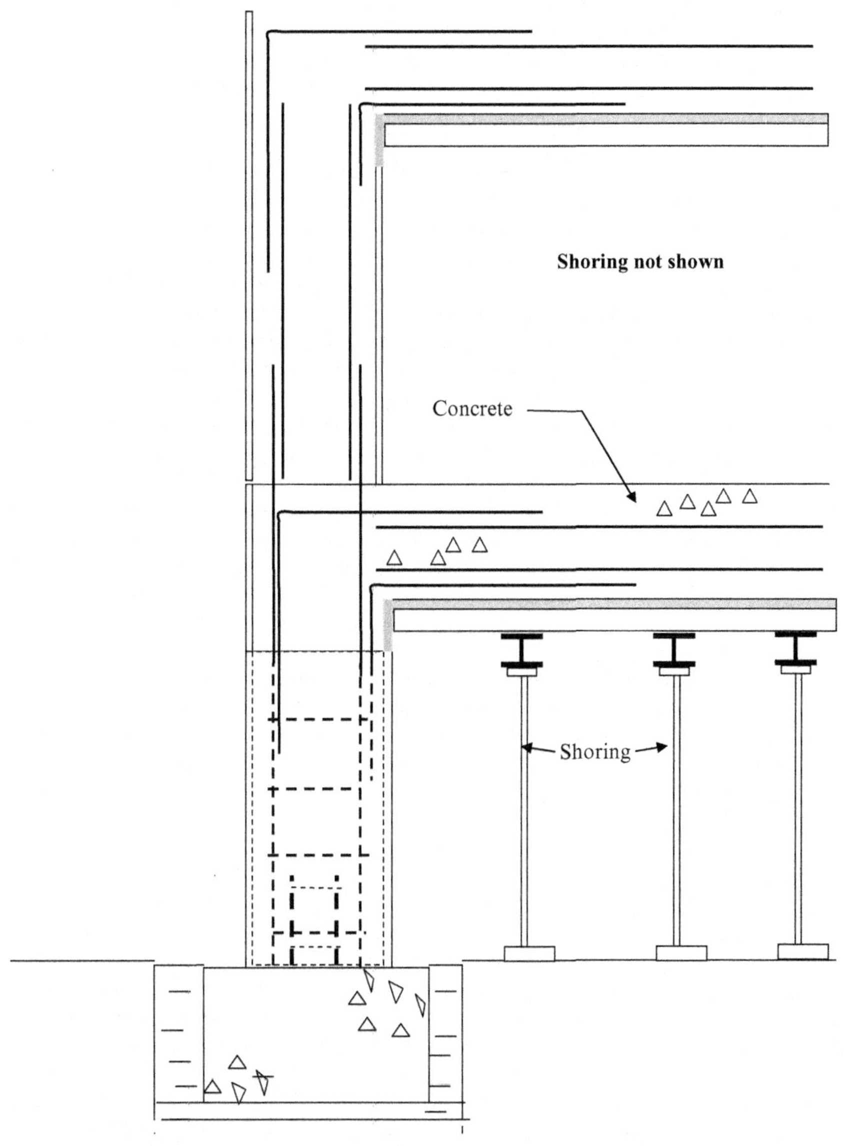

Fig. 26.53 Install formwork and rebars for second floor.

Rebars are inserted into the locking mechanism and then tightened.

STEP 12: *Concrete the deck:* Once the rebars are installed, the deck is concreted (Fig. 26.52).

STEP 13: Formwork and rebars for columns for second floor are installed (Fig. 26.53):

The process can continue for upper floors.

Some of the problems encountered during construction:

- *Concrete pumping during winter:* The pumping of concrete during winter was a challenge. Aggregates and sand start to segregate due to cold. In addition, blocked pipes can delay concrete work.
- *Rebar connections at column—beam intersections:* Rebar connections at column—beam intersections required special attention. Many rebars have to be properly installed. After the installation of rebars, a rebar inspector had to approve the installation. Rebars are checked for the correct size, cover requirements, and splice length.

INDEX

Note: Page numbers followed by *b* indicate boxes, *f* indicate figures and *t* indicate tables.